T0295362

BASICS OF WILDLIFE HEALTH CARE AND MANAGEMENT

The book on **Basics of Wildlife Healthcare and Management** provides the basic information about the introduction of wildlife course by the Veterinary Council of India for veterinary graduate. The book has been planned looking its theme and basic understanding of the various topics which will be useful for biologist, zoologist, veterinarians working in forest ,zoos or at field level where they use to get wild animals for post mortem or for treatment. It also provides helpful information to the forest officers, zoo managers and protected area managers for critical care management and for doing needful things before approaching a veterinarian to save the life of animal or to collect biological material useful for diagnosis.

Rajesh Jani: He is having experience of more than 25 years in academic and research field. Received FIVE awards in academics and research, including Hari oum Ashram Award and Professional Excellency Award in Wildlife Health management. As a member of All India Wildlife and zoo veterinarian association he acted as a secretary for the western region and is an active member. He has published 64 clinical and research articles in National and International Journals, six books and presented lead articles in several seminars and symposiums. As a nominee by the government, during his U.S.A. academic tour visited many wildlife refuses and veterinary schools. He is an active member in 15 different organizations such as Zoo Outreach Organization (ZOO), Captive Breeding Specialist Group (CBSG), Founder member of Society for Promotion of history of Zoos and Natural History in India, (SPOHZ*NHI), Asian Regional Network of International Zoo Educators (ARNIZE), Association of Indian Zoo and Wildlife Veterinarians (AIZWV) and wildlife health advisory member for Gujarat state and coordinator for western region for Indian Wildlife Health Cooperative programme as well as for WHIP.

BASICS OF WILDLIFE HEALTH CARE AND MANAGEMENT

by :

Rajesh Jani

Ex. Coordinator, Wildlife Health

College of Veterinary Science & Animal Husbandry

ANAND-388001

NARENDRA PUBLISHING HOUSE

DELHI (INDIA)

First published 2021
by CRC Press
2 Park Square, Milton Park, Abingdon, Oxon, OX14 4RN

and by CRC Press
6000 Broken Sound Parkway NW, Suite 300, Boca Raton, FL 33487-2742

CRC Press is an imprint of Informa UK Limited

British Library Cataloguing-in-Publication Data
A catalogue record for this book is available from the British Library

Library of Congress Cataloging-in-Publication Data
A catalog record has been requested

ISBN: 978-1-032-06528-1 (hbk)
ISBN: 978-1-003-20269-1 (ebk)

by :

Rajesh Jani
Ex. Coordinator, Wildlife Health
College of Veterinary Science & Animal Husbandry
ANAND-388001

Dedicated
to
My family
and
Wildlife workers

Contents

Preface *ix*

Acknowledgements *xi*

References Acknowledged *xiii*

1. Indian Wildlife 1
2. Taxonomical Classification of Wildlife 12
3. Biological Features of Wildlife 17
4. Population and Habitat Management 39
5. Ethology-Animal Behaviour 66
6. Health Monitoring and Evaluation in Free Living Wild Animals 72
7. Physical and Chemical Restrains 76
8. Transportation 92
9. Health Examination and Recording of Vital Signs 96
10. Wildlife Diseases: An Over View 102
11. Major Infectious Diseases of Wild Animals 107
12. Major Non Infectious Diseases of Wild Animals 118
13. Rescue and Critical Care Management 122
14. Post Mortem Examination 127
15. Forensic and Veterolegal Aspects of Crime Investigation 132
16. Zoonotic Diseases 139
17. Disease Control Techniques in Free Living Wild Animals 147
18. Field Approch for Disease Diagnosis from Dead Animals 152
19. Ex-Situ Captive Management in Zoos 154
20. Housing of Captive Animals 165
21. Hygiene and Sanitation 172

22. Collection, Preservation and Dispatch of Biomaterials 177
23. Nutrition of Zoo Animals 185
24. Orphan Management 197
25. Disaster Management 211
26. Surgical Problems and Intervention 217
27. Medical Management of Reptiles 242
28. Exotic Bird Health Care and Management 248
29 Wildlife Disease Surveillance 253
30 Advance Surveillance Technologies 262
31. Wildlife Protection Act 268
32. Scheduled Wildlife of India 286
Appendices 354

Preface

The book of Basics of Wildlife Healthcare and Management was a long thought from my involvement in wildlife field. During study at Wildlife Institute of India, Dehradun and thereafter at USA study tour meeting with many renowned wildlife veterinarians and biologist including Dr.Murry Fowler, Dr.Scott Citino, Dr.F.J.Dein and many more realized me to do little contribution for the Indian wild animal health care and management. Upon introduction of wildlife course for veterinary graduate, the book has been planned looking its theme and basic understanding of the various topics like wildlife, status of animals on taxonomical and IUCN level, biological variations,ethology, their nutrition, ex- situ and insitu conservation, restrain, handling, orphan management, captive management, zoonotic diseases, forensic and legal understanding, surgical interventions and conditions, reproductive base line information, rescue operation, critical care management, various forms required for legal procedure, post mortem techniques and collection,preservation and dispatch of biomaterials, medical management of reptiles, exotic birds health care and management have been covered along with preparedness of disaster management with a view of holistic approach.

I am highly thankful to several authors from them I have compiled some of the matter. The basic purpose of this book is to have awareness on healthcare and management for the undergraduate, post graduate, teachers, wildlife workers, foresters, zoo veterinarians and many more who loves wild animals and do little for their conservations. I am very grateful to all who directly or indirectly supported me for this outcome.

Dr. Rajesh Jani

Coordinator
Wildlife Health
Western Regional Center
Vet.College, Anand
Gujarat

Acknowledgements

On the execution of the book on **Basics of Wildlife Health Care and Management**, I express my gratitude to all those who have helped me in my field of wildlife, taught and directed me something or just generally made the going easier and gave me the prospect to complete this outcome.

I feel privileged to acknowledge the founder mentor Dr.K.N.Vyas, retired Dean and the then Director of Campus of Gujarat Agricultural University, Anand who initiated me for wildlife along with the founder supporter Dr. Pradeep Malik, Scientist,Wildlife Institute of India, Dehradun and my overseas mentor Dr. F.Joshua Dein, Scientist, National Wildlife Health Center, Madison and Faculty of Wisconsin University, USA for wildlife study tour. The advice and nurturing from Dr.B.M.Arora for encouragement to work in wildlife is highly appreciated. I owe him immense gratitude for all support and boosting my career. It was indeed a valuable opportunity for me to pursue my PG Diploma in Wildlife Management under meticulous, affectionate and brilliant personalities and the faculties and scientists of Institute from Dehradun as well as from overeas is duly acknowledged.

A few lines are too short to account for my deep gratitude towards my colleagues Dr.A.B.Srivastava, Dr.Naseer Ahmed, Dr.M.G.Jayathangaraj, Dr.S.K.Mishra, Dr.Mark Drew, Dr.Jim Sikarski, Dr.Julia, Dr.Scott Citino, Dr.Tracy, Dr.D.Swarup and many more who encouraged guided and nurture my developing period in wildlife. I thank them for their kind support, guidance, indispensable suggestions, untiring help and splendid assistance. Their extensive discussions, company around my work and interesting explorations have been of great value in this field. I also thank my guide, Guru and Ex.Professor and Head Dr. P.R.Patel.

I express my heart-felt gratitude to Several wild lifers and officers like Ashwin Parmar, Late Mr.P.P.Raval, Dr..H.S.Singh, Bharat Pathak, Mr. D.K.Tipre, ,Mr. B.S.Bonal, Ex.Member Secretary, CZA, MR.K.K. Gupta, CZA, Mr. R.D.Katara, Mr.V.J.Rana, Dr.J.H.Desai, Mr. D.C. Mangarola, Zoo Veterianry officers like Dr.C.N.Bhuva, Dr. B.M. Bhadesiya, Dr.R.K. Sabapara, Dr.M.G.Maradia, Dr.R.K.Sahu, Dr.C.B.Patel, Dr.P.C.Mehta, Dr.R.H.Hirpara,,late Dr.T.K. Telang for their cordial support and encouragement, guidance and ever willing help. I extend my sincere thanks to Mr. Rajesh Gopal, for allowing and providing me to use some technical guidance.

I to a great extent thank to my distant invaluable friends Dr.J.L.Singh, Dr. A.K.Mathur, Dr. Parag Nigam, Dr.Paneer Selvam, Mr.Sameer Sinha and all my diploma batch mates of WII for their well wishes and blessings.I wish to express my sincere thanks to my parent departmental staffs for the timely support .The technical support and help from Department of Surgery is also acknowledged.

My heartiest thanks to my junior colleague's Drs. Dhaval Fefar, Jatin Patel, Dhara Patel, Rafi Mathakia, Sarita Devi, Roon, Sunetra, Chitra, Satish, Seemanthini and many more for their support and help.

My parents,wife, son and daughter along with othe family members especially Jayesh, Minal, Vidhi, Yagna, Om, Hiru, and David boon - Raxa are also appreciated for their cordial love.

The special thanks to Narendra Publishers without their periodical reminders this output may not be in your hand. Finally, I will appreciate any suggestions, modification and updates from your feedback for this book.

Anand **(Rajesh Jani)**

Date: 07 / 06 /2020

References Acknowledged

I am highly enriched by referring the following book and used some of the matter from the following books. I respect all the authors and appreciate their contribution in the field of wildlife management.

References

Arnell, C. and Keymer, I.F. (1975). **Bird Diseases**, Neptune, NJTFH Publication.

Boever, W.J. (1979). Restraints of Non domestic pets. Vet.Clin.North Am.J.9: 391-405.

Fowler, M.E.(1986).**Zoo and wild animal Medicine**.2nd edn., W.B.Saunders Co. Philadelphia.

Giles.R.H.(1978) Wildlife Management,1st Ed. ,WH Freeman and Company,San Franscisco.

Joshi,B.P.(1991)Wild Animal Nmedicne,1st edn.Oxford & IBH Publishing co.Pvt.Ltd.,New Delhi-1.

Prater,S.H.(1993)The Book on Indian Animals,1st.ed.BNHS,Bombay.

Rajesh Gopal(1978) Fundamentals of Wildlife Management,1st Ed.Justice House,Allahabad.

Wallach, J.D. and Boever, W.J. (1983). **Diseases of Exotic animals**. 1st edn.

W.B.Saunders Co. Philadelphia.

CHAPTER 1

INDIAN WILDLIFE

India's biodiversity is both rich and varied. The heritage of vivid wildlife of India attracts the world tourist and wildlifers for its magnificient flora and fauna. The country is one of the 12 mega diversity areas in the world, in terms of animal. India is unique in the richness and diversity of its vegetation and wildlife. Almost 340 plus mammal species, over a thousand and two hundred species of birds in nearly 2100 forms and more than 30,000 species of insects - provide evidence to the wealthiness of wildlife in India. Besides, there are a number of species of fish, amphibians and reptiles.With over 4.5% of its geographical area covered by more than 89 national parks and 489 sanctuaries, the range and diversity of India's wildlife heritage matches the grandeur and magnificence of her civilization.

India's national parks and wild life sanctuaries from Laddakh in Himalayas to Southern tip of Tamilnadu. These parks, reserves, sanctuaries and forests are vital to the conservation of endangered species, such as Bengal tiger, the Asiatic Elephant, Lion, the Snow Leopard and Saras Crane. India's first national park, the Corbett was established in the foothills of Himalayas. It supports a great variety of mammals and over 585 species of birds. The wild elephant population is on the increase and both tiger and leopard are regularly seen. Kanha National Park is the largest of the original tiger reserves. The park is noted for its local herd of swamp deer. Also a species of the swamp deer found in Kanha A third subspecies (and the largest population) of swamp deer is at the Dudhwa National Park in the northeastern UP. The magnificent bird sanctuary at Bharatpur Provides a vast breeding area for the native water birds.

During the winters (November-March) migratory birds arrive in large numbers, including the Siberian Crane. In the Indian deserts, the most discussed bird is the Great Indian bustard. In western Himalayas, one can see birds like Himalayan monal pheasant, western tragopan, white crested khalij cheer pleasant and griffon vultures, In the Andaman and Nicobar region, about 250 species and sub species of birds are found, such as rare Narcondum hornbill, Nicobar pigeon and megapode. Here are also other birds like white bellied sea eagle, white breasted swiftlet and several fruit pigeons. All these could be observed in Andaman's six national parks and over ninety wildlife sanctuaries.

The Himalayas (foothills) are known for big mammals like elephant, sambar, swamp, deer, cheetal, wild boar tiger, panther, hyena, black bear, sloth bear, porcupine, Great Indian one horned rhinoceros, wild buffalo, gangetic gharial, golden langur. Wild ass, sheep, deers, smaller mammals, snow leopards, wolf, cats and brown bears are in plenty in the western Himalayas. While the national park and sanctuaries of northern and central India are better known, there are quite a few parks and sanctuaries in South India, too, e.g., Madumalai in Tamil Nadu and Bandipur Tiger Reserve and Nagarhole National Park in Karnataka.

A glimpsis of western zone thrills us with countries richest wild carnivores such as the king of beast Asiatic lion in Greater Gir, tigers in Ranthambhore and Sariska of Rajasthan, leopard, wolf, hyena, sloth bears,wild ass,great Indian bustard,saras crane and the visit of flamingo city of Kutch region of Great Gujarat remind us the richness of biodiversity tour of Indian wildlife sanctuaries and national parks as a favours travelmasti.

WHAT IS WILDLIFE?

Wildlife is a term that does not enjoy a precise or a universally accepted definition. The term implies all things that are living outside direct human control and therefore includes those plants and animals that are **not cultivated or domesticated**. In its fullest meaning, wildlife encompasses insects and fungi, frogs and wild flowers, as well as doves, deer, and trees. Nonetheless, organizations concerned with wildlife generally favor the so-called *higher* forms of animal life.

WHAT IS WILDLIFE MANAGEMENT?

Management means judicious use of the available resources. In general, **wildlife management** is the application of ecological knowledge to populations of vertebrate animals and their plant and animal association in a manner that strikes a balance between the needs of those populations and the needs of people, until the 1960s wildlife management was primarily game management, the husbandry and regulation of populations of birds and mammals hunted for sport. Wildlife management is changing, but its past remains relevant to the present and future. The practice of wildlife management is rooted in the intermingling of human ethics, culture, perceptions, and legal concepts.

IMPORTANCE OF WILDLIFE

1. Ecological value
2. Scientific value
3. Economic value

4. Genetic resources (Conservation)
5. Biological diversity
6. Pleasure value
7. Asthetic Value
8. Game values
9. Ethical value

1. **Ecological value :** Conservation of life maintains a balance of nature through biogeochemical cycles, food chains, population controls by positive and negative feedback. If a species is lost in long run, it may upset the natural balance and as a consequence makes the system vulnerable. **A network involving in the interactions of living and nonliving elements in a manner that sustains life- is called an ecosystem;** Living organisms borrow oxygen, carbon dioxide, and nutrients from the ecosystem and then return these materials through the processes of respiration, excretion, and decomposition. The living part of an ecosystem, at any given times and places, is known as the **biotic community,** or more simply the **community**. The tiger , lions can be considered as at apex and as a top predator who depends on the base of the pyramid.
2. **Scientific and educational value :** to study the normal anatomy, physiology and body related data, efficacy and administration of drugs, genetically for population modelling etc.
3. **Aesthetic and ethical value :** On welfare base and for aesthetic and ethical value one should conserve the wild life.
4. **Conservation:** To propagate a specific species and to translocate, relocate and reintroduction programme.
5. **Recreational and economical value :** By keeping wild animals in captivity the revenue can be generated for fund raising and maintaining the zoos and can be also use as study material for conservation of a specific species.

HOW TO CONSERVE THE WILDLIFE?

Conservation (Maintenance and propagation of a viable population of a species) can be done either by

1. **In situ :** means keeping the wild animals in the nature either in National Park, wildlife sanctuaries or in wild status in any protected areas.
2. ***Ex situ*** **:** means keeping the wild animals either in captivity in close observation in a specific enclosure or in a wider area of captivity , safari park or in advanced form of *frozen zoos*.

WILDLIFE ECOSYSTEMS AND NATURAL COMMUNITIES

Civilization is a state of mutual and interdependent cooperation between human animals, other animals, plants and soil, which may be disrupted at any moment by the failure of any of them. Each of the species we call wildlife participates in a vast network of life- a system in which nonliving elements are brought into the tissues of living organisms. These elements then undergo exchanges between plants and animals and finally again enter the physical environment. Such **'a network involving in the interactions of** living and nonliving elements in a manner that sustains life- is called an ecosystem, Living organisms borrow oxygen, carbon, dioxide, and nutrients from the ecosystem and then return these materials through the processes of respiration, excretion, and decomposition. The living part of an ecosystem, at any given time and place, is known as the **biotic community,** or more simply the **community.** An ecosystem usually consists of several communities, each having distinctive groups of plants and animals. Nonetheless, whether dominated by bacteria, trees, amoebas. or whales, communities are identifiable association of plants and animals living in a finite physical environment.

TYPES OF ECOSYSTEMS

A. Natural Ecosystems :

1. **Terrestrial :** Forest, grassland, desert etc. These operate under natural conditions without any major interference by man.
2. **Aquatic :** Fresh water
   a. **Lotic :** Running water as spring, stream or rivers.
   b. **Lentic :** Standing water as lake, pond, pools, puddles,ditch, swamp, etc.
   c. **Marine** : Deep bodies **as ocean or shallow ones** as seas **or** an estuary, etc.

B. Artificial (Man - engineered) Ecosystems : Man-engineered ecosystems: Crop, Urban, Cropland, Spacecraft etc.

Microecosystem : The complete ecological system of an area, including the plants, animals and the environmental factors is known as ecosystem. The ecosystem as that approach, in which habitat, plants and animals are all considered as one interacting unit; materials and energies of one passing in and out of the others.

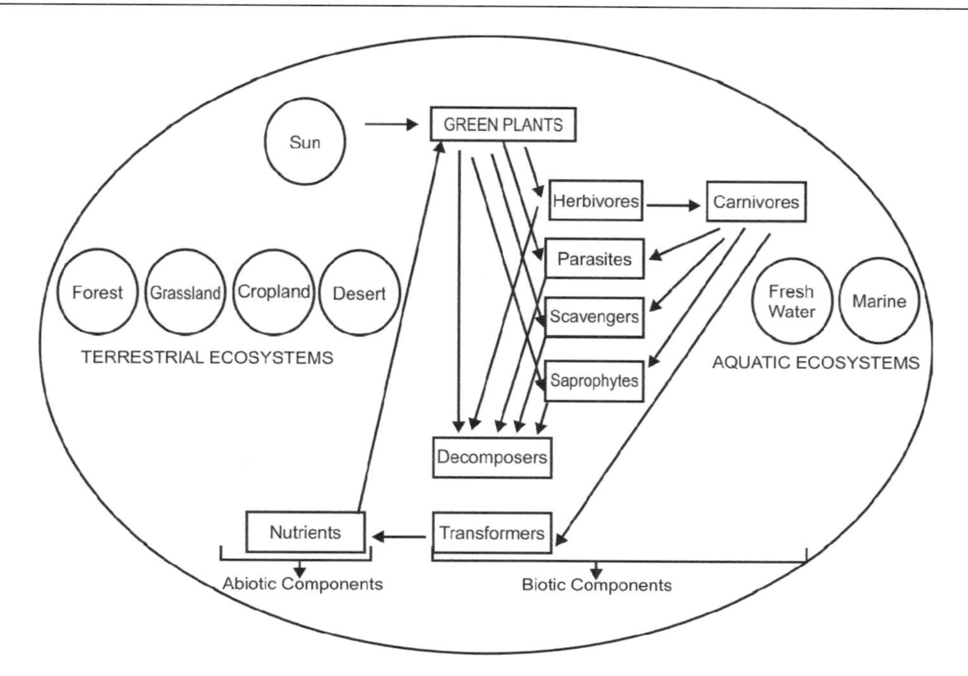

Diagrammatic Representation of Biosphere

Table 1. Biogeographic Classification of India

	Biogeographic Zone	Biotic Province	Some of major wild fauna
1.	Trans-Himalayan	Tibetan	Snow leopard, Musk deer,
2.	Himalayan	North West, West, Central and East Himalaya	Elephant ,leopard, goral, clouded leopard, rhinoceros Manipur deer, wild buffaloes, pheasant birds and reptiles
3.	Desert	Kutch, Thar, Rajasthan	Wolf, foxes, jackals, hyaena, tiger, leopards, jungle cats
4.	Semi- arid	Punjab, Gujarat-rajwada,	Cheetal, leopards, nilgai, wild cats
5.	Western Ghat	Malabar Coast, Mountains of Western Ghat	Elephant, gaur, tiger, leopard, lion-tail macaque
6.	Deccan Peninsula	Southern , Central and Eastern Deccan Plateau, Central high lands, Chhota-Nagpur area	Tiger, leopard, lion(confined to Gir),wild ass, wild dogs, sambhar, barasingha
7.	Gengetic Plain	Upper and lower Gengetic plain	Sambhar, tiger, leopard, peafowl, jungle fowl
8.	North-East India	Brahmaputra Valley, Assam hills	Elephant, rhinocerous, gibbons, flying squirrels
9.	Islands	Andaman, Nicobar and Lakshadweep islands	Marine mammals, local and migratory birds, Andaman pig
10.	Coasts	West and East Coast	Marine creatures, turtles

Conservation area: Conservation Areas are areas of notable environmental or historical interest or importance which is protected by law against undesirable changes. These areas are conserved by varying levels of legal protectionwhich are given by the policies formulated by the government or global conventions.

Following are the major Conservation Areas in India:

Table 2. Important Conservation Sites in India (as on February, 2019)

Reserves/ Sites	**Numbers**	**Total area (in Sq.Kms.)**
Tiger Reserves	50	71027.10
Elephant Reserves	32	69,582.80
Biosphere Reserves	18	87491.6
RAMSAR Wetland Sites	27	11121.31
Natural World Heritage Sites	07	11755.84
Cultural World Heritage Sites	28	—
Mixed World Heritage Sites	01	1784.00
Important Coastal and Marine Biodiversity Areas	107	10773.07
Marine Protected Areas	131	9801.13
Important Bird Areas	467	—
Potential Important Bird Areas	96	—
Key Biodiversity Areas	531	—
Biodiversity Heritage Sites	9	—

Geographical Area of India (http://knowindia.gov.in/) = 32,87,263 km^2

Forest cover of India (FSI, 2017) = 7,08,273 km^2

Percentage Area under Forest cover = 21.54 % of Geographical Area of India

(**Source:** Forest Survey of India)

Table 3. Protected Areas of India (July, 2019)

	Total Area (km^2)	**Coverage % of Country**	
National Parks (NPs)	104	40501.13	1.23
Wildlife Sanctuaries (WLSs)	551	119775.80	3.64
Conservation Reserves (CRs)	88	4356.49	0.13
Community Reserves	127	525.22	0.02
Protected Areas (PAs)	870	165158.54	5.02

What is a National Park?

An area, whether within a sanctuary or not, can be notified by the state government to be constituted as a National Park, by reason of its ecological, faunal, floral, geomorphological, or zoological association or importance, needed to for protecting & propagating or developing wildlife therein or its environment.

No human activity is permitted inside the national park except for the ones permitted by the Chief Wildlife Warden of the state under the conditions given in CHAPTER IV, WPA 1972.

There are 104 existing national parks in India covering an area of 40501.13 km2, which is 1.23% of the geographical area of the country (National Wildlife Database, May, 2019).

Table 4. National Parks of India

State	S.No.	Name of State/ Protected Area	Year of Establishment	Area (km^2)
Andaman & Nicobar Islands	1	Campbell Bay NP	1992	426.23
Andaman & Nicobar Islands	2	Galathea Bay NP	1992	110
Andaman & Nicobar Islands	3	Mahatama Gandhi Marine (Wandoor) NP	1983	281.5
Andaman & Nicobar Islands	4	Middle Button Island NP	1987	0.44
Andaman & Nicobar Islands	5	Mount Harriett NP	1987	46.62
Andaman & Nicobar Islands	6	North Button Island NP	1987	0.44
Andaman & Nicobar Islands	7	Rani Jhansi Marine NP	1996	256.14
Andaman & Nicobar Islands	8	Saddle Peak NP	1987	32.54
Andaman & Nicobar Islands	9	South Button Island NP	1987	0.03
Andhra Pradesh	1	Papikonda NP	2008	1012.86
Andhra Pradesh	2	Rajiv Gandhi (Rameswaram) NP	2005	2.4
Andhra Pradesh	3	Sri Venkateswara NP	1989	353.62
Arunachal Pradesh	1	Mouling NP	1986	483
Arunachal Pradesh	2	Namdapha NP	1983	1807.82
Assam	1	Dibru-Saikhowa NP	1999	340
Assam	2	Kaziranga NP	1974	858.98
Assam	3	Manas NP	1990	500
Assam	4	Nameri NP	1998	200
Assam	5	Rajiv Gandhi Orang NP	1999	78.81
Bihar	1	Valmiki NP	1989	335.65
Chhattisgarh	1	Guru Ghasidas (Sanjay) NP	1981	1440.705
Chhattisgarh	2	Indravati (Kutru) NP	1982	1258.37
Chhattisgarh	3	Kanger Valley NP	1982	200
Goa	1	Mollem NP	1992	107
Gujarat	1	Vansda NP	1979	23.99
Gujarat	2	Blackbuck (Velavadar) NP	1976	34.53
Gujarat	3	Gir NP	1975	258.71
Gujarat	4	Marine (Gulf of Kachchh) NP	1982	162.89

[Table Contd.

Contd. Table]

State	S.No.	Name of State/ Protected Area	Year of Establishment	Area (km^2)
Haryana	1	Kalesar NP	2003	46.82
Haryana	2	Sultanpur NP	1989	1.43
Himachal Pradesh	1	Great Himalayan NP	1984	754.4
Himachal Pradesh	2	Inderkilla NP	2010	104
Himachal Pradesh	3	Khirganga NP	2010	710
Himachal Pradesh	4	Pin Valley NP	1987	675
Himachal Pradesh	5	Simbalbara NP	2010	27.88
Jammu & Kashmir	1	City Forest (Salim Ali) NP	1992	9
Jammu & Kashmir	2	Dachigam NP	1981	141
Jammu & Kashmir	3	Hemis NP	1981	3350
Jammu & Kashmir	4	Kishtwar NP	1981	425
Jharkhand	1	Betla NP	1986	226.33
Karnataka	1	Anshi NP	1987	417.34
Karnataka	2	Bandipur NP	1974	874.2
Karnataka	3	Bannerghatta NP	1974	260.51
Karnataka	4	Kudremukh NP	1987	600.32
Karnataka	5	Nagarahole (Rajiv Gandhi) NP	1988	643.39
Kerala	1	Anamudi Shola NP	2003	7.5
Kerala	2	Eravikulam NP	1978	97
Kerala	3	Mathikettan Shola NP	2003	12.82
Kerala	4	Pambadum Shola NP	2003	1.318
Kerala	5	Periyar NP	1982	350
Kerala	6	Silent Valley NP	1984	89.52
Madhya Pradesh	1	Bandhavgarh NP	1968	448.85
Madhya Pradesh	2	Dinosaur Fossils NP	2011	0.8974
Madhya Pradesh	3	Fossil NP	1983	0.27
Madhya Pradesh	4	Indira Priyadarshini Pench NP	1975	292.85
Madhya Pradesh	5	Kanha NP	1955	940
Madhya Pradesh	6	Madhav NP	1959	375.22
Madhya Pradesh	7	Panna NP	1981	542.67
Madhya Pradesh	8	Sanjay NP	1981	466.88
Madhya Pradesh	9	Satpura NP	1981	585.17
Madhya Pradesh	10	Van Vihar NP	1979	4.45
Maharashtra	1	Chandoli NP	2004	317.67
Maharashtra	2	Gugamal NP	1975	361.28

[Table Contd.

Contd. Table]

State	S.No.	Name of State/ Protected Area	Year of Establishment	Area (km^2)
Maharashtra	3	Nawegaon NP	1975	133.88
Maharashtra	4	Pench (Jawaharlal Nehru) NP	1975	257.26
Maharashtra	5	Sanjay Gandhi (Borivilli) NP	1983	86.96
Maharashtra	6	Tadoba NP	1955	116.55
Manipur	1	Keibul-Lamjao NP	1977	40
Meghalaya	1	Balphakram NP	1985	220
Meghalaya	2	Nokrek Ridge NP	1986	47.48
Mizoram	1	Murlen NP	1991	100
Mizoram	2	Phawngpui Blue Mountain NP	1992	50
Nagaland	1	Intanki NP	1993	202.02
Odisha	1	Bhitarkanika NP	1988	145
Odisha	2	Simlipal NP	1980	845.7
Rajasthan	1	Desert NP	1992	3162
Rajasthan	2	Keoladeo Ghana NP	1981	28.73
Rajasthan	3	Mukundra Hills NP	2006	200.54
Rajasthan	4	Ranthambhore NP	1980	282
Rajasthan	5	Sariska NP	1992	273.8
Sikkim	1	Khangchendzonga NP	1977	1784
Tamil Nadu	1	Guindy NP	1976	2.82
Tamil Nadu	2	Gulf of Mannar Marine NP	1980	6.23
Tamil Nadu	3	Indira Gandhi (Annamalai) NP	1989	117.1
Tamil Nadu	4	Mudumalai NP	1990	103.23
Tamil Nadu	5	Mukurthi NP	1990	78.46
Telangana	1	Kasu Brahmananda Reddy NP	1994	1.43
Telangana	2	Mahaveer Harina Vanasthali NP	1994	14.59
Telangana	3	Mrugavani NP	1994	3.6
Tripura	1	Clouded Leopard NP	2007	5.08
Tripura	2	Bison (Rajbari) NP	2007	31.63
Uttar Pradesh	1	Dudhwa NP	1977	490
Uttarakhand	1	Corbett NP	1936	520.82
Uttarakhand	2	Gangotri NP	1989	2390.02
Uttarakhand	3	Govind NP	1990	472.08
Uttarakhand	4	Nanda Devi NP	1982	624.6
Uttarakhand	5	Rajaji NP	1983	820
Uttarakhand	6	Valley of Flowers NP	1982	87.5

[Table Contd.

Contd. Table]

State	S.No.	Name of State/ Protected Area	Year of Establishment	Area (km^2)
West Bengal	1	Buxa NP	1992	117.1
West Bengal	2	Gorumara NP	1992	79.45
West Bengal	3	Jaldapara NP	2014	216.51
West Bengal	4	Neora Valley NP	1986	159.89
West Bengal	5	Singalila NP	1986	78.6
West Bengal	6	Sunderban NP	1984	1330.1

(**Source:** National Wildlife Database, Wildlife Institute of India)

What is Wildlife Sanctuary ?

Any area other than area comprised with any reserve forest or the territorial waters can be notified by the State Government to constitute as a sanctuary if such area is of adequate ecological, faunal, floral, geomorphological, natural. or zoological significance, for the purpose of protecting, propagating or developing wildlife or its environment. Some restricted human activities are allowed inside the Sanctuary area details of which are given in CHAPTER IV, WPA 1972.

There are 551 existing wildlife sanctuaries in India covering an area of 119775.80 km2, which is 3.64 % of the geographical area of the country (National Wildlife Database, May, 2019).

Table 5. Protected Areas of India

State	National Parks	Wildlife Sanctuaries	State	National Parks	Wildlife Sanctuarie
Andaman & Nicobarlslands	9	96	Maharashtra	6	42
Andhra Pradesh	3	13	Madhya Pradesh	10	25
Arunachal Pradesh	2	11	Manipur	1	2
Assam	5	18	Meghalaya	2	4
Bihar	1	12	Mizoram	2	8
Chandigarh	–	2	Nagaland	1	3
Chhattishgarh	3	11	Odisha	2	19
Delhi	–	1	Pondicherry	-	1
Daman,Div	–	1	Punjab	–	13
Goa	1	6	Rajasthan	5	25
Gujarat	4	23	Sikkim	1	7
Haryana	2	8	Tamil Nadu	5	299

[Table Contd.

Contd. Table]

State	National Parks	Wildlife Sanctuaries	State	National Parks	Wildlife Sanctuarie
Himachal Pradesh	5	28	Telangana	3	9
Jammu & Kashmir	4	15	Tripura	2	4
Jharkhand	1	11	Uttarakhand	6	7
Karnataka	5	30	Uttar Pradesh	1	25
Kerala	6	17	West Bengal	6	15
Lakshadweep	-	1			
National Status*				104	551

Number of species of different group of animals found in India has been documented in several books, which includes as per Table 3.

Table 6. Number of species in different groups

Group	*Number of Species*
Mammals	340+
Birds	1200+
Reptiles	420+
Amphibians	140+
Fishes	2000+
Insects	50,000+
Moluscs	4000+
Other invertebrates	20,000+

CHAPTER 2

TAXONOMICAL CLASSIFICATION OF WILD LIFE

The word animal is ordinarily used to warm-blooded homoiothermic furred creatures belonging to the group known as mammals. Mammals are a major group or class of animal kingdom. Their main characters are warm blood, backbone and ability to suckle the young one. The great Swedish naturalist (Carolus Linnaeus, 1707¬1778) who laid the foundation to the modern system of classification divided the class mammalia into three subclasses.

All these species are widely distributed in different biogeographic zones of the country. The important species of Indian wild animals are mentioned taxonomically here for the basic information.

MAMMALS

As mentioned in table 3 (Chapter 1), about 300 plus species of mammals under class mammalia (major three subclass; monotreme (spiny ant eater), Marsupialia (e.g.Kangaroo, koala bears) and eutheria (total 16 orders) out of them, 12 different orders have been reported in India (Prater, 1957).

CLASS - MAMMALIA
Oviparous Mammals (Egg laying)
Order : Monotremata
Duckbills and echidnas
Viviparous Mammals (Produce living young)
Pouched Mammals
Order : Marsupialia
Opossums, kangaroos, wallaroos, wallabies, tasmanian wolf, tasmanial devil, wombat koala, phalangers
PLACENTAL MAMMALS: Eutheria

The following are the orders

1. Edentata: sloths, armadillos, ant-eaters
2. Pholidota: pangolins

3. Insectivora: insect eaters
4. Lagomorpha: pikas, rabbits, hares
5. Rodentia: rodents
6. Chiroptera: bats
7. Cetacea: whales, dolphins, porpoises
8. Carnivora: dogs, weasles, lions
9. Primates: tree shrews, lemurs, monkeys, apes, man
10. Artiodactyla: pigs, camels, deer, giraffes, antelopes.
11. Perissodactyla: horses, tapirs, rhinoceroses
12. Proboscidea: elephants
13. Pinnipedia: seals, sea-lions, walrus
14. Tubulidentata: aardvark
15. Hyracoidea: hyraxes
16. Dermoptera: flying lemurs or colugos
17. Sirenia: manatees, dugong

1. **Edentata :** They have long snout and claws. E.g. sloth.
2. **Pholidata :** They do not have teeth but have a five-clawed digit. e.g. scaly ant eater (Old world) has only one genus (*Manis*). **e.g. Pangolins:** They usually prefer burrows, found in low hills and plains of India. Indian pangolin (*Manis crassicaudata)* and Chinese pangolin (*Manis pentadactyla)* are species of pangolin reported in India.
3. **Insectivores :** They are terrestrial or nocturnal creatures having small pointed teeth. Indian tree shrew, Himalayan tree shrew, long eared hedge hog, pale hadge hog, Indian short tailed mole, white tailed mole, grey musk shrew found in forest areas of country.
4. **Lagomorpha :** They have canine teeth, long hind legs. e.g. Hares,(Black napped, rofous tailed, desert and Himalayan mouse hare are found in India).
5. **Rodentia :** The order Rodentia comprises 32 families and 352 genera. The basic anatomical difference between lagomorpha and rodentia are 1. Os penis present in rodents, absent in lagomorphs, 2. Scrotum posterior to penis in rodents and anterior to penis in lagomorphas. Capybaras (*Hydrochoerus hydrochaeris*) of South America are the largest of rodents, being 1 to 1.3 meters in head and body length and 36 to 50 kg in body weight. In India, the major one includes flying squirrels (Kashmir wooly, large brown, Hodgkin's, grey headed, small travencore, lesser giant, large brown, parti-coloured, hairy footed etc.), giant squirrel (Indian, Malayan, grizzled), orange bellied and hoary bellied Himalayan squirrel, five, dusky, Himalayan and three striped palm squirrel, marmot, gerbil, rat, bandicoot, mouse, vole, porcupine are grouped under this categories.

6. **Chiroptara :** Bats are only mammals capable of true flight. The major species of bats found in India are flying fox, fruit bat, Indian false vampire, Great eastern horse shoe bat, serotine, Indian pipistrelle, short nose fruit bats, common yellow bat, tickells bat, and painted bat are some of bats well distributed in many part of country.
7. **Cetacea :** see marine mammals.
8. **Carnivora :** They are terrestrial, aquatic or arboreal. The major felidae and canidae under this order includes (I). **Felines:** Includes Asiatic lions, tiger, leopard or panther, snow leopard, clouded leopard, caracal, lynx, pallas cat, fishing cat, leopard cat, jungle cat, desert cat. (II).**Civets:** They live in and around the dense jungles. Large Indian civet, small Indian civet, common palm civet, bear civet (Binturong), tiger civet are some of the examples of civets of India. (III). **Mongoose:** Mongooses are well distributed from Kashmir to Kanyakumari, The other species includes small Indian, striped necked, crab eating and brown mongoose. (IV**). Pandas**: red panda live in the temperate forests of central and eastern Himalaya. They are nocturnal. (V). **Bears**: They are distributed from Himalaya to Kanyakumari. Major bears of India include sloth bear, brown bear and Himalayan black bear. They come under Ursidae family. (VI).**Hyaenas**: they are the scavengers of forest. Striped hyena found all over India. (VII).**Canines**: Wolf, jackal, foxes (Red, Indian, Hill, White footed), dhole (Indian wild dog) are well distributed in several part of India. (VIII). **Weasel**; Otter (Common, Smooth Indian, claw - less), marten (stone, yellow throated, nilgiri), badger (Chinese, Burmese, Hog, honey), weasel (Himalayan, pale, yellow bellied, striped back) are some of the examples of these family.
9. **Primates :** The order primate comprises of 11 families and 60 genera. Indian monkeys are belongs to two subfamily of **cercopithecidae** family. The **cercopithecinae** (macaques) and **Colobinae** (langurs). Major species of primates which are found in India are hoolock gibbon or white browed gibbon (north-east), bonnet macaque (Indian peninsula), rhesus macaque (north-east, northern, central India), assamese macaque (foot hill of Himalayas, north-east, sundarban delta), stump-tailed and pig tail macaque (north -eat), lion -tailed macaque (western ghat), common langur (common in India), capped langur (north-east), golden langur (Indo-Bhutan border), nilgiri langur (western ghat), slow loris (north-east) and slender loris (South India).
10. **Artiodactyla :** This diverse order contains all even-toed hoofed mammals, including the swine, peccaries, hippopotamuses and ruminants. Families of ruminants of world with in Artiodactyla includes camelidae, tragulidae (mouse deer), cervidae (true deer), giraffidae (giraffes and okapi), antilocapridae and bovidae (wild buffaloes, bison, wild goats and sheep). The bovidae differs from cervides on the following points..1. *Upper canines* are not present in bovides where as it is present in cervides, 2. *Gall bladder*; bovides posses gall bladder, where is it is absent in all cervides except musk deer. The major species of India includes (a). **Pigs:** Usually found almost all over the country, they prefer to live in grasslands, scrub forest and at times dense forest

also. Viz. Indian wild boar and pigmy hog.(b). **Antelopes**: Chiru, chinkara, black buck, chawsingha and nilgai or blue bull are examples of antelopes found in India. (c). **Deer**: Out of 53 species of deer in world, Indian chevrotain (Mouse deer), musk deer, barking deer, cheetal (spotted deer), hog deer, sambhar, barasingha (swamp deer),thamin (brow antlered deer) and hungul (Kashmir stag) are the native deer species of our country. The sambhar is taller and larger deer. (d). **Bovidae**: Several groups of bovine includes Gaur or Indian bison, bentang, yak, wild buffaloes, urial, nayan, bharal, ibex, markhor, tahr (Himalayan, nilgiri), serow, goral, takin are goat - antelopes.

11. **Perissodactyla :** Perissodactyla is a diverse order containing odd toed hoofed mammals (Burton, 1962). Major wild equine species of India includes Asiatic wild ass of little rann of Kutchh in Gujarat, Tibetan wild ass of trans - Himalayan cold desert and great Indian one horned rhinoceros of swampy wet land areas of north east India.
12. **Proboscida :** Indian elephant (*Elephas maximus*) is relatively smaller than the African elephant (*Loxodonta africans*) along with other several distinguishing features. In past it was distributed in many parts of our country but at present its distribution is confined to Himalayan foot hills, bihar, up, W.Bengal, north east India, orrisa, western ghat and south of Mysore.The great Indian rhinocerous also supports the biodiversity of India.

MARINE MAMMALS

The Cetacea (whale, dolphins and porpoises), Sirenia (sea cows) and Pinnipedia (seals, sea lions and walruses), pinnipedia are not found in Indian seas. Of the eastern and western coastal area of our country Blue whale, finner whale, sea whale, hump backed whale, sperm whale, pigmy sperm hale, common dolphin, red sea bottle nosed dolphine, little Indian porpoise, dugong (sea cow), gangetic dolphine are some of the examples of biodiversity of marine.

BIRDS

In India, 1200 plus species of birds have been documented, which comprises of major 20 orders listed in our country. Most of the birds have keels for attachments of flight muscles known as carinates, but is absent in ratites (Kiwis, ostriches, rheas, cassowaries and emus).

REPTILES

They are poikilotherms, largely oviparous and mostly aquatic in habit. In western countries they are now commonly kept as pets (turtle, terrapins, iguanas, lizards, chameleons, snakes, alligators, geckos, crocodiles). Temperature, humidity, light and variety of food are imporant factors in their life.

The major orders includes chelonia (turtle, tortoises, terrapins), crocodilia (crocodiles, gharials) and squamata, having suborder of serpentes or ophidia and lacertlia (lizards). The important venomous snakes are cobra, Russell's viper, krait and saw scaled viper.

FISHES

In India fishes occur in the rivers, streams, swamps, marshes, lakes and marine habitats from cold Kashmir to Indian Ocean. The major ones include are katla, mirgala, trout, labeo etc. The management includes hygiene, temeperature, pH and O_2 concentration of water.

CHAPTER 3

BIOLOGICAL FEATURES OF WILD ANIMALS

The classifications of the mammals into different groups indicate their relationship with each other genealogically. Their capacity to adapt to their environment is most important. The environment includes the surroundings, the conditions that influence the body forms and the habitat of the animal that it supports. It is characterised most easily by the vegetation. In an enlarged definition environment includes both physical or abiotic and living or biotic environment. The abiotic environment includes the medium of life and the climate. This medium and the climatic conditions regulate and considerably affect the behaviour of the organism. Climatic conditions like the temperature, rainfall, day length, soil, topography all exert influence.

STATUS OF WILD ANIMALS

The wild animals according to their population and habitat, as specified by International Union of Nature and Natural resources (IUCN), based red data book are categories as

(a) extinct (no reasonable doubt that its last individual has died),

(b) extinct in the wild (survive in captivity),

(c) critically endangered (facing extremely high risk of extinction in the wild in the immediate future),

(d) endangered (it is not critically endangered but is facing a high risk of extinction in the wild in the near future),

(e) vulnerable (is facing a high risk of extinction in the wild in the medium term future),

(f) conservation dependent , low risk data deficient and not evaluated.

Salient Features

(a) Prototheria, (Gr. Pratos, first, ther, wild beast), as the name indicates, they are the most primitive of the mammals. Prototheria has only one order the Monotremata (Gr. monos, single, trema atos, a hole). This is the lowest order of mammals having a single opening for both the genital and the digestive organs.

(b) Monotremata is represented by the echidnas (spiny ant-eater) and the duck-billed platypus. They show some reptilian characters and are egg layers. When the young ones hatch the mother suckles them. The milk glands do not have a teat or nipple. The milk exudes from the pores on the skin that are the forerunners of the nipple.

(c) Marsupialia or pouched mammals, represent a further development stage of mammalian evolution. They belong to metatheria. The young ones are born in a very immature stage and find their way into the mother's pouch and remain there until the development is complete. The pouch contains teats for suckling. This order includes kangaroos, wombats, wallabies, opossums and pouched mice.

(d) The Eutheria (Gr. eu, well, ther, beast) includes the placental mammals. As the name indicates they are in the most advanced stage in evolutionary terms. The young ones are retained in the uterus till they reach an advanced stage of development. Placenta provides the nutrition and the oxygen. Placental mammals are considered as success in the evolutionary ladder and reflect diversity and have no less than nineteen orders.

(e) Order Insectivora comprises of small primitive creatures like hedgehogs, shrews and moles. They feed mostly on insects.

(f) Chiroptera (Gr. cheir, hand, pteron, wing) popularly known as bats, are either insect eaters or fruit eaters.

(g) The order Dermoptera ("winged skin") has only two species, the flying lemurs and they are leaf eaters.

(h) Primates are characterised by their ability to grab the objects. They may grab the objects with hand, legs and even with tail. E.g., monkeys of the new world. Their diets are usually mixed: The monkeys, apes, lemurs and human beings and possibly the tree shrews are all primates.

(i) The sloths, armadillos and termite eating ant-eaters, make up the order Edentata. They are seen in Central and South America.

(j) Pholidota, which are known, as pangolins are similar in many ways to armadillos but unrelated and are seen in Africa and Asia.

(k) Rabbits, hares and pikas are considered as rodents and included in Lagomorpha and are mainly herbivores.

(l) Rodentia consists of gnawing animals and is the largest of all the mammalian orders. They include a great variety of animals like mice, rats, guinea pigs, hamsters, porcupines, squirrels and beavers.

(m) Carnivora as the name indicates are flesh-eating animals. They are endowed with claws and the common carnivores are dogs, cats, lion, leopard, tiger, weasels, badgers otters and bears. The aardvark, which is a termite eater, is quite distinct in its anatomy and hence given an order onts own, Tubulidentata.

(n) The elephants both Asian and African with their distinctive trunk and tusk are included in the order Proboscidea. Rock climbing hyrax, once closely related to the elephants is placed in a separate order of its own.

(o) The hoofed animals that are known as ungulates are herbivores.

(p) Those with odd number of toes come under Perissodactyla (Gr. perissos, odd, daktylos, finger/toe). They include the horses, tapirs and rhinoceroses.

(q) The rest of the large herbivores are even toed and called Artiodactyla.(toes; even in number). It includes the cattle, sheep, deer, antelopes, pigs, giraffes, camels, hippopotami and goats.

(r) Sea mammals belong to three marine orders, Pinnipedia (Seals, Sea-lions and Walruses), Cetacea (Whales, Dolphins and Porpoises) and Sirenia (Dugongs and Manatees). They show considerable adaptation to the aquatic habitat.

ORDER PHOLIDOTA

Pangolins or Scaly Ant-eaters of the old world belong to the order Pholidota with only one genus, Manis. Formerly they were classed under Edentata, meaning without teeth. They are seen in Africa and Southeast Asia and have pointed heads with small eyes, long and broad tail, long tongue and no teeth.

ORDER INSECTIVORA

Insectivores are the most primitive placental mammals. 345 species are recognised. They have small narrow pointed snout and eat insects and other invertebrates. They include the Tree Shrews, Hedgehogs, Moles and Ground Shrews. As the name indicates Tree Shrews climb the trees. Moles are adapted for living and finding their food from underground. Hedgehogs and Ground Shrews are mostly terrestrial. Common species found in India are

Indian Tree Shrew (*Anathana ellioti*) Malay Tree Shrew (*Tupaia glis*)

Long Eared Hedgehog (*Hemiechinus auritus*) Eastern Mole (*Talpa micrura*)

Grey Musk Shrew (*Suncus murinus*) .

ORDER CHIROPTERA

It has 951 species. The name is derived from cheir meaning hand and pteron meaning wing. They are the only mammals capable of sustained flight. Usually they rest with their head hanging down. Bats in cold climates are found to hibernate. They are nocturnal and live on night flying insects. Sub-order megachiroptera include all frugivorous bats and that of microchiroptera eat insects. Some bats eat fish, frogs, birds and even other bats. Bats use echolocation to detect their prey and to sense obstacles. Echolocation means the perception of the objects using reflected sound waves, usually high frequency sounds. They use it for orientation and prey location. Their nose and ears are complex in shape. Fruit bats or Flying foxes eat fruits and leaves, food is detected by smell. They are the largest of the bats and have large eyes and dog like head, small ears and a long muzzle. They have better vision than other bats and few use echolocation. Nearly a quarter of the living mammals belong to the group of bats.

The species found in India are Flying Fox (*Pteropus giganteus*), Fulvous Fruit Bat (*Rousettus leschenaulti*), Short Nosed Fruit Bat (*Cynopterussphinx*), Bearded Sheath Tailed Bat (*Taphozous melanopogon*), Indian False Vampire (*Megaderma lyra*), Great Eastern Horse Shoe Bat (*Rhinolophus luctus*), Common Yellow Bat (*Scotophilus heathi*) and Painted Bat (*Kerivoula picta*).

ORDER PRIMATES

Primates are the highest order of mammals, including lemurs, monkeys, anthropid apes and man. This classification probably gives a pride of place for man in the animal kingdom. Physiologically there is nothing superior in primates when compared to the other living organisms. We can call it superior development of the brain and associated higher intelligence. However, intelligence wise lemurs and some monkeys are not much better than some of the lower mammals.

The major distinctive character in primates is the structure of their hands and feet. They are designed with the purpose of grasping objects. This is an adaptation to the particular habits and mode of life of these creatures. The hands of apes, monkeys and lemurs are similar to human beings, but the thumb is opposable to the other fingers. This helps the primates to hand pick and hold objects. Unlike man, hands are their primary organs of locomotion for tree climbing and arboreal movements. Many apes have no thumb at all and in some they are small and useless. This adaptation helps in rapid movement, quick hooking and instant release. Quick progression through the branches may injure a protruded thumb. However unlike apes and monkeys, all lemurs have well developed thumbs and in some, the index finger is poorly developed. Double bones in the fore arm, which are equally developed and free, provide perfect movement for the wrist. The wrist can be turned upward, downward and rotated. The foot is provided with the same facility. The feet of primates have almost the same design as that of hand. The toes are long and flexible. The big toe is highly developed like the thumb and can oppose the other digits for grasping objects. In man the grasping power of feet is lost. Gibbon has an extensively long arm, powerful chest and shoulders and a weak hindquarter, which is well adapted to its type of progression through the trees. In langur arms are not excessively long, legs are longer than the arms and loins and thighs are well developed. They move fast, springing from one branch to another and from tree to tree.

Tail helps them to balance while moving in leaps and bounds. Tails have a variable feature in different primates. In the new world monkeys they are used as an organ of prehension. Apes have no tail and they maintain balance with the help of outstretched arms.

While walking on the ground, or along a branch, gibbon walk erect on the soles of the feet and keep balance with stretched arms. Langurs and monkeys walk and run like any other quadrupeds similar to dogs. The whole palm is pressed to the ground, but not the entire sole, the heel is raised above the surface. Monkeys in general are good swimmers, especially macaques. They swim vigorously in breaststroke style.

Apes, monkeys and lemurs eat, flowers, leaves and fruits. The teeth of these herbivores can grind tough vegetable matter. Most lemurs thrust out their snout for food. Apes and monkeys use hands as prehensile organs to take food to the mouth. Some have large pouch in their cheeks to which they cram food that they cannot immediately eat. They continue to eat even when the pouch is full. Baboons and macaques possess these pouches, but it is not seen in langurs. Stomach of langur is compartmentalised into three pouches, somewhat similar to ruminants. While langurs are herbivores and macaques are omnivores, they eat grubs, spiders and insects. Some even eat lizards and frogs and one of the tribes even eat crabs. Lemurs are nocturnal, but monkeys feed only during the day. They get along very well with other animals. Some ungulates prefer to forage underneath the trees on which monkeys are feeding. Monkeys at times will drop wastefully and intentionally fruits and leaves. Ungulates feeding on the ground in turn eat them.

Main predators of monkeys apart from man are large cats, especially the panther, large snakes and crocodiles. They escape from the predators with well-developed vision, hearing, extreme alertness and agility. Hiding behind the natural cover or concealment by deliberately drawing branches together is a common habit. Most common impulse is escaping by fleeing from danger. Interestingly sometimes they slide down to escape from the predator. Why these arboreal animals come down exposing themselves more to danger and sometimes get killed is yet to be explained. They get protection by living in collective groups, the troop. A threatened attack on any member of the troop draws aggressive reaction from other members. Alarm calls of langurs and macaques, when a large cat is on the prowl are famous. The hunters often notice this for the presence of a tiger or a leopard. An alarm call from anyone of the troop members will send the entire members to bolt without even finding out the reason for the threat. They never use tools in self-defence or to attack animals. They fight to protect themselves or their young ones or to establish dominance over other males for females. The monkeys usually live in the tropical climate; however some langurs and some like Assamese macaques have extended territories in the very cold regions of Himalayas. They are adapted with special winter coats for this purpose. Seasonal movements are influenced by the availability for food.

Monkeys cannot talk, not for want of intelligence but because of the anatomical peculiarity of their voice box. Many birds are capable of imitating human words. They vocalise several communications like pleasure, anger, fear, warning and calls to come together and show distinct facial expressions corresponding to different emotions.

Fur picking is not like the popular belief for hunting lice or ticks.

It is universal and is a form of amotive caress or courtship. Repeated indulgence in fur picking suggests a powerful bond and means of social communication between the members of the troop.

Each troop does not range all over the habitat, but often confine to a specific territory. They may marginally overlap and fight for territory. This is rare in the wild. In urban environment, it is very much seen, often due to the shortage of food. Rhesus monkey is the common example for this type of behaviour. Males dominate the troop and are ranked in a linear manner similar to the pecking order in poultry. Different sub-groups in a certain tribe are established.

All primates give birth to their young ones singly. Immediately after birth the young one cling to its mother's body sucking her teat. It is able to hold fast on to the mother even during quick movements and jumps. While sitting, the mother supports the young one by holding the baby with its arms. Long tailed lemur supports its baby to her body with her tail. When the baby is grown up to crawl on its own, it is carried on the back of the mother. A similar method of carrying the young ones is seen in bats, sloth and armadillos. Needless to say, marsupials carry the young ones in their pouch. Apes and monkeys suckle their young ones for a long time. A baby gibbon is suckled for nearly two years. The mothers do not ordinarily gather food for the young ones. It is interesting to note that the intelligent animals look after the young ones for a considerably longer period of time and have long period of maturity. Great apes and bigger monkeys take anything from six to twelve years for mental development and to become independent. Smaller monkeys take 3-5 years and little lemurs that are physically smaller but with poorly developed intelligence take only two to three years to achieve full-grown status. It is reasoned that intelligent animals take longer period to reach adulthood, because they have to learn the trick of the trade of the adult intelligent world.

A dominant monkey takes tremendous risks to protect its followers. A nursing mother protects the young ones even at the cost of her life. It is noteworthy that a mother continues to carry the dead body of the young one in some instances. This may be due to the fact that it is not able to recognise death and carry around the young one purely on an instinct Indian Primates.

There are apes, monkeys and lemurs in India. None of the great apes are seen in India. Gorilla and Chimpanzee are seen in Africa and Orang-utan in the forests of Borneo and Sumatra. Hoolock Gibbon is the only tribe of ape seen in India in the forests of Assam and Chittagong. As mentioned earlier they are tail-less with well-developed arms that are longer than the legs. Indian monkeys belong to one Family viz. Cercopithecidae and two sub-families Cercopithecinae (Macaques) and Colobinae (Langurs). Macaques are sturdy solid and squat, while the langurs are slim and with long tail. Macaques have cheek pouch while the langurs have a pouched stomach. Among lemurs, only one Family is found in India the Lorisidae or the Lorises.

Hoolock Gibbon (***Hylobates hoolock***) Males are black in colour with white eyebrows. Females are brown. They feed on fruits, leaves and also on insect grubs and spiders. They travel along the top galleries of forest foliage and live in small groups. Young ones are covered with yellow tinted greyish white hair. They stand erect and are more than one metre in height and weigh 6.8 kg on an average.

Primate Groups of South Asia

Among macaques, Rhesus Macaque ***(Macaca mulatta)*** and Bonnet Monkey ***(Macaca radiata)*** are common. The human blood group classification is based on studies on rhesus monkey. They are common in North and Central India. Bonnet monkey has a longer tail compared to that of Rhesus. The temple monkeys of south India are bonnet monkeys and are seen very commonly.

Other group of monkeys are langurs. Common Langur ***(Semnopithecus entellus***, Old name ***Presbytis entellus)*** is distributed all over India and worshipped as Hanuman of Ramayana. However for some people Hanuman is from Deccan and should be Bonnet monkey.

Lion Tailed Macaque ***(Macaca silenus)*** is found only in Kerala, Tamil Nadu and Karnataka. It is endangered and protected in the Silent Valley National Park. It eats lizards, snakes, and insects and also fruits and leaves. Assamese Macaque ***(Macaca assamensis***), Stump Tailed Macaque ***(Macaca speciosa)*** and Pig Tailed Macaque ***(Macaca nemestrina)***.

Other langurs are, Nilgiri Langur (***Trachypithecus johni***, Old name ***Presbytis johni***) as the name implies is seen in Nilgiris in Kerala, Tamil Nadu and Karnataka, Capped

Langur or Leaf Monkey (***Trachypithecus pileatus***, Old name ***Presbytis pileatus***) and Golden Langur ***(Trachipithecus geei***, Old name ***Presbytis geei***).

Golden langur **Capped langur**

The golden *langur,* is endemic to north-east India. The species was discovered in 1955-56. It is confined to a small area of reserve forest between the Manas and Sankosh river along the foot of Bhutan Himalayas. On the north its range further extends into the central Bhutan region. The surviving population of the species comprises around 1,000 individuals, living in herbs of 7 to 15. The species is threatened mainly due to destruction of habitat. Golden *Langur* are distributed in various parts of India. Data of various sanctuaries reveal that they cover total 2, 18,229: Ripu = 60,529, Kachugaon = 21,446, Chirang = 59,254, Manas = 77,000. The body coat of golden *langur* looks uniform deep cream in dull light. The species is endangered due to indiscriminate felling of the trees in the forests of North Assam and north-eastern States has caused the monkeys homeless and subsequently perished in large numbers. Golden langur bas been included in Schedule I (the Entry 10) of the Wildlife (Protection) Act,1972 and also in Appendix I of the CITES.

EXOTIC APES

Apes are anthropoid primates and comprises of lesser apes (Gibbon) and great apes (Orang-utan, Gorilla and Chimpanzee). They differ from the monkeys in not having a tail and in using their arms to swing through the trees.

Orang-utan (*Pango pygmaeus*) is native to forests of Sumatra and Borneo. Height measures about 1.5 m and weighs about 90 kg. Body is covered with sparse long shaggy red brown hair. The arm span is up to 2.25m. In adult males large naked fatty folds form a collar around the face. It is the largest anthropoid ape of Asia. Legs are short and bowed, with knees turned out and feet in. Feeds mainly on the fruits and buds of plants.

Gorilla ***(Gorilla gorilla)*** There are two varieties, the low land one of West Africa and Cameroon and the mountain variety of the Eastern Congo Basin. Gorilla is the largest primate and grows to a height up to 1.8 metres, weighs about 200 kg and has a massive and muscular body. Usually walks on all fours. Adult males have a marked crest and are black except in old males (Silver blacks), which have a silvery grey torso. Gorilla roams the forest during the day in small family groups for fruits and plants and spends the night on the trees.

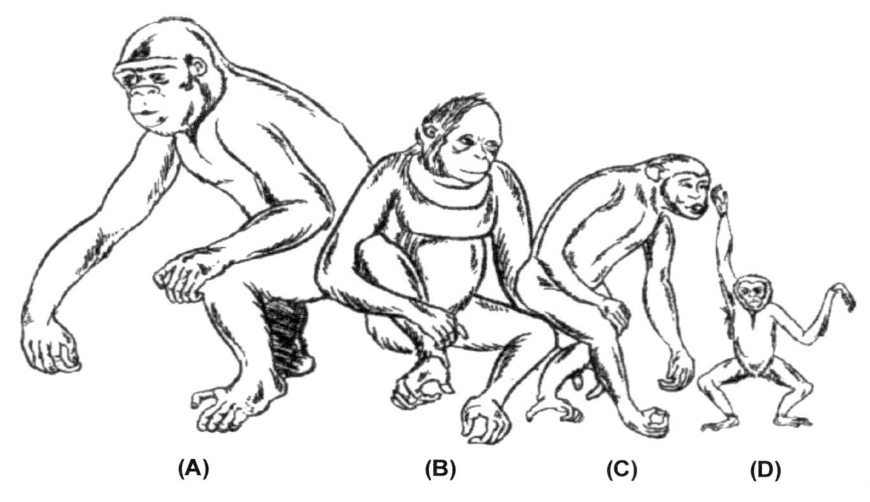

(A) Gorilla, (B) Orang-utan, (C) Chimpanzee, (D) Gibbon

Chimpanzee is native to equatorial Africa, and they are believed to be the closest living relatives to human beings. Height 1-1.7 m. Black coat, hair on head is parted or directed backwards, skin pale and dark with face and ears naked. They are territorial and feed on plants and insects and use small twigs to get food. Two species are identified; Chimpanzee (***Pan troglodytes***) and the smaller black faced Pigmy Chimpanzee or Bonobo (***Pan paniscus***). They produce single ones at birth, may live up to an age of 35 years and are the most intelligent of apes.

Slow Loris

Among lemurs, the two types seen in India are Slow Loris (***Nycticebus coucang***) and Slender Loris (***Loris tardigradus***). They have large eyes and are hence hunted to make products allegedly to improve the eyesight. Slender Loris is seen in south India and Slow Loris in northeastern India and is on the verge of extinction.

ORDER DERMOPTERA

This order includes a primitive primate of Madagascar with large eyes and pointed snout. Most species have a long tail. 27 species in 3 families Lemur (Lemuridae), Mouse or Dwarf Lemur (Cheirogalidae) and Leaping Lemur (lndridae) are described. Colugo or

Flying Lemur is often regarded as a taxonomic puzzle. The head shows features of both lemurs and insectivores. Body is about 40 cm long and 'wings' of skin is stretched between the fore and hind limbs extending up to the shoulders. They are arboreal creatures and more efficient gliders than Flying Squirrels.

ORDER CARNIVORA

It carried cats.overs lion, tiger,leopard,hunting cheetah,clouded leopard,snow lopard,bears and varied cats.The major species are as:

Indian Lion (*Panthera leo persica)*

The Indian (or Asiatic) lion is now almost extinct from Asia except in the Gir Forest (Gujarat). The population of the lion in the Gir forest declined from 290 in 1955 to 177 in 1969. The Gir Lion project was started in 1972. The 15th Asiatic Lion Census could not be conducted in 2020, as scheduled, because of the COVID-19 pandemic. In June 2020, an estimation exercise counted **674 Asiatic lions** in the Gir forest region, an increase of 29 per cent over the 2015 census figure. The Greater Gir Lion National Park which now covered the Girnar sanctuary is well-managed protected areas in the country that provides good facilities for visitors to see these unique animals in their natural habitat.

Characteristics of Indian Lion

Body length 260 cm to 295 (male); 200 cm to 275 cm (female); Tail 30cm; Height 95cm Body weight 250 kg. Colour: The coat of the lion is light brown in colour.The male unlike female has a mane which grows on the head, neck and shoulders, making the lion only cat exhibiting sexual dimorphism. The lion lives in a small group (pride). It consists of up to 20 members and has one or more mature lions and a number of lioness with juveniles or cubs. The African lion (*Panthera leo leo*) differs phenotypically with more weight,height and length without abdominal belly fold which is unique in Asiatic lion(*Panthera leo persica).* The Asiatic lions in most cases about 90 % the belly fold are prominent.Asiatic lion were given special protection by the Newab of Jungadh Sakkar baug Zoo was established in 1863 by the Nawab especially to assist in the conservation of the Gir Lions by collection of ill and injured animals and giving them medical treatment in the zoo. After for breeding lion to supply the world's zoo. This may be the visit zoo in the world to be established with a specific and practical Conservation programme.

Lion prefer family life unlike other large carnivorous which are often solitary. The male Lion is a defender of the prides. Territory which the female hunts and rears the cubs. Gestation period for lions is between 105 and 115 days. A normal from 2 to 4 cubs. The record measurement of an Asiatic Lion is 2.92 meters. While that of African lion 3.23 preferred habitat for lions open scrab forest.

Indian Tiger *(Panthera tigris)*

It is the magnificent and best known animal of Indian wildlife. The population of tigers reduced from 40,000 at the turn of century to 1827 in 1972 due to hunting, habitat destruction by deforestation and taming of the rivers for human needs. It is described as endangered in Red data book. Fit its conservation *Project Tiger* was started on April 1,1973. At present, the project is operating in 27 tiger reserves located in the 14 States of the country. The latest count indicates rise from 2,226 in 2014 to 2,967 in 2018.India is "now one of the biggest and most secure habitats of the tiger". India is now estimated to be home to around 70% of the world's tigers.

Leopard (Panther)- (*Panthera pardus*)

What is the difference between Panther and leopard?

Some call it a panther (*Panthera pardus*), some leopard. Both are right, may call it a Cheetah, that is surely a mistake.It is a different group. The leopard is more often, it not always mistaken foe a Cheetah. The two are quite different, while the coat of a leopard is marked with small close rosettes, that of a Cheetah is oattwerned with solid black spots. Now extinct in India. The Cheetah is slimmer of the two. The leopard has shorter legs and a longer tail. The colour much lighter, the distinct gace of the Cheetah marked by "Tear stripes" running from the corner of its eyes, down the sides of the nose. Differentiates it from a leopard. The leopard is found in closed country, While the Cheetah is an open country, Status is Endangered.The normal weight is 60 kg. (Equivalent to one person). Gestation period is around 90 to 95 Days.There are 13 races of leopards as per Prater.The western region, specifically Gujrat is having highest number of leopards.Their ususal diet is: Antelope, Deer, Rodents, Birds, Monkey's and Fish.

Hunting Cheeta (*Acinonyx jubatas*)

We have had cheetahs before 1951,now extinct from the natural habitat of our country,however in 2009 the Sakkarbaug zoo,Junagadh(Gujarat) brought two pair of cheetah in captivity.They differ from true cats like leopard phenotypically with small black spots on body coat,with long slender angular body, small round skull,long tail,tear canal at inner canthus of eye favouring the fastest runner on the earth with presence of non retractile claws,which is retractile in lion,tiger and in leopards.

ORDER PROBOSCIDA

Indian Elephant *(Elephas maximus)*

They are widely distributed in the Himalayan Terai in northern India and in Kamataka, Kerala and Tamilnadu is southern India. They live in dense tropical and subtropical forests near water. In the country, elephants have been domesticated and trained for hunting, transportations, processions, traveling, circuses and several heavy works. Though the population of Indian Elephant is increasing in past decade (about 8 to 10 per cent). In India wild elephant population is about 18,000 to 20,000. They need protection due to habitat destruction by the local people for firewood, grazing land, forest produce, encroachments and poorly planned development activities. Efforts have been made and Project Elephant was launched recently by the Government of India which will be executed in 11 States of India. It has been notified as endangered species in Schedule I (the Entry 12-B) of the Wildlife (Protection) Act of 1972 and also in Appendix I of the CITES.

The elephants, pachyderms are having maternal higherchy.The basic difference of African and Indian-Asiatic elephants are as

Table 1. Difference of Asiatic and African elephants

Character	Asian Elephant	African Elephant
Biological name	*Elephas maximus*	*Loxodonta africans*
General Appearance	Shorter and round shape	Taller and slimmer body
Height	Average 2 feet shorter (9-10 feet)	Average 2 feet taller (11-12 feet)
Weight	Average 2000 pounds lighter	Average 2000 pounds heavier
Back	Rounded with slight hump in the middle	Have a dip in the middle
Ears	Smaller	Larger and in the shape of the
Head	Consists of two domes	Consists of a single dome
Tusks	Developed only in males.	Developed in both male and female Female tusks too small to be visible outside
Legs	Shorter and stockier	Longer
Trunk	Has one “finger” at the tip	Has two “fingers” at the tip

Indian Rhino *(Rhinoceros unicornis)*

Several species of rhinoceros were surviving on the earth million years ago, and out of them only five of them are surviving today. Rhinos has its economic value. Its flesh and blood are considered highly acceptable food in Nepal. Urine is considered as antiseptic. However, rhinos are mostly killed for their horn which is believed to be a strong aphrodisiac and an antidote for poison.

The great Indian one-horned Rhion *(Rhinoceros unicornis)* at present they are confined to a few pockets of suitable habitat in the Brahamputra valley, and small sanctuaries of North Bengal and the Royal Chitwan National Park in Nepal. The number is reducing due to destruction of the natural habitat and wanton killing that eliminated the species from most parts of its former range. The surviving population is less than 2,000 and is scattered and is on the verge of extinction. These species have been listed as endangered in Schedule I (the Entry 30) of the Wildlife (Protection) Act, 1972 and in Appendix I of the CITES.

Pigmy Hog *(Sus salvanius)*

Pigmy hog is distributed along the Himalayan foot hills from Assam to the terai areas of Uttar Pradesh. The pigmy hog is about 25 cm in height. They live in small herd to 5 to 25 individuals. They are shy and nocturnal. The habitat consists mainly of grasslands with interspersed patches of trees in the moist deciduous forests. However, it has disappeared from most parts of its earlier range as a result of habitat destruction for human settlement and agriculture development. At present, the pigmy hogs are restricted to a few pockets along Assam's border with Bhutan and Arunachal Pradesh. The only areas where viable population of Pigmy hog is found are Manas and Bamadi wildlife sanctuaries in Assam.'The total surviving population of Pigmy hog in the aforementioned two sanctuaries would be around 500 individuals. Destruction of grasslands by converting them to agricultural fields in the terai region coupled with plantation of exotic species are the main causes of disappearance of the species. They have now been included in the first IUCN/WWF 1984 list as one of the twelve most endangered animal species. It is also included as endangered species in Schedule I (the Entry 29) of Wildlife (Protection) Act, 1972 and in Appendix I of the CITES. The species was thought to be extinct until its rediscovery in 1952. Breeding efforts have been made to save the species in the field. World's smallest pig was set up at Basishta under a Pigmy Hog Conservation Programme in 1995. At the Centre, the population of the nearly extinct species has increased by 666 per cent in just 18 months, i.e. 43 pigmy hogs bred in capacity from the original stock of 6 captured from the Manas National Park. The research and breeding programme at the Centre was begun in **1966.**

Hispid Hare (*Caprolagus hispidus*)

The hispid hare shares the same habitat as that of the pigmy hog. The species has disappeared from most parts of its earlier range of distribution as a result of destruction of its natural habitat. It has already been listed in IUCN Red data book. It is an endangered species and has been included in Schedule I (the Entry 11) of the Wildlife (Protection) Act, 1972 and in Appendix I of the CITES. The surviving population of the species is known to exist in Manas and Borandi wildlife sanctuaries and some other suitable habitats in the region. Only 600 individuals of the hispid are available.

ORDER ARTIODACTYLA

This diverse order contains all even-toed hoofed mammals, including the swine, peccaries, hippopotamuses and ruminants. Families of ruminants of world with in Artiodactyla includes camelidae, tragulidae (mouse deer), cervidae (true deer), giraffidae (giraffes and okapi), antilocapridae and bovidae (wild buffaloes, bison, wild goats and sheep).The major species details are as for

Brow-antlered Deer *(Cervus eldi)*

The brow-antlered deer or dancing deer of Manipur is locally called *Sasigai.* Brow-antlered Deer is the endangered species and is perhaps the rarest and most localized mammal sub-species in the world. In Manipur, the species is confined to a small area in Keibul Lamjao National Park. Until 1950, the species was extinct, however, after declaration of Keibul Lamjao sanctuary in 1966 and later as National park (1977) the species could be conserve. The present surviving population of the species in the wild has been reported to be 76 as per census (April 5,1990). About 29 males, 35 females and 12 fawns were reported. About 100 individuals are now available in 14 zoos of India. It has been included in the list of endangered species(Schedule I, the Entry 3 of the Wildlife Protection Act. 1972.

FOUR HORNED ANTELOPE *(Tretacerus Quadricarnis)* Chowshingha

Distribution	:	Peninsular India, South of the Himalauas
Size	:	Soulder height 65 cm.
Weight	:	About 20 kg
Horn	:	Anterior 1 cm to 25 Cms., Posterior 5 cm to 10 cm.
Colour	:	The colour of its course coat is dull red- brown above and white below-old back are yellowish.
Habitat	:	Undulating or hilly area and shelters in tall grass and open jungle.
Gestation	:	8 to 8.5 months.
Characters	:	Chowshingha the only member of this group with two horn pairs of horns the front pair always the shorter. The presence of pair of a well developed glands between the false hooves of the hind legs in both male and female. Horn are not ringed.

Barking Deer *(Muntjac muntiac- Zimmerman)*

The Muntjac or Barking deer

Distribution	:	Indo- Malayan contries China and Japan.
Size	:	Soulder height 50 to 75 Cms.
Weight	:	About 22 kg
Horn	:	5 cm to 17.8 Cms.
Habitat	:	Thicky Wooden hills in Himalayas and S. India. Its occurs up to level of 1500 mt. Tp 2450 mt.
Gestation	:	6 months.

Food : Various leaves, grasses and wild fruits.

Characters : The call from a distance sound much like bark of a dog when alarmed give out a series of short cacking barks. the young usually one some times two are born at the begning of the rains. Horn are shed during May of June. The upper vanines of the male are well developed and are used by the animal in self defence.

Wild Buffalo (*Bubalus bubalis*)

The wild buffalo (commonly known as *Van Bhainsa, is* riverain and grassland species. The animal lives in small herbs in the swampy grasslands. They were once associated with the large rivers and their tributaries in the Gangetic plains, the north-east, and the eastern peninsular India, extending from the riverain grasslands of Rohilkand terai to Assam in the east and Godvari in the sough-east. It is now extinct in the terai of Uttar Pradesh, Bihar and West Bengal, and has become very rare in Orissa and Madhya Pradesh. (The only stronghold of the species at present is the Brahmputra valley in Assam. The wild buffalo is not much different in general appearance from the domesticated one, except for the former it is heavier and more robust in looking. The height of a bull at the shoulder level is around 170 cm, weighing 900 kg. The colour, like the domestic one, is slaty black. They have two types of horns (i) covered upwards in a semi-circle and the tips remain separated by a small gap; and (ii) spread outwards horizontally, slightly curving near the tips. The species has been drastically reduced in numbers, therefore, they have been included in the list of endangered species [Schedule I (the Entry 41) of the Wildlife (Protection) Act, 1972.The surviving population of the species in the area is about 3000 and out of which the major population (about 1000) is found in Manas and Kaziranga National parks. The Kaziranga NP is the most suited location for conservation of this species. The major reason for its disappearance is the non-availability and shrinkage of riverain and grassland habitats. There is no natural enemy of the species except tiger. Incidence of poaching is rare, though some local inhabitants relish its meat.

Indian Crocodiles

There are 22 species of crocodiles in the World. In India, three types of crocodiles are available: the freshwater crocodile or marsh crocodile or mugger (croco*dylus palustris),* saltwater or estuarine crocodile (croco*dylus poiosus)* and gharial *(Gavialis gangeticus).* In India, Orissa State is the home for all the three species of Crocodiles found in India. The commonest of all the three species is th e broad snouted mugger or marsh crocodile, which inhabits all kind of the fresh water habitats. In South India, muggers are found even in tidal influenced saline water. The salt-water crocodile in India is confined to its highly adopted mangrove habitats found in Sunderbans in West Bengal, in Bhitarkanika in Orissa, and Andaman & Nicobar Islands in the Bay of Bengal. The gharial, with a long and slender jaw, once found in the Ganges, Indus and Brahmaputra and in the Mahanadi river system, now it survives in the Chambal river in Madhya Pradesh/Uttar Pradesh/Rajasthan. These species were once abundant in all the major rivers and even ponds, however, they are now threatened animals, therefore, they have been listed in Schedule I, Pt. II (the Entry 1-D and 2) of the Wildlife (Protection) Act of 1972). The decline in population is due to uncontrolled and all season hunting for skin, flesh and sport. Loss of habitat due to construction of dams, diversion of rivers and human interference are some other factors. Efforts have been made to save these species. The Government of India has launched the 'Crocodile Breeding Project' in 1975.

SNAKES

There are over 200 kinds of snakes in India and over 2,500 in the world.In India there are only four common poisonous snakes Cobra, Krait, Russells viper, and saw scaled viper learning to identify these is important.Snakes are less choosy about foot then we are. They are opportunistic feeders. Mostly bigger bakes eat Rats, and smaller snakes eats Mice, Frogs, Birds, Lizards. They rarely take eggs and never drink Milk, some snakes also eat other snakes.The Krait is said to be the most toxic snake in India of all the snakes in the world. The Australian Tiger snake is the most toxic.Only males and female of the same species mate, some like the cobras and kraits lay eggs, others like sand boas, vine snakes and all viper give birth to young ones.Snakes generally live about 10 to 20 years, occasionally to over 30 years.

SOME IMPORTANT AVIFAUNA

White-winged Wood Duck *(Cairina scutalata)*

They are the resident bird of north-east India. They are also distributed in Burma, Thailand, Indonesia, and through Malaysia to Sumatra and Java. In India, the species are found in Buri Dining, Upper Dihing, Joypur, Doom Dooma, Dangari, Saikhowa and Dibru reserve forests in Assam and adjoining forests of Arunachal Pradesh. The species has become drastically reduced due to hunting, collection of eggs and destruction of habitat as a result of large scale clearance for cultivation of tea and for oil exploration. For maintaining the viable population of the species, Dibru and Saikhowa reserve forests with some other adjoining areas have been declared as a sanctuary in 1986. White-winged Wood Duck is listed as endangered species, and has been included in Schedule I, Pt. ID (the Entry 18) of the Wildlife (Protection) Act, 1972.

Hornbills

Hombills belong to Family-Bucerotidae. Hombills are large birds with broad curved bills. They nest in suitable trees. Deforestation poses serious problem to their survival because of scarcity of suitable nesting trees. They occur in India. The common available hombills in India are: grey hornbill *(Tochus birostris)* Malabar grey honrbill (*Tochus griseus).*

White-throated brown hombill *(Ptilolaemus tickellil)*. Rufous-necked hombill *(Aceros nipalensis),* Wrethed hombill *(Rhyticeros unduletus),* Narcondam hombill *((Rhyticeros unduletus),* Giant hombill (Buceros *bicornis),* Indian pied hombill (*Anthracoceros malabaricus),* Malabar pied hombill (*Anthracoceros coronatus).* The area of their distribution are North India in semi-open forests, Western Ghats, evergreen forests; 300-1500 m altitude, deciduous evergreen forests of Sikkim, Arunachal Pradesh, Manipur and Mizoram, evergreen forests along Himalayan foothills sough of Brahmaputra, Narcondam Island of Andaman group, Narcondam Island of Andaman group, heavy forests of Terai upto 2,000 m altitude, Western Ghats, Terai and lower foot hill forests and Plains to 300 m altitude forests of South India. Tribals kill them for the alleged medicinal value of their flesh and bones. Hornbills *(Ptiloaemus tickellil austeni, Aceros nipalensis, Rhyticeros unduletus-ticehursti)* are included as endangered in Schedule I, Pt III (the Entry 4-C) of Wildlife (Protection) Act of 1972.

The Great Indian Bustard *(Choriotis* or *Ardeotis nigriceps)*

The Great Indian bustard (popularly known as *Godawanj* belongs to Family- Otididae. It is one of the largest flying birds in the world. The cock is polygamous. The hen lays one (rarely two) egg per clutch in a shallow depression at the base of a bush. It is a great friend of farmers as it feeds on locust, mice, grasshoppers, beetles, centipeds, small snakes and lizards. Sometimes, it raids field and eats grains and tender shoots of crop plants. The Great Indian bustard is one of the rare species of the birds that wear majestic look. The bird is larger in size, four feet tall with a long white neck and legs. It is a very heavy bird, probably the heaviest Indian bird to be able to fly. It can weigh 20 kg and is sandy-brown and white bird with a black head. Its wings span in almost two and half metres. Earlier it was abundant in the zones of India, as far as south of the Malabar coast and Sri Lanka, is now rare and restricted mainly to the deserts of Rajasthan. The males is larger than female. It is found singly or in pairs or threes in semi-arid regions of Gujarat and Rajasthan; lives in scattered droves of 20 to 30. It is estimated 1,000 Great Indian Bustards in the world and 75 per cent of them are found in Rajasthan (Thar desert, specially in Jaisalmer, Jodhpur and Bikaner. In Rajasthan, large areas of Desert National Park have been fenced so that bustard may not be disturbed during breeding season (July-October). To save the bird in Rajasthan, 'Save Godawan Campaign' has been launched in the desert district of Jaisalmer by the Bombay Natural History Society with assistance from the Salim Alt Centre for Ornithology and Natural History, and WWF. In Madhya Pradesh, the bird is mainly found in Karera sanctuary (202 tan^2) in Shivpuri district. The destructive human activities (hunting) has threatens the species. Its population has declined due to habitat description and therefore, it is considered as endangered. It has been included in IUCN Red data book and in Schedule I, Pt III (the Entry 3) of the Wildlife (Protection) Act, 1972. This is .clearly evident from the growth pattern of the population of the species in Desert National Park, Jaisalmer.

Bengal Florican *(Houbaropsis bengalensis)*

It belongs to the group of bustards and floricans (Family Otididae) of the old world. Its habitat is characterised by *Phyragmites-Sacchamm-Imperata* type of grassland. They are found in the terai grasslands of Nepal, Uttar Pradesh, West Bengal and the Assam valley. Its population has declined sharply in recent years because of destruction of grassland habitat. In the north-east India, the species can still be seen in the grassland habitats scattered between the S ankosh river in the west and upto the Dihing river in the east. In the protected areas of India, it is found in Dudhwa N.P. (UP), Jaldapara (WB), and Manas, Kaziranga, Orang and Pabitora (Assam). A few individuals may be surviving in Barandi and Sonai Rupai sanctuaries. The surviving population of the birds in the protected areas could be 200-300. It has become extinct in Bangladesh but survives in Nepal. As an endangered species, it is included in Schedule I, Pt. Ill (the Entry 1-C) of the Wildlife (Protection) Act, 1972. The Bengal florican, has been notified by the IUCN as being one of the 12 most endangered species of the world.

RED DATA BOOK

IUCN was formed the Survival Service commission, which sought to supply upto date information about every species of wildlife in danger of extinction. The Red Data Book provides information about endangered species. According to an estimate, in last about 2,000 years the world has lost about 160 species of mammals and 88 species of birds, by way of extinction. In India 134 plant species have been declared, threatened as against 10 mammals, 22 reptiles and amphibians and 41 birds.

Endangered species of mammals, amphibians, reptiles and birds

MAMMALS

1. Andaman wild pig *(Sus scrofa & S. amanensis)*
2. Bharal *(Ovis nahura)*
3. Binturong *(Arctictis binturong)*
4. Bison or gaur or mithun *(Bos gaurus)*
5. Black buck *(Antelope cervicapra)*
6. Blue whale *(Balaenoptera musculus)*
7. Brown-antlered deer or Thamin *(Cervus eldi)*
8. Capped langur *(Presbytis pileatus)*
9. Caracal *(Felis caracal)*
10. Cetecean species
11. Cheetah *(Acinonyx jubatus)*
12. Chinese pangolin *(Manis pentadactyla)*

13. Chinkara or Indian gazelle *(Gazella gazelta bennetti)*
14. Chital *(Axis axis)*
15. Clouded leopard *(Neofelis nebulosa)*
16. Crab-eating macaque *(Macaca irus umbrosa)*
17. Desert cat *(Felis libyca)*
18. Dugong *(Dugong dugong)*
19. Fishing cat *(Felis viverrina)*
20. Flying squirrels (all species of genera *Petaurista, Eupetaurus. Belomys, Hylopetes)*
21. Four-horned antelope *(Tetraceros quadricornis)*
22. Gangetic dolphin *(Platanisfa gangetica)*
23. Giant squirrels *(Ratufa macroura. R. indica. R. bicoloi)*
24. Golden cat *(Felis temmincki)*
25. Golden langur *(Presbytis geel)*
26. Gorals *(Nemorhaedus goral, N. hodgsoni)*
27. Himalayan black bear *(Selenarctos thibetanus)*
28. Himalayan brown bear *(Ursus arctos)*
29. Himalayan crestless porcupine *(Hystrix hodgsoni)*
30. Himalayan ibex *(Capra ibex)*
31. Himalayan tahr *(Hemitragus jemlahicus)*
32. Hispid hare *(Caprolagus hispidus)*
33. Hog badger *(Arctonyx collaris)*
34. Hoolock or gibbon *(Hytobates nootock)*
35. Hump-backed whale *(Megaptera nevaeangliae)*
36. Hyaena *(Hyaena hyaena)*
37. Indian elephant *(Elephas maximus)*
38. Indian lion *(Panthera leo persica)*
39. Indian wild ass *(Equus hemionus khur)*
40. Indian wolf *(Canis lupus pallipes)*
41. Kashmir stag or hangul *(Genus elephus langul)*
42. Leaf monkey *(Presbytis mefatopnos)*
43. Leopard cat *(Felis bengalensis)*
44. Leopard or panther *(Panthera pardus)*
45. Lesser or red panda *(Wilurus fulgens)*
46. Lion-tailed macaque *(Macaca silenus)*

47. Loris *(Loris cardigradus)*
48. Lynx *(Felix lynx isabeflinus)*
49. Malabar civet *(Viverra megaspifa)*
50. Malay or sun bear *(Helarctos malayanus)*
51. Marbled cat *(Felis mormorare)*
52. Markhor *(Capra falconeri)*
53. Mouse deer *(Tragulus memi'nna)*
54. Musk deer *(Moscnus moscriiferus)*
55. Nilgai *(Boselaphus tragocamefus)*
56. Nilgiri langur *(Presbytis Johni)*
57. Nilgiri thar *(llemitragus hytocrius)*
58. Otters *(Lutra lutra. L. perspicilata. Aonyx cinerea)*
59. Ovis ammon or nyan *(Ovis ammon riodgsoni)*
60. Pallasts cat *(Felis manuf)*
61. Pangolin *(Manis crassicaudata)*
62. Pig-tailed macaque *(Macaca memestrina)*
63. Pygmy hog *(Sus sufvamus)*
64. Rate! *(Mellivora capensis)*
65. Red fox *(Vulpes vulpes}*
66. Rhinoceros *(Rhinoceros unicornis)*
67. Rusty-spotted cat *(Feffis rubiginosa)*
68. Sambar *(Cervus unicolor)*
69. Serow *(Capricornis sumatraensis)*
70. Sloth bear *(Melursus ursinus)*
71. Sloth bear *(Melursus ursinus)*
72. Slow loris *(Nycticebus coucang)*
73. Small travancora flying squirrel *(Pteromys fuscocapillus)*
74. Snow leopard *(Panthera uncia)*
75. Spotted linsang *(Prionodan pardicolot)*
76. Swamp deer or Gond *(Cervus duvauceli,* all species*)*
77. Takin or mishmitakin *(Budorcas taxicoloi)*
78. Tibetan antelope or Chiru *(Panthelopes hodgesoni)*
79. Tibetan gazelle *(Procapra picticaudata)*
80. Tibetan wild ass *(Equus hemionus kiang)*

81. Tiger *(Panthera tigris)*
82. Urial or shapu *(Ovis vignei)*
83. Wild buffalo *(Bubalus bubalis)*
84. Wild dog or dhole *(Cuon alpinus)*
85. Wild yak (Bos *grunniens)*

REPTILES AND AMPHIBIANS

1. Agra monitor lizard *(Varanus griseus)*
2. Atlantic ridley turtle or Kemp's ridley turtle *(Lepidochelys kempii)*
3. Barred, oval or yellow monitor lizard (*Varanus flaescens)*
4. Estuarine crocodile *(Crocodylus porosus)* Salt water
5. Crocodile *(Crocodylus palustris)*
6. Gharial *(Gavialis gangeticus)*
7. Ganges soft-shelled turtle *(Trionyx gangeticus)*
8. Green sea turtle *(Chelonia mydas)*
9. Hawksbill turtle *(Eretmochelys imbricata inlscata)*
10. Himalayan newt or salamander *(Tryletotriton verrucosus)*
11. Indian egg-eating snake *(Elachistodon westermanni)*
12. Indian soft-shelled turtle *(Lissemys punctata puncfata)*
13. Indian tent turtle *(Kachuga tecta tecta)*
14. Large Bengal minitor lizard (*Varanus bengalesis)*
15. Leathery turtle *(Dermochelys coriacea)*
16. Loggerhead turtle *(Caretta caretta)*
17. Olive-Back loggerhead turtle *(Lepidochelys oivacea)*
18. Tortoise (genera of families of *Testudinidae, Trionvchidae)*
19. Viviparous toads *(Nectophyrynoides* spp.)
20. Water lizard (*Varanus salvatoi)*

BIRDS

1. Andaman teal *(Anas gibberifrons albogularis)*
2. Bazas *(Aviceda jeordoni and A. leuphotes)*
3. Bengal florican *(Eupodotis bengalensis)*
4. Black necked crane *(Grus nigricollis)*
5. *Blood* pheasants *(Ithaginis cruentus tibetanus, Ithaginis cruentus kuseii)*

6. Brown headed gull *(Larus brunnicephalus)*
7. Cheer pheasant *(Catreus wallichii)*
8. Comb duck *(Sarkidiornis melanotos)*
9. Forest spotted owlet *(Athene blewitti)*
10. Great Indian bustard *(Choriotis nigriceps)*
11. Great Indian hornbill *(Buceros bicornis)*
12. Hooded crane *(Grus monacha)*
13. Hornbills *(Ptilotaemus tickelli austeni, Aceros nipalensis. Rhyticeros undulatus ticehursti)*
14. Houbara bustard *(Chlamydotis undulata)*
15. Humes bar-backed pheasant *(Syrmaticus humiae)*
16. Indian pied hombill *(Anthracoceros malabahcus)*
17. Jerdon courser *(Cursor/us bitorquatus)*
18. Lammergeier *{Gypaetus barbatus)*
19. Large falcons *(Faico peregrinus, F. biarmicus and F. Chicquera)*
20. *Large* whistling teal *(Dendrocygna bicoloi)*
21. Monal pheasants *(Lophophorus impejanus)*
22. Mountain quail *(Ophrysia superciliosa)*
23. Narcondom hombill *(Rhyticeros undulatus narcondami)*
24. *Nicobar* megapode *(Megapodius freycinet)*
25. Nicobar pigeon *(Caloenas nicobarica pelewensis)*
26. *Peacock* pheasant *(Polyplectron bicalcaratum)*
27. Peafowl *(Pavo cristatus)*
28. Pinkheaded duck *(Rhodonessa caryophyllacea)*
29. Sclater's monal *(Lophophorus sc/ateri)*
30. Siberian white crane *(Grus leucogeranus)*
31. Spurfowl *(Gallaperdix spo.)*
32. Tibetan snow-cock *(Tetraogallus tibetanus)*
33. Tragopan pheasants *(Tragopan melanocephalus, T. biythil*, T. satyra, T. temminckii)
34. White bellied sea eagle *(Haliaeetus leucogastel)*
35. White-eared pheasant *(Crossoptilon crossoptilon)*
36. White spoonbill *(Platalea leucorodia)*
37. *White-winged wood* duck *(Cairina scutulata)*

CHAPTER 4

POPULATIONS AND HABITAT MANAGEMENT

Ecological importance of habitat, cover and population dynamics are basic parameters to be considered for wildlife survival in free living area. The objectives of wildlife management for a given area may be achieved either by manipulating or protecting the habitat or by manipulating the population itself. At times it may become necessary to increase a particular endangered species to make the population viable; on the other hand, increase in number of certain species may be necessary for the purpose of harvest also. Stabilizing a wildlife population at an optimum level is important so that the animals do not exceed the carrying capacity of the given habitat. Thus a wildlife manager 'has to deal with animal populations and this call for understanding some basic principles.

Over growth of particular wild species is equally considered as a problematic issue in the ecosystem. Although improvement in the population of wild animal is a good sign. It is unfortunate, that some of the wild herbivores and other species unwantedly increasing in free range as well as in captivity condition, which is creating public menace. For maintaining balance in the food pyramid, population management places an important role. Therefore, population limits depends on the following factors :

Habitat and Carrying Capacity

The abundance of all wildlife is directly related to the *amount, quality,* and *availability* of wildlife habitat. As a wildlife population increases, it uses more resources. No limited-size area of land can provide an inexhaustible supply of habitat for an ever-increasing number of animals. One area can support only a limited number of animals using similar resources.

This limit is called **carrying capacity**. If the number of animals in a habitat exceeds the carrying capacity, they degrade the habitat by eating available food and eliminating cover, reducing the carrying capacity for that species. Increased disease, lower reproduction, and/ or starvation decrease the number of animals. The population fluxes around the carrying capacity.

Each habitat supports a certain amount of wildlife. There is only so much food, water, shelter, and space available. If more animals are added, they will not find enough food, water, shelter, or space. Either the surplus animals must move to new habitat or they will die from predation, disease, or starvation. Carrying capacity of a habitat may change from season to season. Typically, it is highest in the growing season when resources are plentiful and lowest in winter when resources are most restricted.

Many factors may affect the carrying capacity of any habitat. Shortage of any of the basic needs (food, water, shelter, and space) is a limiting factor. Limiting factors are usually habitat-related, based on the quality and quantity of available resources. Human influences, such as destruction of habitat or disturbance of nesting or brooding sites, also may limit carrying capacity of an area for certain animals.

Mortality Factors

Many factors contribute to the death of wild animals and reduce wildlife populations. Mortality factors (causes of death) are related to climate, diseases, parasites, starvation, weather, predation, and hunting. They usually affect the overflow, or surplus, animals. It is normal for a certain number of animals to die each year. If the habitat remains healthy, wildlife will make up for the loss of individual animals by producing more young ones. Mortality factors help balance wildlife populations with their habitat.

Hunting is an important wildlife management tool for some wildlife populations. One example is Nilgai wild boar. Without wolves and other large predators, deer numbers can increase above their habitat's carrying capacity. Deer licenses are issued for certain numbers and sexes in different zones of the state. Most hunted wildlife species are examples of good conservation. Hunters take annual surplus animals that would otherwise die from natural causes, avoiding over population problems.

FORMS OF WILDLIFE MANAGEMENT

Wildlife management techniques are used to increase, maintain, or reduce wildlife populations.

Habitat Restoration and Management

Habitat restoration/management is a primary tool wildlife biologists use to manage, protect, and enhance wildlife populations.

Increased wildlife diversity in an area may be a wildlife management goal. It is difficult to develop strategies for managing each species separately because there are hundreds of species of birds, mammals, fish, amphibians, reptiles, and invertebrates, each with different needs. Several wildlife species can benefit when a complete habitat type or ecosystem is improved, created, or preserved intact. Managers often restore/ manage habitats to meet the needs of threatened or endangered species, or groups of species (e.g., grassland birds).

Restoring wetland areas has many benefits. Wetlands cleanse water and improve water quality as it flows through to nearby streams and rivers, improving these habitats for fish and other aquatic species. They provide nesting and escape cover for waterfowl and other game birds and mammals. They also supply food, shelter, and denning and nesting sites for dozens of species ranging from snails, dragonflies, and turtles to rails, muskrats, and mink.

Managers may restore wetlands by removing or plugging tile lines, or create new ones. Often, wetland plants return once the basin fills, but a new wetland may also be seeded. Managers may manipulate water levels to increase the plant growth for food and cover. They also plant surrounding areas to native grasses to provide nesting sites for some wetland birds and to protect water quality.

Managers may enhance grassland areas by clearing brush (prescribed burning, cutting, herbicides) and removing trees, as well as over-planting them with native prairie species. This helps reduce cover used by edge predators (skunks, raccoons, red-tailed hawks) and improves the quality of the habitat for grassland animals.

Biologists plant food plots (corn, sunflowers, legumes) and grasslands to provide winter food and spring nesting and brood rearing cover for upland game birds (e.g., ring-necked pheasants). Success or failure of spring nesting and rearing of young often has the greatest impact on populations. Harsh winters (with long periods of snow cover and icy conditions) and very wet springs can reduce nest success and increase mortality of young.

Harvest

Management goals are dictated by the success or failure of rearing young. Changes in weather conditions over several years can have severe impacts on wildlife populations. Adjusting the harvest may be the best way to maintain certain game populations. For example, when major areas used by ducks for nesting experienced several years of drought, the number of ducks hunters could shoot was decreased until the wetlands refilled and duck numbers recovered.

Managers may strive to reduce or maintain populations so animals conflict less with human activities. For example, macaque is abundant in urban areas. This presents challenges for wildlife managers because hunting with firearms is not allowed. Trapping and relocating deer are expensive, time-consuming, and do not provide a long-term solution. Relocated deer do not survive well in unfamiliar areas that probably are at carrying capacity for deer already. The most effective solution has been controlled hunts.

Another example is the trend in the buck:doe deer harvest ratio. The percentage of bucks harvested is increasing. This may create an imbalance and lead to an unhealthy deer population. Regulations that encourage harvest of more does in areas with many deer may result in more mature bucks and a healthier deer population.

Endangered Species Management

Endangered or threatened species require intensive management. Critical habitat and locations of existing populations must be identified so they can be managed successfully. Numbers of individuals and survival rates in existing populations are tracked. Specific habitat types may be created. Existing areas where endangered species are found are protected and/ or managed.

An animal species is considered endangered when its numbers become so low that experts think it may become extinct unless action is taken to save it.

Species Reintroduction

Another wildlife management goal may be to re-establish species in suitable habitat. Eastern wild turkeys, white-tailed deer, peregrine falcons, barn owls, river otters, beaver, sharp-tailed grouse, giant Canada geese, greater prairie chickens, sand hill cranes, and trumpeter swans once were extirpated (entirely gone from Iowa) due to loss of habitat, unregulated hunting, and/ or persistent pesticides in the environment. They are found in the state once again as a result of IDNR reintroduction programs and management efforts.

These are directed through the Wildlife Bureau (including the Management, Research, and Wildlife Diversity sections). Most programs have been very successful. Biologists consider several factors before initiating a reintroduction effort: availability of appropriate habitat, concerns the public genetically suitable individuals of the species to be reintroduced, and depending on the situation.

Conservation and Preservation

Wildlife conservation helps ensure future generations can enjoy our resources.

Conservation can include consumptive (involve taking or harvesting natural resources) activities such as hunting, fishing, trapping, and harvesting timber as well as non-consumptive (do not involve taking or harvesting) activities such as bird watching, photography, and hiking. Both affect wildlife issues between populations.

Conservation must balance issues between wildlife and human populations.

Conservation of wildlife implies insuring threatened and endangered species receive special management to protect their presence in the future.

Conservation may include preservation (protection of natural resources that emphasizes non-consumptive activities). A habitat or ecosystem can be preserved by manipulation (e.g., managing a prairie with fire to eliminate woody species and preserve prairie plants). An area also may be managed by doing nothing at all. For example, when a forest is allowed to mature without any human manipulation such as timber harvest, grazing, or tree planting.

Monitoring Wildlife Populations

Department of forest and environment is legally charged with responsibility for the protection, enhancement, management, and preservation of Iowa's wildlife resources. Wildlife biologists use many techniques to monitor wildlife and gather information that helps determine wildlife management policies and practices. Biologists attempt to standardize information-gathering techniques so data can be shared and compared with other agencies.

Biologists conduct surveys, inventory existing populations, evaluate habitat, and do research. These monitoring techniques provide information about distribution, abundance, or needs of wildlife species ranging from salamanders and frogs, to deer and pheasants, to songbirds and eagles. This information is used to make management decisions.

Biologists use some surveys to estimate the number of a particular wildlife species in a specific area at a given time. Others track general population trends that allow biologists to correlate wildlife abundance with weather, habitat influences, hunting pressure, and more over time. Small representative tracts of habitat are surveyed for the animal being studied. Biologists can estimate how many animals might be found in similar habitats using information from these sample areas. They can then estimate that species' total population for the region.

Biologists can determine birth and death rates by studying animals. This may involve systematically capturing animals, marking them, and releasing them for tracking. Information from a recaptured, marked animal can be compared to data collected at its initial capture. Biologists use this information to learn more about the animal's age, growth, health, and range of habitat use.

Similarly, other methods like aerial surveys, roadside survey, post card survey, non-game survey could be used. However, for important species photographs is also a good option. Dr. Ullas Karanth has devised an alternate methodology to determine the tiger population, called the camera-trap method it identifies a high prey- density area in a forest method, it identifies a high prey-density area in a forest and then a few cameras are positioned there. At night, the stalking tigers trip the cameras, which capture the profiles of the felines. Because the body strips of each tiger are different, one can determine the number of tigers in a specific area more accurately in this method than in most other methods. However, this method has its own pitfalls. The high prices of the imported cameras render this method very expensive. A combination of methodologies has to be used for a more accurate estimate.

Research

Wildlife research involves scientific study of animal species. For example, the effects of habitat fragmentation and degradation on geotropically migrant birds (e.g., hummingbirds, vireos, warblers) have been researched. These species require large, unbroken tracts of undisturbed forests in order to nest successfully.

Endangered animals have been studied to determine their specific habitat needs and survivability in Iowa. Documentation and publication of research helps wildlife agencies manage these species in neighboring states.

Radio telemetry is an important tool biologists use to track movements of wild animals. Individual animals are fitted with radio transmitters so their movements can be monitored. Death/birth rates and migratory patterns also can be determined using telemetry.

Research also includes monitoring demographics and opinions. Iowans' attitudes and opinions affect our wildlife management programs. Public opinion may affect laws passed by the Legislature, which in turn, may greatly influence Iowa's wildlife.

Regulations

It is illegal to harm, harass, possess, or kill most wildlife species. Federal and state laws protect them. Bird nests, feathers, and eggs also are protected. Endangered or threatened mammals and those in taxonomic families that include game animals have legal protection. They do not belong to a family group are protected and cannot be harmed unless they are in dwellings where people live.

Hunting regulations help managers maintain wildlife populations. They: 1) control the number of each game species taken; 2) provide a more even distribution of game taken among hunters; 3) protect wildlife species that are more vulnerable to hunting pressure; and in some cases 4) protect or regulate the harvest of females to· increase or decrease the potential number of young produced.

Habitat Enhancements

Many things can be done to increase wildlife populations. Landscaping, nest boxes, and reintroduction programs for some of Iowa's native species have been very beneficial for all of Iowa's wildlife.

Wood ducks and bluebirds are cavity nesters. They use holes in snags (standing dead trees) to nest. Nest boxes, which simulate cavities in trees, have helped increase their numbers. American kestrel boxes have been placed on interstate road signs.

Landscaping for wildlife is very popular, Hummingbirds and butterflies are attracted to gardens with nectar producing flowers, Trees provide food and shelter for wildlife while reducing heating and cooling bills. Small ponds, added to any backyard, provide water for birds, chipmunks, frogs, and toads. Bat houses in suitable areas attract these mosquito-eating animals.

People purchase land to enjoy outdoor activities such as camping, hunting, viewing wildlife, or prairie restoration. Current farm programs (traditionally oriented to saving soil) now include wildlife and habitat conservation/ enhancement as important objectives.

The following outline is intended to suggest a means of organizing efforts. The objective, of course, is to identify those factors that are most responsible for renting the further growth of the population. These factors include, births, increase deaths, or both.

A Extrinsic Factors

1. Density-independent (primarily weather conditions)

a. Cause of direct mortality?

b. Center or periphery of species range?

c. Does weather have a substantial influence on food quantity-which are density-dependent factors?

2. Density dependent

a. Food

(1) Quality: Are necessary nutrients present?

(2) Quantity: Is enough food available?

b. Cover

(1) Shelter from elements: Are quality and quantity sufficient?

(2) Escape or hiding cover (for predators or from predators), are quality and quantity sufficient?

c. Refugia available : Are there patches of habitat in the range of the population in which animals have a high likelihood of escaping various mortality factors, such as predators, hunters, parasites, and disease?

d. Competitors : Is there competition for resources by other species?

e. Diseases and parasites : Are these factors influencing birth and death rates?

f. Predators : Are predators controlling the population?

g. Buffer species : Are other prey species present that absorb some of the impact of predation, particularly when the species being considered is at low densities?

h. Hunting harvest : Is harvest toll replaced by the next season's production of huntable animals?

i. Interactions among various factors : What interactions occur? Food supply-disease? Food supply-predation? Food supply-competition? Cover-predation? Buffer species-predation?

B. Intrinsic Factors

1. Genetically stable factors

a. Litter or brood sizes. What is the inherent potential of the species to reproduce?

b. Longevity. How long can individuals live?

c. Habitat selection for breeding, feeding, resting. Is it available according to inherent needs of the species?

d. Self-limiting factors. Does the species possess self-limiting behavior such as territorial spacing or restricted breeding among selected members of a group?

e. Dispersal. Is there an opportunity for immigration and emigration?

f. Interaction. How do inherent features interact, such as territorial behavior-food supply, dispersal-food supply, birth rates-food supply?

2. Genetically variable factors

a. Birth rates. Within the physiological limits of the species, does the population show varying birth rates?

b. Survival rater. Does a population differ genetically from time to time in the ability of individuals to withstand stress, or has there been a response to a strong selective factor such as disease or biocides?

The answer of these questions about a population is little bit difficult. The most fruitful and simplest approach is usually to examine the more obvious factors first, such as weather, food, cover, and the behavioral nature of the animals. Should such and approach fail to provide satisfactory answers, then more subtle interactions must be examined.

METAPOPULATIONS

Metapopulations-groups of local populations of a species have recently been studied by ecologists of modem age. Each group occupies separate patches of habitat, which are often connected by corridors through which dispersal and exchange may occur. Because each group is relatively isolation, any one of the groups within a metapopulation in prone to extirpation (often called "local extinction"), but when this happens, the vacant patch also has the capacity to be re-colonized by individuals from the other patches.

COMPUTER MODELING

Using computer modeling, simulated the distribution, reproduction, and survival of cottontails in a hypothetical metapopulation, a configuration of 30 patches shown below. They then tested four scenarios, each populated with 200 cottontails. The first scenario consisted of the entire 30-patch landscape. Scenarios 2 to 4 consisted, respectively, of the following:

1. 10 patches of equal size with 20 cottontails per patch,
2. 5 patches of equal size with 20 cottontails per patch, and
3. 2 large patches with 100 cottontails per patch.

Variables included fluctuating temperatures, show cover, and, based on recent trends in development, 4 percent annual loss habitat. Because of the large proportion of patches with small populations, the 30-patch landscape was the least likely to sustain its cottontail population (i.e., the probability of extirpation at the smaller sites was too high). In contrast, each of the three other scenarios maintained relatively secure populations. These results suggest that cottontails, which favor early successional forests. might best be managed by rotational cutting of adjacent 5 - to 10-ha patches, with one-third of these cut every 10 years, thereby perpetually producing three age-classes of forests. Such a manipulation allows cottontails to find appropriate habitat without having to travel across large areas of unfriendly terrain. Elsewhere, habitat for ruffed grouse *(Bonasa umbellus)* is currently managed using a similar practice.

POPULATION MODELS

When wildlife managers predict how many mallard ducks will be present in the fall population based on sample counts of breeding ducks and the number of prairie ponds in the spring, they are using a model. Similarly, an estimate of the number of deer present obtained by sampling and counting pellet groups (droppings) left by the deer is another example of the use of model. Computer models have been developed to simulate wolf (Coni's *lupus]-moose {Alces alces)* populations on Isle Royale and the projected recovery of whooping cranes (*Grus omericona*).

A population normally exists in a habitat within the carrying capacity, which is determined by certain habitat conditions relevant for the given species. When wildlife populations are fairly low with respect to the carrying capacity, the birth rate may be high and the population would increase in number. On the other hand, those populations which are already existing at the level of carrying capacity in a habitat may show relatively lower birth rate but a higher death rate. Both births as well as death rates thus differ as per the population density.

Population is derived from the latin word ‘popular’, which essentially means people. In the biological sense, it can be considered as an assemblage of organisms, -plants or animals-which may belong to a single species or to several closely related species occupying a definite area. It is a self regulating system. There are basically two types of populations, viz. the ‘monospecific populations’ having individuals of the same species, and the ‘mixed or poly-specific populations’ having individuals of several species.

A population has many characteristics or attributes which are typical; these function for the entire group and do not pertain to an individual. The major factors which governs the population of wild animals are viz. population density, birth rate, death rate, age distribution, growth form, population equilibrium, biotic potential, dispersal, dispersion and population fluctuations. Change in population is affected by changes in these parameters.

Population density : This can be understood as the number of individuals occupying a unit area or volume of the habitat. The population density units may vary as per situations. Usually, there are two ways of expressing the population density, viz. the crude density and the ecological density; crude density is the total number of individuals or biomass per unit of total area or total volume; on the other hand, ecological density can be understood as the number of individuals or the biomass per unit of that area or volume which is actually occupied by the animals.

Natality or birth Rate : The actual rate of birth is usually known as 'fecundity', and is measured as the number of live births per female over a given period of time, usually one year. It can also be defined as the average number of new individuals produced by a given population in a unit time. The 'maximum birth rate' or 'potential natality' is the maximum number of new individuals which can be produced per unit of time, under a given set of ideal ecological conditions. Species differ widely in their fecundity; some animals may reproduce in large numbers, whereas others may restrict their off-springs to a few only. The size of a litter is governed by various physiological and morphological characteristics of a given species.

As far as mammals are concerned, the uterus size of the females, the body cavity and also the number of mammary glands serve as limiting factors for the number of off-springs that are produced at a time. The maximum birth rate is known as the **'potential natality'** for a given population; the 'actual' or 'ecological birth rate' or **'realized natality'** is the actual number of new individuals which are added to a population in a given time. This is considerably less when compared to the maximum birth rate, since all the animals may not be equally fertile, and also all the eggs that are produced may not hatch into young ones and attain adulthood.

The ecological birth rate or the natality rate of a population is usually expressed by the formula :- B = Nn/t , where 'B' = natality rate per unit of time, Nn = number of new individuals that are added to the population by natality and 't' is the unit of time. The natality rate per individual 'b' in the population per unit of time can also be expressed by the formula :- b = Nn/N t. The typical types of reproductive rates are such as:-

1. Primary natality: fecundity = in utero young per total female adults.
2. Net natality: total female young per total female adults.
3. Refined natality: total young per total female adults.
4. Gross natality: total female young per total adults.
5. Crude natality: total young per total adults.
6. General natality: pregnancies (mean litter size) (total adults).

First three types of natality are female specific; usually the precision also decreases in the various types indicated. Sex ratios affect the reproductive age. A male: female: young ratio has been used by many people for combining certain information on sex ratios with an expression of natality.

Mortality or death Rate : Essentially, mortality rate of a population means the number of individuals which die per unit of time; it may be expressed as 'potential mortality' or 'realized mortality'. Potential mortality or the minimum mortality is understood as the number of deaths which would occur under ideal conditions due to various physiological changes pertaining to old age. For a given population this value is constant. On the other hand, realized mortality means the actual death rate, which depends upon certain physical factors and also on the composition, density and size of the population. The incidences of death in a population can be expressed by a survivorship curve, where the number of survivors in a given population are plotted against time; usually there are three basic types of such curves.

Diagonal curve : Here the species show a fairly constant mortality rate at all ages which results in a straight diagonal line; this is typical for certain lower organisms like the hydra and many species of birds such as the song sparrow.

Convex curve : Many animals live out their full potential life span and die only during old age, as a result of which their curve is highly convex. It is almost horizontal until the span is reached and later drops down sharply; this are typical of man, rabbit, deer and many other larger mammals.

Concave curve : This curve is typical of those animals where the mortality rate is very high in the early life, characterised by frequent deaths during infancy; many invertebrates, fishes and oysters show this curve. Survivorship curves Indicate when a particular species was most vulnerable, and how the increase or decrease in mortality, during this vulnerable period, affects the future size of the population. We can determine the life expectancy from survival data; life expectancy is the probable years of life to be lived by individuals in a particular age class.

Fertility Schedules : Animals differ in their fecundity and accordingly the rates used for expressing them also differ. The time of measurement of fertility rate should be carefully recorded, since more often than not, this depends on the convenience of data collection. Therefore, fecundity schedules are usually made from placental scar counts or adult-female embryo counts. Schedules of fertility usually are sex-specific. In case the life table pertains to female population, then only female births are recorded; if the table corresponds to male, then only male births are noted. It is not possible to ascertain the age of males contributing to fertilization and therefore, very rarely

Stable age distribution : This pertains to a population where the proportion of the individuals in each age class remains constant through time. This is possible only if the fertility and survivorship schedules are almost constant. It is likely that a population may be constant, increasing or decreasing but never fluctuating.

Stationary age distribution : This refers to a population in which the number of individuals pertaining to each age class is constant, thereby making the total population density also constant. A population having a stationary age distribution will always have a stable age distribution but not vice-versa.

Age composition : The number of animals or the percentage of individuals in a population in different age groups is known as the age distribution of a population. Populations vary widely in the relative numbers of young and old and thus they show different natality and mortality. Usually three age groups are recognized in a population, viz. pre-reproductive, reproductive and post-reproductive.

Sex ratio : It is said that a sex ratio of 50 : 50 at birth appears to be a general rule for many species of vertebrates. If this ratio is not maintained in a population because of variations in mortality amongst species, then the population dynamics gets affected in varying degrees. For a proper management of a wildlife area, one must know the sex ratio of various wildlife populations. It indicates the procreation potential. It has an effect on the main population body weight and consequently the requirements of animals. If the food supply in a habitat is fixed, then it win support more young females than adult males. As per convention, sex ratio is usually expressed as males per 100 females.

SEX RATIO COMPUTATION:

MALES: 100 FEMALES.

From field data:

1647 males 3921 females

1647/ 3921 = X/100

3921 x == 164700

x == 42

Therefore, 42 : 100 is the conventional sex ratio.

From female percentage: (e.g. 64% females)

64/100 =100/x+100

(x + 100) 64 == 10000, :. x == 56

Therefore, 56 : 100 is the conventional sex ratio.

From male percentage: (e.g. 64% males)

100-64/100 = 100/X+100

x = 177

Therefore, 177 : 100 is the conventional sex ratio.

Percentage :-

Percent females from field data-

1647 males, 3921 females.

3921/1647+3921= 0.704

Therefore, the percentage is 70% females.

Percent males from field data: 1647 bulls, 3921 cows.

1647/1647+3921 =0.295

Therefore, the percentage is 30% males.

Percent females from sex ratio: (e.g. 64 : 100)

100/64 + 100 = 0.609

The breeding behaviour of a population is also important. Some species are monogamous, i.e. there is a pairing of a single male and a single female for one complete breeding season, which may be extended for a further period of time; in case this sex ratio is disturbed, then the population of monogamous species declines, irrespective of the fact whether the males or females suffer mortality. On the other hand, there are polygamous species where each male will be mating with more than one female; here also, the maintenance of a viable sex ratio is important; abnormal changes in the ratio can have drastic effects on the growth of the population. As far as polygamous species are concerned, the rate of reproduction becomes a function of the number of breeding age females which are present in the population.

Usually in wildlife management, a population analyst refers to a cohort of animals; this represents a standard number of one thousand or ten thousand animals; when a population is described in tenants of cohort, ready comparisons are facilitated with other populations. Thus, a cohort with a sex ratio of 40 : 100 will have a productive component of 714, viz. [(100/40 + 100) x 1000]. If the males are harvested in a population, it is said that the productive component gets increased. If welfare factors in a particular habitat are of limited quantity, and in case an increase in the population is desired, then the proportion of females should be kept high when compared to the males. This becomes an important managerial strategy, because keeping the females high would mean that the population has a chance to increase in future. However, this is reversed in case we have to reduce the population. In general, the average female birth rates when multiplied by the number of females give us the annual production of young ones.

Dispersal : The movement of an animal from its birth or release site, to the place where it reproduces or would have reproduced had it been alive, is known as dispersal. Dispersals are essential in nature since the conditions of a habitat may change with time. Dispersals prevent a population from becoming extinct since the local habitat conditions may not be in a position to sustain it for long. The outward movement of animals is usually called as 'emigration', whereas the inward movements are referred as 'immigrations'. Thus emigrants of a particular habitat may become the immigrants in another region. In case a given population is fairly stable and is almost at the level of carrying capacity, then such dispersals may not have much effect on the population density. However, immigrations initiated by overcrowding or by the shonage of welfare factors may certainly reduce the population, thereby influencing the age structure and the rate of production. Many vertebrates show dispersal: the armadillo- ***Dasypus novencinctus*** is a classical example of emigration: Dispersals are also seen amongst fox, squirrel, and lemming. Both immigration and emigration involve the one way movement of animals.

There are periodic or seasonal dispersals within animal populations; such periodic dispersals are called as 'migrations'. Many mammals, birds, fishes and also some invertebrates show migrations which are two-way movements.

Internal distribution patterns of populations : The mode of distribution of individuals in a given population is known as the 'internal distribution pattern'. It affects the internal structure and the other attributes of a given population. Usually, three patterns of distribution are seen, viz. uniform, random and clumped.

Uniform distribution : In this, the individuals are roughly or almost evenly spaced with respect to one another; this is seen only in those populations where competition between various individuals is not acute or in cases where there is a positive antagonism.

Random distribution : In random distribution, the individuals of a population do not group together to form clumps nor they are spaced evenly but are irregularly scattered. Such a distribution is seldom seen in populations. It may be present only in such situations where the environmental conditions are uniform throughout the habitat, which avoids the clumping tendency in animals.

Clumped distribution : Clumped distribution is the "aggregation" which is most common. Various degrees of clumping are seen in animal populations but these groups are of specific size. Three categories of clumped distribution are usually encountered, viz. random clumping, uniform clumping and aggregated clumping. Many factors are responsible for clumping such as the differences in the habitat conditions, weather, the reproductive processes and the social attraction between individuals. Clumping or population. In general, the average female birth rate when multiplied by the number of females gives us the annual production of young ones.

Population fluctuations : All populations fluctuate to some degree; some of these fluctuations are very distinct. In general, initially fluctuations are of two types, viz. the 'cyclical fluctuations' and 'irruptions'. Cycles are major population changes which occur at regular intervals, usually in a 3 to 4 years cycle, typical of certain rodents in the arctic regions. It may also occur in 9 to 1 year cycle in certain populations of lynx, ruffed grouse and snow shoe hare. Longer cycles are typical of boreal forests. The 9 to 10 year cycles were initially described by the noted British ecologist Charles Elton, and since then many ecologists have accepted this as a valid biological phenomenon met within populations under certain conditions, However, causes for such cycles have not been understood and the issue still remains controversial. Two types of causes have been postulated, viz. the extrinsic and the intrinsic ones. Extrinsic causes are those which occur due to forces operating outside the population, whereas intrinsic ones are those which occur as a result of forces which work within the populations themselves. Amongst the extrinsic factors, various phenomena such as climatic cycles. diseases, decline in food quantity and quality. Predation and the like can be included; intrinsic factors, on the other hand, may consist of certain physiological stresses due to over-crowding, changes in social system between animals and also genetic changes within the population themselves.

Irruption, the other type of fluctuation, is more important to a wildlife manger. They occur at irregular intervals and are common to both temperate as well as the tropical regions. Irruptions result in considerable increase in the population which may have for reaching consequences. In one observation between 1920 and 1940 in the United States, many cervid populations increased sharply causing concern to wild lifers. At that time it was considered that the elimination of certain natural predators, such as the gray wolf and the cougar, might have caused these irruptions. However, recently these views have been challenged by many who believe that these increases were the normal oscillations due to the interaction of those herbivore populations with their food supply.

Carrying capacity, biotic potential and environmental resistance : The growth rate of a population in an unrestricted manner is usually known as the biotic potential. It is unlikely that the full biotic potential of any species would be reached, because, certain inborn limitations in the environment counter this. This resistance, as indicated earlier, is known as the environmental resistance, and is seen during the shortage of welfare factors such as food, water or mates; environmental resistance brings down the population level and prevents the biotic potential from being reached for a particular population; thus, in a way it defines the carrying capacity of an environment. When a population approaches the carrying capacity then the numbers start leveling. Sometimes, a population may overshoot the level of the carrying capacity; if this happens, then the growth rate falls off abruptly and picks up again when the numbers are below the carrying capacity; this process continues. In stable populations, oscillations of numbers closely near the carrying capacity level are frequently seen. Theoretically, it must be understood that if a population severely overshoots the carrying capacity, then it may result in a complete depletion leading to a 'crash'. For many species, such crashes are a normal feature in their cycle. Thus, many insects and plants, which are annual, crash every year but re-appear again and carryon the process of reproduction during the favorable months. Some crashes, however, are permanent in nature. If the environmental resistance on a group of animals is removed by eliminating the limiting factors such as predators, diseases, food and water shortage, then, the animals may go on increasing to a point at which they stand devastating their own habitat. Thus, an elephant population, beyond a certain size may spoil its habitat completely as a result of which they may get eliminated from the environment subsequently. In the normal course, the habitat would have sufficed for many generations. Such a devastation of habitat, beyond the carrying capacity, spoils the habitat for many years by bringing down the carrying capacity.

Patterns of population : Natural selection has shaped the various population growth patterns in nature. From a theoretical study of island biogeography, a useful distinction was made between two main kinds of population strategies. According to this, the first animals which reach uncolonized islands have abundant resources due to which they expand rapidly. Natural selection, during this period, undoubtedly favours reproduction, so that the animal species can take advantage of the available resources in the area; this evolution strategy has been termed as **"r-selection"**. As time passes, the island gets crowded and consequently the available resources also decline, as a result of which the rapid growth of the' r-selection' phase is replaced by a more slower growth pattern which is better suited for the increased competition and the limited number of resources. This conservative strategy of slow growth is called as **"K-selection"**. This concept of rand K selection has been applied to species that live on continents as well as those present on islands. From the wildlife management point of view, it is important to appreciate that certain population· characteristics are linked with one another; thus the 'r selected' species show rapid reproduction because of favourable conditions of the environment and hence they can withstand high rates of mortality; usually, the 'r selected' ones are very well adapted to early and mid-successional stages of the habitat. On the other hand, the 'K selected' species have lower rates of reproduction and show considerable age at the' first breeding also; thus, they are not in a position to withstand high mortality. These animals are usually large with comparatively lower metabolic rates. They .have long life spans and spend more time and energy to bring up the young ones .However, the distinction between the r and K selected species is not absolute in nature.Many species in nature fall somewhere in between these two extremes.'

It must be understood here that the management strategies for 'r' and 'K' selected species vary considerably. Species showing extreme 'r selection' exhibit tremendous build up of population which may cause severe damage to nearby agriculture crop and property, due to which they may quickly come into conflict with human beings. A wildlife manager thus, is confronted with the problem of controlling these damages either by manipulating the habitat suitably or by controlling the population itself. Extreme 'K selected' species are usually the endangered ones; they do n6t exist in very high numbers and generally lack the capacity to increase rapidly, even under improved conditions of the habitat. It has also been pointed out that 'K selected' species are usually adapted to a stable, late successional habitat, sensitive to human or any other biotic disturbances. Many species of water fowl are 'r selected'; they show quick increase if the habitat conditions are improved. As far as Indian conditions are concerned, the spotted deer (***Axis axis***) can be consideed as a 'r selected' species, whereas the hard ground Barasinga (***Cervus duvauceli brandert)*** is an example of K selection.

Population genetics : A wildlife manager, apart from increasing the animal numbers, must also evolve a strategy so that the perpetuation of wild animals is ensured for the future. Thus, the importance of gene pool should be understood and appreciated in the proper perspective. A gene pool can be considered as the sum total of the genetic diversity which is contained within a population or within an entire species. Gene pools are different from populations and can increase only through mutations or by the process of out breeding. Mutations do not occur very frequently and thus, we cannot depend on them for the immediate replenishment of the genetic pool. Various conservation efforts can only bring back the numbers and restore the animal population, but they cannot restore the losses in the gene pool. In the near future, wildlife managers may be faced with various problems pertaining to population genetics. A common problem usually encountered with the endangered species, is the drastic reduction in the size of the gene pool; another one, which is equally severe is the interbreeding of the wildlife species with another. Such interbreedings contaminate the gene pool of both the parent species. When one parent species is much more numerous than the other, then, the less numerous one may be faced with the threat of extinction. Various human activities may aggravate both these problems in future. Reduction in a gene pool results mainly due to the decline in numbers of a particular species combined with the shrinkage of wilderness; as the numbers diminish, the unique genes, typical for that animal species, are lost. These genes may be adapted variants (ecotypes) and by losing them we lose the diversity. Because of such reduction in the gene pool and the shrinkage of original range, the species ultimately remains isolated in a small patch, as a result of which, interactions within the population at the genetic level are drastically reduced, which enhance the process of genetic loss. The smaller the population is in its isolated patch, greater will be the rate of loss of the genetic diversity due to chance recombinations; this process is called as the' genetic drift'. The obvious effects of in-breeding and genetic drift are the reduction in fecundity and survival rates; such problems are usually called as 'in-breeding depressions'. Many long term effects include certain random changes in the phenotypes and gradual depletion of genetic variants (Franklin, 1980). Many phenotypic characters in the population may change, which may not be due to the various natural selective forces, but may be the result of certain random processes of genetic drift, which further exclude some more part of the already much depleted gene pool. Genetic drift depletes the gene pool and weakens the ability of a population to recolonize the former portions of its range; this may happen even if the animals within the population increase in numbers substantially. Over a long period of time, all these animals may not be in a position to survive and thus would stand out of the evolutionary process.

The contamination of a gene pool due to in-breeding with closely related species is an example of 'genetic contamination' or 'genetic swamping'. Various human disturbances cause hybridization between species which are closely related; the disturbed habitat appears congenial to the hybridized offspring, which may be unsuitable for the parents. If the hybrids survive and maintain their fertility, then the risk of genetic contamination by back crossing with either of the parent species always exists.

Certain genetic problems can also arise due to restocking, when members of one sub-species are introduced into the natural range of another. This results in in-breeding and production of offsprings which may be suitable for the local habitat conditions than the original ones, but the genetic identity between the two sub-species is lost. In certain cases, where the native population of a species is completely absent, then restocking may be the only alternative. With the help of population genetics, animals with greater genetic variability should be selected and introduced in such areas so that they are least related to one another in the environment. This will enlarge the gene pool and the animals may have greater chances of survival in the near future.

Regulation of population: density stabilizing factors : Various factors in nature affect the levels of population. Factors which are outside the population are known as 'density independent factors', whereas, those acting within the population are termed as 'density dependent factors'. The density dependent factors usually stabilize a population at the level of asymptote, determined by the carrying capacity of the habitat: The density independent factors indicate the potential level which a population may attain in a given habitat and they are physical in nature. They do not have the stabilizing effect and their intensity is independent of the population size. All these factors, taken together, are collectively considered as the environmental resistance. No factor acts alone; the various limiting and stabilizing factors act in an 'inter compensatory' manner *(Kendeigh, 1974).*

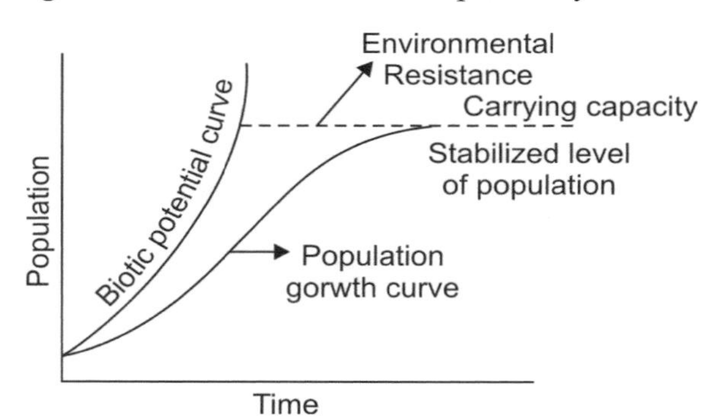

Density independent factors : The density independent or environmental factors set the ultimate limit or carrying capacity, which cannot be permanently exceeded; the density dependent or biotic factors usually stabilize populations at that level or slightly below it. The changes in the availability of space, cover, favorable weather conditions, food and other welfare factors are independent of the density of population, but they can drastically affect the animal abundance in a given habitat. Shortage of these welfare factors may result in the mortality of animals. Species differ in their demand for various welfare factors such as space, food, and shelter; consequently, these factors affect the levels of population which are attained in a given area. Smaller species may require less space when compared to the larger ones. As the population of a particular species goes on increasing, the welfare factors necessary for its survival gradually diminish, and a stage may come when the habitat may not be in a position to support the population. thus indicating that these factors are also responsive to the size of the population indirectly. Since animals have their own tolerance limits, the availability of various welfare factors sets the upper limit for the population size.

Density dependent factors : The intensity of density dependent factors increases with a rise in population density; it decreases when the population level comes down. The important density dependent factors are competition,fecundity,survival of young ones, predation, emigration, disease and physiological status. It must be remembered that not all density dependent factors have a stabilizing effect on animal populations.

Competition : In nature, competitions are common. It helps to stabilize a population at a particular level. Usually, competition is seen for space, cover and food which constitute important welfare factors for wild animals. The competition for space is fairly seen in territorial animals. Competition is also seen for the best possible place available in a habitat with respect to cover and protection. Competition, thus, is an important factor which brings about the regulation of population; it ensures food supply, space and other welfare factors which are essential for survival. For those species in which competition results in the establishment of hierarchies or territories, the population is stabilized at a level below the one at which the welfare factors become limiting or exhausted; this level ensures an adequate rate of reproduction. Those animals which fail to secure their territory or a place in the social hierarchy, are forced to leave the habitat and seek refuge elsewhere. Even if they survive, they may not be in a position to reproduce in the original area. However, when such factors do not fully bring down the population size then competition acts in an indirect way by exposing the vulnerable individuals to predation, reducing the reproductivity, and at times causing the emigration of extra individuals to other places. According to the 'competitive exclusion principle', whenever two species compete for the same resources, one of them will emerge as the superior competitor and the other will be pushed out.

Competition usually occurs between individuals of the same species (intraspecific competition) or members belonging to different species (interspecific competition). However, competition between the same species appears to be more strong, because members of the same species fiercely compete with one another for the welfare factors such as food, shelter, or mates. Competition between different species is also interesting; there is a balance between the intraspecific as well as interspecific competitions which occur in a given habitat. As far as intraspecific competition is concerned, it is said that all members of a given population face the same level of competition; this is caned as the 'principle of equal opportunity'; this results in the broadening of resource utilization. Interspecific competition, on the other hand, restricts the resource range available to a population. Competition is significantly reduced if there is no overlap between the geographical ranges of animal species; this condition is known as 'allopatry'. However, in nature, many a times, the allopatric species have come together and have ended up as competitors. They can be considered as 'potential competitors'. Sympatric populations are those which co-inhabit a particular geographical area. In case species occupy adjoining areas, the condition is known as 'parapatry' and there is no significant overlap. Such parapatric populations, perhaps, can co-exist only in situations where two ecotypes have been sharply demarcated; various influences are required to ensure such ecological separations in nature. Any separation of niche between animals will be clearly reflected in the morphology. Sometimes such differences may be present in a more subtle manner.

Predation : The animal who depends on the prey species and the changes in the population level of predators closely follow variations in the population of prey animals; however, it is not certain whether the number of predators available in an area is simply dependent on the abundance of prey animals which constitute their food, or, whether the predators, by their feeding, regulate the abundance of prey animals. Under certain conditions, predators largely affect the number of prey species on which they feed. In the predator prey relationship, both functional as well as numerical responses may be seen. A predator may respond numerically to the increase in prey base by increasing in number; it may also show a functional response by increasing the intake of prey animals. Predation in a habitat is dependent upon the availability of various potential prey species, it is also affected by the number and kinds of predators available in a habitat apart from the density and vulnerability of the prey species.

Predation is a very controversial subject in wildlife management and has been discussed by many workers. It is believed by some that unchecked predation may result in the eradication of natural prey animals from a habitat; on the other hand, there is a school of thought which maintains that predation, in reality, may have no effect, because it removes only the vulnerable prey animals which would have died naturally due to some other causes. Some workers believe that prey populations are benefited by the process of predation because it involves selective killing of unfit individuals in a population. In nature, a definite prey-predator relationship may exist, but sometimes this gets altered due to the introduction of various exotic animals.

Like competition, predation favours species diversity and helps in the stabilization of populations. If certain factors like the availability of food or weather conditions affect a prey population apart from predation, then the role of predation may be to serve as a partial check on the number of prey animals. Predation which regulates a prey population is less rigid than the predation which limits a population.

Predations vary as per situation, and at times it may involve selection of different age classes of prey. sometimes, different members, belonging to various age classes, may be randomly consumed by predators and this may not alter the population structure of the prey animals; but many a times, predation occurs is proportionately and either very young or old classes in a population are affected, which results in a higher percentage of animals belonging to the middle age class. Such predations are very common in nature as far as the carnivores of Africa are concerned. The reproductive value is dependant upon age, and different age groups have their own reproductive values. The older age groups in a population may not be in a position to reproduce in nature and hence have lower reproductive values. Animals which survive through the rigorous juvenile period of high mortality, later mature into adults and enter into an age class which can produce a large number of offsprings. In case thc predators consume disproportionately more from the less reproductive group, then they may not alter the population at all; sometimes, this may enhance the fecundity of the population because the percentage of age classes with higher reproductive values may be more. Various examples of predators, selecting different age classes from thc prey animals, are available from the studies of predators which feed on large ungulates. Predators like cheetah, wild dog and the gray wolf prefer large proportions of old or very young animals, since these age classes are not in a position to run during a long chase by the predators. Animals which stalk or ambush are not highly selective; their killing, more or less, resemble the age composition of the population as such.

Various population characteristics of the prey animals also affect predation. It has been observed that the population of predators increase along with the prey populations and the predation pressure also builds up gradually. If the prey animal exhibits a population trait such as irruption at periodic intervals, then various responses can be seen in the predators. Because of the extra availability of the prey animals, the predators may change their feeding pattern. They may prefer to kill more members of the irrupting population. This is called as a 'functional shift' or response shown by the predators to the increasing numbers of prey. If the irruption is of a localized nature, it may attract predators from outside, and consequently, the predation pressure in the habitat may further increase. On the other hand, if the irruption continues and persists for a greater length of time, then the predator population may also go on increasing because of the food availability arid this may become a numerical response.

Territoriality prevailing in the prey animals also affects predation.

There are certain population characteristics of the predator which deserve consideration. In case the population of a predator is affected by factors other than the availability of food, then it may not show a numerical response to the increase in the numbers of prey animals, due to which it may not affect the prey population. The social behaviour of predators affects their population growth. Thus the pack order, hierarchy or other social set up within a group are important. In the gray wolf packs, the alpha male as well as the alpha female restrict other members of the pack from breeding. George Schaller observed that the male lions often kill the young ones sired by the mates in the neighbouring prides; this curbs the lion population growth. It is believed that such a social behaviour may act as the upper limit for the predator population before it reaches the limit on account of available food. Territoriality amongst the prey animals as well as the predators are important; predators are sometimes highly territorial which may have an influence on the predation pattern.

Many predators depend upon more than one prey species, and the effects of predation on any particular prey animal is greatly influenced by the availability of other prey animals which act as 'buffer species'. Predation intensity is density dependent; the predation pressure increases along with the prey population. Because of the buffering effect, predators exert more pressure on the prey animals which are available· in ,abundance rather than on those which are scarce in a habitat. The buffering effect results in species diversity amongst the prey populations. Predation pressures are essential on the more numerous prey animals, otherwise, they may out number those animals which are less abundant in a habitat. Many ecologists believe that in the process of evolution, predation favours an increase in species diversity by allowing more overlapping of niches between competing species.

Other prey animals which act as 'buffer species'. Predation intensity is density dependent; the predation pressure increases along with the prey population. Because of the buffering effect, predators exert more pressure on the prey animals which are available in ,abundance rather than on those which are scarce in a habitat. The buffering effect results in species diversity amongst the prey populations. Predation pressures are essential on the more numerous prey animals, otherwise, they may outnumber those animals which are less abundant in a habitat. Many ecologists believe that in the process of evolution, predation favours an increase in species diversity by allowing more overlapping of niches between competing species.

By using simulation models, four major "tactics" for population stability, pertaining to both prey and predator, have been evolved

1. The 'system stabilizing predator' is one which is capable of altering its own physiology in such a way that even at low prey densities, it can reproduce.
2. Some predators are' wasters' in nature and they may kill all the prey animals present in a single location. On the other hand, those predators which hunt only such animals which are found in aggregates, kill only one animal at a time and often return to the carcass until it is fully consumed; tigers and lions exhibit this, and quite often it is a familiar sight to find a tiger guarding its kill. Under such situations, stability will result only if the predator is able to find its specific prey type.
3. If a primary predator is affected by a secondary predator, then the oscillations of both primary predator as well as its prey may be dampened by the effect of the secondary predator on the primary one.
4. There are certain predators which do not attack randomly but prefer to choose a certain type of prey from a given population; this results in multiple attacks on a single prey species which may stabilize prey oscillations.

Predation and Foraging strategies : Various attempts have been made to bring out certain uniform principles relating to the foraging behaviour of different animals. Foraging and predation involve considerable expenditure of energy and time and it would be worthwhile to examine whether a particular strategy is beneficial and economical for an animal. By foraging, an animal gains certain amount of matter and energy, which may be considered as profits. These profits are utilized for growth, maintenance and reproduction by the animals. Foraging involves considerable costs, and at times these may be very high. By resorting to a particular foraging strategy, an animal may be taking a risk by exposing itself to all sorts of attacks. An 'optimal strategy' increases the profits and reduces the costs or risks. A close look at the predator-prey relationship shows how efficiently animals develop their foraging strategy to avoid wasting. It is important that an optimal predator should not waste much time in catching very small animals which contain lesser amount of energy, because, the energy spent in chasing small prey may be more than that would be secured after predating them. An optimal predator or forager would also avoid searching certain prey animals or food at the wrong time and at the wrong places. Optimal situations may vary according to habitat conditions. Animals differ from one another as far as spatio-temporal utilization of habitat is concerned. Thus foraging patterns and strategies adopted for predation vary between animals. It may not be optimal for a cheetah to chase a black buck, at full speed, inside a dense cover; on the other hand, the predator may be successful in chasing a prey in the grasslands.

Predation and patchiness : Environments change at different places and they are not uniform throughout. It is said that these changes are rather abrupt usually, environments assume the form of a mosaic which are called "patchy". The type of patchiness which affects a given animal differs as per its habits and size. The environment can be considered as composed of both coarse grained as well as fine grained components. Animals, whether big or small, are affected by the mosaic pattern of the habitat. This is a very broad and flexible concept and can be defined in various ways. Broadly, it can be stated that in a fine grained environment, animals spend lesser time in searching their food, and this is more so if there is a chance of welfare factors getting repeated at predictable frequencies in the habitat. Patchiness is also affected by the behaviour of animals. If an organism does not make any selection amongst the resources available in a habitat) and uses them in the same proportion in which they occur, then, it can be said to utilize the habitat, in a fine grained manner. On the other hand, those animals which spend considerable amount of time in selecting certain patches of the environment can be said to use them rather in a coarse grained way. Firstly, there are certain species which tend to be "viscous"; these species do not move around considerably and they produce offsprings which may continue to inhabit the parental patch type. In due course of time, the offsprings get used to the patch. On the other hand, there are "vagile" species which move around frequently among patches, but rather prefer to rear their young ones in certain specific places. In case a particular patch is not favourable, then they prefer to move on to an adjoining patch. There is another response shown by animals to a coarse grained environment; this is seen in viscous species where the local populations disperse their gametes randomly so that their offsprings can thrive in such areas where they are able to settle down. Obviously, this would mean that those offsprings which land on unfavourable patches will soon get eliminated. Patchiness, broadly, is dependent upon the behavioural attributes of animals which occupy a given habitat apart from the various physical characteristics of the habitat itself. There is a gradual integration or blending of various patches.

Mac Arthur has reviewed several aspects of 'optimal foraging theory'; certain basic assumptions were made by him, which are as follows :-

1. The environment is basically repeatable (patchy) with a certain , probability of finding a particular welfare factor.
2. There is a continuous and "normal" curve for the spectrum of food items, intermediate types being most common.
3. Similar animals forage in the same manner, and any animal can exploit the food sources in the best possible way if it falls between those that are optimal for two neighbouring species.
4. No single phenotype can be efficient to the maximum extent on all prey types. Increase of efficiency on anyone prey type reduces the efficiency in foraging the other.

5. A species normally tends to maximize its total food intake. Apart from these assumptions, Mac Arthur divided foraging into four phases:

    Deciding where to search

    searching for palatable food item

    taking a decision whether to pursue a located food item and

    pursuing it

An efficient system of energy and time budgeting is essential for a predator to survive in nature. When prey animals are less in number, the predator may succumb; therefore it is essential to understand the prey size. The 'optimal prey size' for a given type of predator should be such that it should meet the energy requirements. Large prey animals obviously yield more energy, which may compensate for the time and energy spent by the predators to capture them. Small prey animals are usually more abundant in a habitat when compared to the larger ones; therefore, many animals consume a large number of smaller prey than a larger one. Small predators thriving on smaller prey have a better chance of securing their food, when compared to the larger predators which require larger prey. Therefore, large predators maintain a wide range of prey animals. The social system of predator is important since it determines the range of prey size for a predator; a wild dog is capable of killing a small fawn at ease; however, since the animal is adapted to pack hunting, it prefers to bring down a large stag. We can also compare the feeding patterns seen amongst herbivores with those of the predators. A herbivore consumes a large amount of plant material as food; since these are available readily in an environment and are not mobile, a herbivore does not spend any energy in chasing its food material unlike a predator. However, we must remember that the food ingested by a herbivore is of inferior quality when compared to the animal protein consumed by a predator. The value, as far as the nutrients are concerned, is more in the food of a predator; hence, a predator can afford to spend considerable energy in getting its food which is of a better quality unlike those of the herbivores.

General principles :

1. When predation starts affecting the population density of prey animals it means that the habitat has weakened either the predator or the prey.
2. The effect of predation on population densities is more significant when the predator is comparatively a more prolific species than the prey.
3. The numerical ratio of game animals to their natural predators may be disturbed when human hunters enter the field in large numbers.

Future of Wildlife Management

Wildlife management involves political, social, and biological factors. Biologists must monitor wildlife through surveys and research to effectively manage all wildlife species. Citizens also play an important role in wildlife management by supporting conservation programs and legislation and creating wildlife habitat. Wildlife management has restored wildlife populations, including Manipur deer vultures and more. Additional wildlife species will benefit from future management efforts.

Providing quality habitat areas, large enough to support specialized wildlife species, is a challenge facing wildlife managers. Urban sprawl, intensive agriculture, confinement livestock operations, and industrial development often decrease wildlife habitat and threaten water quality.NGO have emerged as leaders in conservation. These groups work cooperatively with government agencies to accomplish conservation goals.

Management of Human- Animal Conflict

This conflict has become a common event these days. Veterinarians are supposed to play vital role right from the capture, post capture management and post mortem which is the general sequel of the unmanaged cases. Generally animals are presented in maximally exhausted state thereby making the capture operation more risky as far as life of animal is concerned. So many causes have been attributed to this conflict like deforestation, habitat destruction, extensive cultivation, industrialization, shrinkage of prey base and disability. But a vet is concerned with the well being of animal irrespective of the causes responsible for the conflict. It is important to have the basic knowledge of the behaviour of animal for efficient management of conflict. Basic points in immobilization are:-

1. Safe handling of animal
2. Safety of the people involved.
3. Transportation to rescue centre.
4. Post capture medical attendance.

Wild animals are presented in following situations -

Trap Cage : Trap cage are positioned at various strategic locations in a conflict area with a suitable live or dead bait. The trapped felid is easy to immobilize with blow-pipe. Self inflected cage wounds due to struggle in need proper care.

Wire Trap : This is an indigenous device made out of vehicle clutch wire with a noose at one end while other end of the wire is tied to the stem of firm vegetation or an iron peg. The noose gets tightened in a abdominal area and causes incised would as the animal struggles. The forelimbs and teeth may also get injured due to struggle. Capture in this stage is risky. All necessary precautions must be taken and darting gun should be used.

Snap Trap : This is also an indigenous device made of iron in which mostly fore limb of animal gets entangled and injured due to struggle. Other limbs and body parts may also be get injured during struggle. Darting gun should be preferred in this situation.

Open Bait : When a felids is expected to come to the kill again or planted bait of same animal like dog or goat is kept, the position is taken at a safe distance hide from where dart can be fired.

Rescue Centre : Once the animal reaches the rescue centre, immobilization for subsequent examination and medical attendance can be accomplished by using blow pipe.

IMPORTANT CONSIDERATION

First Hand Information : One must seek on the spot information like distance of the spot from main road, on foot distance, type of vegetation, type of trap, terrain, position of animals weather upward, downward or parallel, nearest approachable safe distance. These are very good indicators for planning and immobilization operation. A responsible forest official must be deputed to gather and disseminate this information.

Coordination : One must have ability to coordinate with forest officers who are there as protected areas or park managers. After receiving information, meticulous planning has to be done for moving manpower, machinery and equipment to and fro the spot up to rescue centre.

People's participation : Involvement of peoples representatives of the conflict area is very important in seeking local help like materials, volunteers for carrying immobilized animal to the cage and vehicle, crowd control and other arrangements in case the operation is postponed. Help of a local shooter can also be sought as a felid in a wire or snap trap can get free and attack the people around.

Safety Measures : Safety of the people involved in the operation, local people who gather in large number and leopard as well as of prime importance. Help of law and order machinery can be sought from S.D.O.(Civil) for deploying police personnel for controlling the crowd and also the shooters in case of suspected attack. Crowd management is very important aspect of conflict management. Crowd must be controlled around the open trapped felid by fixing security tape in a radius of about 25-30 meters and nobody should be allowed to touch and cross it. Public address system may also be used to control the crowd. The dart should be fired from a distance beyond the striking range of a leopard in a particular situation. It has been observed that leopard attacks back with full force in the direction from where the dart strikes. Thus the trap may break resulting in attack by the leopard. To cope up with such situation services of an experienced shooter and few persons with sticks and prongs can be utilized. Heroic display by any person to approach leopard bare handed should not be allowed at all.

Escape Route : As a safety precaution, a felid should not be approached from the side which can serve as escape route in case the wire or snap trap gives away. Generally a passage covered with vegetation leading to wildness be kept for this purpose.

Instructions to the spot : The forest official deployed initially on the spot should be instructed not to allow the people to disturb the animal till the rescue team reaches the spot. It has been observed that people tend to tease and make animal excited and aggressive resulting in myopathy and injuries which could prove fatal after chemical immobilization. The rescue team must be kept informed about the latest position.

Forensic Aspect : Sometimes a animal is found in precarious condition due to gun shot or malicious poisoning. One should be prepared to send the relevant materials to forensic laboratory. Sealed samples are handed over after receiving written request. It is better to collect the samples in duplicate and retain one part after preserving and sealing them properly.

People's Awareness : It has been observed that people gather in large number and are curious. They can be well educated at this time. Causes of leopard menace and its possible out fall can be explained to them. For example if leopard are killed or trapped there would not be any check on other species like herbivores, wild boar, monkeys and jackals etc. If these species grow in large number can become vermin and pose so many serious problem. Therefore, a harmonious existence between every creature on this earth is very vital. Media should also be briefed from public awareness point of view.

Telemetry : Radio collaring of the animals around the conflict area, particularly felids can also be opted. As the animal approaches the locality, they can be chased away or captured if found disabled.

Capacity Building : Refresher courses and workshops for veterinarians, para-veterinarians and forest staff should be organized from time to time on a common platform for better interaction leading to better coordination. They should also be exposed to the reported conflict area in order to have better understanding.

CHAPTER 5

ETHOLOGY-ANIMAL BEHAVIOUR

Animal behaviour, ethology or animal psychology is a branch of zoology deals with reaction and activities of the animal in its natural environment. It is a fascinating subject which has attracted the attention of biologists. Observations were made by the animal behaviourists on their learning abilities, intelligence etc. Study of animal behaviour is a complex subject difficult to interpret unless efforts are made to understand its meaning in relation to other species, particularly man. Nevertheless, differences in the consciousness of mind among animals and man render the subject more complicated. Animal behaviour is part of biology and is composed of physiology, psychology and sociology and these subjects are an integral part of ecology. Animal behaviour is response to the stimuli the animal receives from the environ-ment and is based on the interpretation of the motives of human mind which is called anthropomorphism (attribution of human qualities like motives and emotions to the behaviour of animals) and is not free from errors. It is suggested that animal behaviour may be studied without regard for mental evolution of man as the animals do not behave in the same manner as human beings.

Ethology thus forms an important speciality for wildlife veteri-narians, for proper health management and care. Greek and early Darwinian biologists who studied develop-ment of instinct and intelligence among animals were unable to answer questions such as, do animals have souls? do animals reason? Have animal sensation? do animals have memory?do animal dreams? do animal counter act with human behaviour etc.

Detailed synthesis on etnology and psychology on tne behaviour of animals has been reviewed with over 400 cross references for detailed specific reading. Only a few aspects are discussed herein for basic understanding of the major parameters.

Aggression : Aggression takes place between two members of the same species which involves fighting, leading to injury or death. Attack of predator on their prey is not considered aggres-sion as it is aimed at obtaining the food. Aggression is directed at the territorial defence found in birds and mammals (stags, monkeys). Aggression is considered as an essential drive of life but the nature of aggression in animals is still a matter of controversy.Mutual grooming is a powerful means of reducing aggresssion in some animals, while raising of tail and errection of hair is a threatening signal in some animals and expansion of wing in some birds which under the autosomic control.

Acclimatization : It is the process by which the animal adjusts to the altered conditions of the environment to ensure the maintenance of normal body functions e.g. animals living in high altitudes become adapted to low levels of oxygen in the atmosphere by increase in the amount of haemoglobin in the erythrocytes. Animals can become acclimatized to the changed environmental temperature. An animal can normally tolerate a particular range of environmental temperature (tolerance zone) flourishes best at a particular temperature (optimum zone) and may die ata higher temperature (lethal zone). If animals are not able to acclimatize, particularly the cold blooded, their meta-bolic rate may either increase or decrease. However, exact mechanism of acclimatization in animals is not fully under-stood, although, anima l's behaviour is influenced by the activity of adrenal glands and anterior pitutary.

Agonistic behaviour : It may also be described as appease-ment behaviour and is characterized by both attack or escape behaviour to its rival. The animal may attack (aggression), flee or compromise. Aggression between two animals of the same species may at times be prevented by appeasement or submis-sive behaviour on the part of one of them. It is antithesis of aggressive behaviour and is necessary for a close and peaceful living. Food begging in birds, dog lying on its back, exposing its underside and allowing its opponent to sniff its body, offer to groom, shrieks and squeals of the defeated animal, sexual presentation in primates (lordosis) and immobility (flight stimulates aggression) are some of the common examples of appease-ment behaviour in animal societies. Dominance-subordination behaviour is commonly observed among animals. Raising of stright flag like tail in alpha male, keeping tail underneth the body as submissive expression in many primates and in carnivores.

Alarm Behaviour : It is the response of the animal to a dan-ger signal and is characterized by immobility, stillness and cryptic colouring which provides complete camouflage. Behavi-our of one alarmed animal is a warning to others. Alarm call is an interesting means of communication which may be under-stood by other species as well e.g. birds may give special calls to indicate the presence of a wandering cat and is probably understood by all jungle fauna. The jungal bablar's chirping and congregation, alarm call of deer species on visualisation of predators,,hanuman langur's call are some of the example of it.

Communication : Communication in animals is an intri-cate mechanism and it is vital that the animals communicate with each other, particularly during the mattng season. These signals are species specific which reduce the possibilities of interspecies mating, keeping a psychological barrier. There are many communication signals in the animal kingdoms. Visual signals;-Plumes in pheasants, peacock dance, light signals produced biochemically by fireflies, acoustic signals- many animals make sounds such as song of birds, touch signals-necking of giraffes while courting,chemical signals:- pheromones attracts the sex in musk deer,civet cats,scat glands in many animals(anal gland,interdigital gland,inner canthus of eye gland etc.)

Competition : Cometition is the process of natural selection.Incompatibilities may evoke conflict.The animals competefor the limited natural resources which leads to the survival of the fittest.The type of cometition is intra specific or inter specific.

Consciousness : It is the alert state of mind when the animal responds to the external stimuli. It is contrasted with the state of unconsciousness when the animal is asleep, anaesthetized, comatosed or hibernating (hibernating animals are semicon-scious or unconscious), which is related with the activity ofthe cerebral cortex. Mostly in carnivores like bears .

Courtship behaviour : Mutual behaviour of animals of dif-ferent sexes which are ready to mate is described as courtship. This behaviour is dependent on the changes in the endocrines which stimulate special postures. Usually the male makes itself conspicuous to attract the female. Sexual mating requires pro-mixity of the two sexes and advertisement that they are ready to fertilize or be fertilized. Different animals differ in their courtship and breeding behaviour, to avoid interspecies matings and prolonged courtship behaviour is probably aimed at it. In-tricacies of courtship behaviour of the animals are interesting (male carnivorous spider is eaten by its female mate; some car-nivorous flies, Empids are known to make courtship presents to the female consort in the form of worms wrapped in a parcel of silk as food and he copulates while she eats.

Breeding behaviour : It is concerned with mating, care of the young and laying of scent trails to attract the mate. Building and maintenance of nests in the birds (weaver birds) and establishment and protec-tion of harems(laking in black buck) in some animals is also considered in breeding behaviour. For reproduction, gametes from opposite sexes must come together to produce fertilized ovum. For this, male and female must come together in physical contact in the act of copulation. For fertilization, synchronization of gametogenesis is essential and courtship enables it. The zygote must be embedded in the uterus of the female or laid as egg with a protective shell in birds, to be hatched later. After birth or hatching, prenatal care is necessary for the survival of the neonate.

Vertebrates : Mating behaviour becomes more complex in higher vertebrates, reaching to its climax in man. Fish like many other aquatic animals send acuostic signals by grinding of teeth etc. Churring and whistling type of breeding songs of toads and frogs are followed bylong period of mounting and fertilisation.In birds, songs and displays are more conspicuous to attract the member of the opposite sex. These signals are so diversified and specific that different species of birds which are otherwise capable of hybridization, breed successfully without breeding catastrophes in nature. Most birds are monogamous, forming pairs for the season or whole life, some are polygamous but polyandry is rarely seen. Brooder parasitism is a phenomena in which incubation of eggs and rearing of the young is foisted on to other species, occurs in about six families of birds (cuckoos).Some bird nest in solitary area where some in colonies(heron,flamingo).The nest may be abandoned after breeding or to be reuse after repair in the successive years.

In most mammals, visual and vocal displays are thought significant but are not elaborate as in aves.Antelopes, deer,monkeys maintain harems of females and the alpha dominant male ususally copulates, based on seasonal cycle.In caninesurine scants excites sexual desires in males.Pack hunting in wolves,aunting behaviour in elephants represnts the social structure in this species.

Insects are ususally unisexual and copulate to breed.They prefer scent signal,visual trails,sound or dance.Scorpion's mating dance is associated with discharge of the sperms on the ground in a capsule called spermatophore and during the dance female lowers its posterior part of the body to pick it up. In molluscs, whether unisexual or hermaphrodite, cross fertilization is a rule. In some species of molluscs, calcarious darts are shot which pierce the body of the other animal, releasing a chemical which brings the snail in mating mood. Male Siamese fish is beautifully coloured and exhibits vital postures. It builds a nest of bubbles at the surface ofthe water, coating them with saliva and welcomes the mating female into the trap for courtship.

BEHAVIOUR OF SOME WILD ANIMALS IN CAPTIVITY

Primates

They are most responsive to formal training sessions amongst captive species. Among the wild species they have social hierarchy. Insecticidal problem is also there. However in captivity, keeper should take care that they should not do some of the activities in front of primates like turn of gas stove, paint, open refrigerators etc. The positive responder should be rewarded during training. The trained primates help a lot for collection of blood, vet. Clinical examination, crowed entertainment etc.

Abnormalities : Hyper grooming, bizarre, stereotype, mal adaptive human imprint effects are common. Cage destruction, paint chewing, masturbation, self mutilation, emesis followed by coprophagia, anorexia, body weight loss, depression are commonly observed psychopathology due to stress and improper introduction to other cage mate or to those primates caught from wilderness.

Ruminanats

Ethological understanding of ruminants is more important to manage their physical and mental health in captivity. Conditioning is of basic part by music, bell, and horns or by human voice as an auditory signal for their gathering, feeding and medication. The concept of adaptation is important for veterinarians for such ruminants. Introduction of herd, group managment seperation of males from females, concept of interaction, space, neighbours, inside furnitures, constructive occupation are vital for their behaviour.

Abnormalities : Hair plucking, coprophagia, fighting, stereotyped behaviour are expressed on account of unoccupied hours except feeding.

Felids

Most of the felids behaves in free living often mask their behaviour in captivity. The solitary nature of tiger, pride and social bonding of lion, wider prey base, hiding habit and preference of shelf by leopard is some of the ethological observation in free living.

Abnormalities : Circle running, figure eights, rolling over, tail and paw chewing, enuresis, encopresis, pica, hyperesthesia, aggression, frustration, stereotyped behaviour have been documented in felids.

Canids

In free living the canids stay in pack. They have some typical behavioural problem in cages which are some what similar to felids.

Ursidae/Viverridae and Hyaenidae

Hiking eggs, anxiety, pacing. anal sac material throwing are common natural behaviour in viverridae and hyaenidae, stereotyped movement, head swaying, to and fro pacing and pacing with head bobbing, paw sucking and dancing are some of the observed unusual behaviour to express their unhappiness by bears.

ERRISODACTYLA, PROBOSCIDAE AND HIPPOPOTAMIDAE

They are social, living in groups and may face problem of stereotyped in captivity as others described earlier.

Rodents and Lagomorphs

They are burrowers and rooters by nature and needs environment compatible with their natural requirement. Constructive occupation (foraging activity) is needed to solve some of the problems.

Abnormalities : If placed isolated can become more aggressive expressing circling, convulsion and "figure-8" walking patterns. Cannibalism and cage mate fighting, hypoglycemia, compulsive water drinking are some of the abnormal behaviour.

Procyonidae

Raccoons, coatimundi, giant panda prefers washing of food prior to eating it. Enrichment of enclosures is helpful to prevent psychopathology.

Mustelidae

Otter, ferret and skunks are hyperactivity, aggressive and prey-seeking behaviour. They release an odoriferous sebaceous material when extreme anxiety occurs.

Aves

They express territorial ,breeding activity based on day light hours. Cannibalism, vent picking, feather picking are some of the behavioural problem.

Reptiles and Amphibians

They do not express psycholopathology if kept with proper caging and environment. Pacing back, ecdiais and shedding are common in snakes. Basking in reptiles and amphibians is often common and one must provide larger area and movement as well as hiding space with darkness also.

CHAPTER 6

HEALTH MONITORING AND EVALUATION IN FREE LIVING WILD ANIMALS

The regular record keeping and feed intake in captivity with close observation helps the veterinarian to understand the distress and diseases in captivity but it is very difficult to assess in free living wild population. The health status of a wild population can be assessed by evaluating the following components:

a. Habitat evaluation
b. Population statistics &
c. Direct animal assessment

Habitat evaluation is essential because long term survival of a population depends to a greater extent on the habitat quality. This evaluation includes estimating vegetation / prey biomass, level of disturbance (over grazing, competition etc.) and other habitat components, perhaps by using HIS models information on population statistics is important because the ultimate measure of a population condition is its reproductive success. This includes various population parameters such as natality, mortality, net recruitment rate etc. Direct animal assessment is also important because individuals of a population act as indicators of the status of their relationship with the environment. This last component is otherwise the health monitoring of animal populations.

Health monitoring procedures for wild animals should be a regular exercise in protected areas. Unfortunately, such programs rarely form a regular part of the management practices because of the wrong notion among managers and biologists that (i) the need for disease monitoring arises only when there is an epidemic and (ii) that diseases occur only when an infective agent is involved. This opinion is not justified because (i) only a few infections are manifested as disease while many are not so immediately obvious and (ii) the most prevalent disease syndrome in wildlife is the malnutrition / starvation complex.

MONITORING CONDITION OF ANIMALS

Condition of an animal responds to the changes in its habitat quality, which is governed by many interrelated component factors which are often seasonal in nature. Through most animals are subjected to many of these factors, their effect on the overall health of a population may be more apparent in some species or in different age or sex groups of the same species.

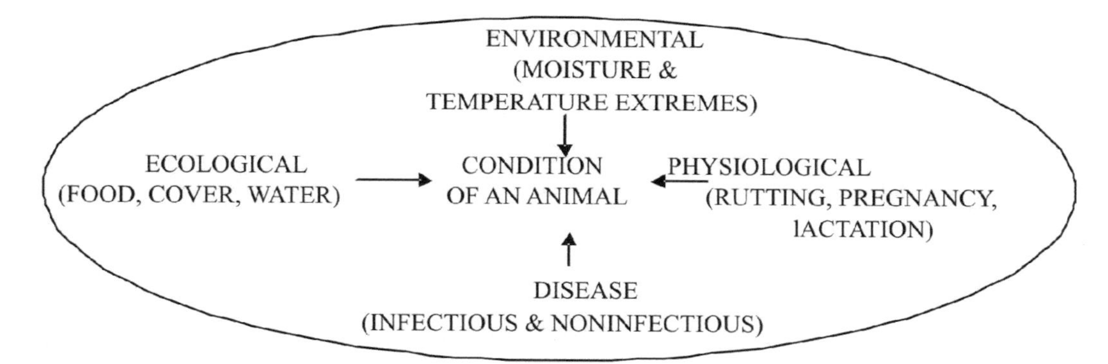

Fig. 1. Condition of an animal is governed by a complex of factors which are interrelated.

The deterioration of condition could be due to nutritional stress chronic diseases resulting in muscle wasting of natural biological reasons (pregnancy / lactation in females & rutting in males). Many non-infectious diseases and biological processes, however, leave a noticeable effect on the body condition of animals. Condition of live animals can be assessed by looking at their appearance or body condition and condition of dead animals can be assessed by estimating the extent of fat deposition in the body.

Monitoring Infection and Disease

Observational methods of monitoring infectious diseases

Laboratory based methods of monitoring infections and diseases

(a) Postmortem Investigation

(b) Serological surveys

(c) Macro-parasitological investigations

(d) Screening for micro parasites

STUDY OF KILLS

Study of kills means post-mortem examination, in the present context. Detailed post-mortem examination is important for the diagnosis of the diseases. It also gives information about the health of the animal.

Diagnosis of the diseases and parasitism require careful and detailed evaluation of dead animal with subsequent collection, preservation and eventual examination of appropriate specimens.

The field investigator should record all pertinent data that may be of importance in making a diagnosis. Any fact that the investigator feels, might have some bearing, should be recorded. The history should include species, sex, age of animal, locality from which the animal came, habitat type, date and time when it was located, if the animal is dead, complete description of the clinical signs of illness, number of other animals of the same species involved and died.

Post mortem examination if at all possible should be conducted in diagnostic laboratory, where they can be necropsied by trained pathologist. Essential tools required for post-mortem examination are rubber gloves, sharp knives, scissors, bone shears and forceps. It is also desired to have a clean, well lighted placed to work, disinfectant, good camera, adequate water supply, 10 % formalin, 70 % alcohol, sterile plastic bags, vials for collection of blood and glass slides are necessary material for collection of the specimens.

The post- mortem examination should be done thoroughly and examined all external characters of the carcass. Open the carcass in such a way that the systematic and system wise post-mortem examination can be made with ease.

BODY CONDITION EVALUATION

Table 1: Generalized description and evaluation of different body parts of ungulates*

	Body part	Point = 0	Point = 1	Point = 2	Score
1.	Flank area	Depression is barely visible. Flank area outline is indistinct	Flank area slightly concave and outline visible	Depression concave and tucked in	
2.	Ribs	Thoracic surface is smooth and ribs are difficult to see	Ribs are visible but not all can be counted with ease	Ribs prominent with distinct inter-coastal depressions	
3.	Pelvic Girdle	Bony projections of pelvic girdle are barely visible	Pelvic girdle outline slightly visible	Bony projections of pelvic girdle are clearly visible	
4.	Vertebral column	When seen laterally, it runs smooth without any breaks. Lumbar processes visible	Lateral processes of lumbar vertebrae are visible but not prominent	Lateral processes of lumbars very prominent. Dorsal processes of vertebrae seen	
5.	Lumbar shelf	No depression in shelf. Appears almost round from behind	Slight depression on either side	Depression deep and concave	

BCI = (Interpretation: 0-4 = Good; 5-7 = Fair; 8-10 = Poor) * Modified from Riney (op. cit.)

Table 2: Description and evaluation of different body parts of Asian elephants*

	Body part	Point = 0	Point = 1	Point = 2	Score
1.	Temporal depression	Flat and frontal ridge vaguely defined	Slightly concave and frontal ridge defined	Deeply concave	
2.	Scapula	Spinous process of the scapula not visible	Spinous process visible	——	
3.	Ribs	Thoracic barrel smooth and ribs barely visible	Ribs visible	Ribs clearly demarcated with pronounced intercostal depressions	
4.	Flank area	Flank are depression barely visible	Sunken flank area Depression visible	——	
5.	Pelvic girdle	External angle of illium not visible	Visible but not pronounced	External angle jutting and pronounced	
6.	Tail	Muscular, not bony	Joints of tail vertebrae seen	Thin and bony, vertebrae can be easily counted	
7.	Lumbar shelf	Shelf not present. Round when seen from the rear	Lumbar shelf appears flat and not round	Pronounced concave shelf on either side of the vertebral column	

BCI = (Interpretation: 0-4 = Good; 5-8 = Fair; 9-12 = Poor) * Based on V. Krishnamurthy

CHAPTER 7

PHYSICAL AND CHEMICAL RESTRAINTS

The wild animals are often required to catch and restraint for various kind of routine operations like detail examination, treatment, shifting and crating of animal.

There are various kinds of equipment and techniques used for capturing and restraining of wild animals. The basic techniques includes (I).Physical restraint and (II) chemical restraint.

It is very much essential to evaluate the physical condition of animals and the justification of capture because each restraint has some effect on behaviour, the life or physiological activities of animal. Some of the animals like black buck, chinkara and deers etc. are very much sensitive to restraint and susceptible to shock, even a simple capturing may result in, injury or death to animal and personal involved.

MAJOR DEFINITIONS

Anesthesia – renders an animal unconscious. Provides analgesia

Amnestic – Most drugs (if not all) wipe out the short-term memory of the animal. The animals do not remember the few minutes leading up to the drugging. This is a great quality and allows you to repeatedly drug the animals without them becoming wary.

Analgesia – Loss of pain and sensation

Tranquilization/sedation – Does not cause loss of consciousness. Animals can be roused with stimulation.

Catalepsy – Stiff rigidity in muscles.

Therapeutic index – The ratio of the amount of the drug that will cause mortality of the animal compared to the amount that will serve as an effective dose for immobilization. Most of the drugs we use in wildlife have a high therapeutic index. It is better to overdose the animal (in most cases) than allow them to run around overheating, which may result in capture myopathy and eventual death.

Shelf-life – Drugs lose their efficacy by sitting on a shelf for too long. Need to keep drugs above 90% efficacy.

Induction time- the time from when the drug is administered to the time the animal goes down and is out.

Thermoregulatory Effect – An increase or decrease in core body temperature

Antagonists – Administered to counter the effects of the primary anesthetic and/or the tranquilizer. It is preferred if they are administered intravenously, but intramuscular injection also works fine – it just takes longer.

PLANNING

Most state wildlife agencies offer in-house courses to their personnel on methods of wildlife capture and immobilization. In addition to attending a complete course on wildlife capture an immobilization, you should seek the advice of more experienced personnel and practice the techniques before using them on animals. Conditions in the field are highly variable, and you should learn to recognize how they may affect a potential capture and immobilization event. Incompetent personnel can lead to death of the animal or even you! Equipment Readiness – Gear should always be available, in good working condition, and current (e.g., drugs should be properly inventoried and not outdated. Outdated drugs may be less effective.) The drug box should be clean and organized. Take personal responsibility for correcting problems with equipment and drugs and do not rely on others in order to always be prepared for a call. Replace equipment immediately after use so you will be prepared for the next call.

Release Sites – Always check your surrounding before releasing an animal. This is for the safety of the animals as well as your own. After all the time and expense you have gone through in capturing an animal, be sure to release it in an area where it has the best chance of survival. Here is the example of coordination system of USA.

Interagency Coordination – Each supervisory district should have a plan or agreement for coordinating response and animals care with the local Police Department, Sheriff's Office, Animal Control, US Forest Service Ranger District, electric company, wildlife veterinarians, and wildlife rehabilitators. If a plan does not exist, MAKE ONE!!! Check with you Area Coordinator and District Supervisor. The plan should list the role of each group. Discuss problems that have occurred, problems you may encounter, ways to improve response, and respective roles of each agency.

ASSESSING THE SITUATION

When you receive a nuisance of depredation complaint, try to contact the complainant directly. Ask for the following information:

Location of the problem

Number of sightings

Damage to property

Aggressiveness of the problem animal

Threat to human safety

The initial contact will help you to decide your response and what equipment to bring. At the scene, consider the following:

1. **Should you capture the animal?** Often nuisance situations cab be resolved without capturing the animal through education, diversions/scare tactics, habitat modifications, etc. Assess the need to capture based on whether the animal has caused damage, whether it threatens human safety, and whether a capture is safe for you and the animal. *Recognize the public often expects you to capture the animal, but rarely understands that capture is the least desirable alternative.*
2. **Physical or chemical restraint?** The rule of thumb is to use physical restraint unless it jeopardizes human or animal safety.

   Educate people Scare off animal Trap or physically restrain animal Dart
3. **Type of terrain and type/condition of animal.** If the area can easily conceal a downed animal, chemical immobilization should be reconsidered. A chemically immobilized animal will have at least 5 minutes travel time to escape. Likewise, carefully plan your strategy if the animal is aggressive (e.g., males in rut or females with young). Excited animals usually require higher doses of the drug. Debilitated animals (diseased, malnourished, dehydrated) are less difficult to chemically to immobilize than healthy animals.
4. **Adequate equipment and assistance?** Based on the information received from the complainant, ideally you would have gathered the right equipment and obtained the assistance required. However, situations may change at the scene or your equipment may be inadequate for the call. If you do not have adequate equipment or assistance, you should reconsider your response for your own safety as well as the animal's well-being. You are no match for wild animals when it comes to strength, quickness, or reactions.
5. **Human Safety.** Human safety is your first priority! Ideally, you should have arranged assistance from police or sheriff officers to maintain crowd and traffic control (see *Interagency Coordination*). If you do not have assistance, you must enforce control or wait until help arrives. Expect chaos if crowd control is not enforced immediately or authoritatively. If you are dart-gunning, the public must remain at a safe distance from you as you are loading drugs, from the animal, and from the dart. Never attempt a shot unless you are sure the public will not be hit if you miss. Account for all missing darts. Do not accept help from unknown or untrained personnel. If you cannot capture the animal without endangering the public of yourself, reconsider your response.

PHYSICAL RESTRAINT (cited from Fowlers)

In order to be successful in working with animals, we must understand their behavioral characteristics and the aspects of their psychological makeup that will allow us to provide for their best interests.

The successful restraint operator must understand and have a working acquaintance with the tools of restraint (including the use of voice) and with manual and chemical restraint techniques. The use of special restraint devices should be mastered.

The tools used in effecting a given degree of restraint vary greatly. Some tools may be desirable for dealing with one species but are contraindicated for used with another. Success in the art of restraint requires both experience and study to know when it is appropriate to use a specific type of restraint. Inappropriate use of certain techniques may be not only unwise but also dangerous to animal or human.

Tools are place into the following groups: (1) psychological tools, which enable more satisfactory manipulation of a given animal because certain biological characteristics are understood; (2) tool that diminish sense perceptions of animals; (3) methods of confinements (4) tools to lend added strength or to extend the arms; (5) physical barriers of tools that protect us or allow closer scrutiny of animals; (6) physical force, a tool used to subdue animals; and (7) chemical agents, which sedate, immobilize, or anesthetize animals.

Psychological Tools

Voice, an important tool, is frequently overlooked by animal handlers because of its simplicity. Emotional states are reflected in the voice. Both domestic and wild animals readily perceive fear or lack of confidence. Timidity in approaching an animal, the way the hands are held, quickness or slowness in use of the hands, and general stance indicate to the animal the presence or absence of confidence.

With wild animals, it is important to recognize that training may involve establishment of dominance over the animal by the trainer. This is a complex behavioral phenomenon, and it is unlikely that a different person who comes in to manipulate that animal can acquire such dominance in a short period of time.

The successful restrainer must acquire detailed knowledge of the anatomy and physiology of the species to be manipulated, including the distance the limbs can reach to kick or strike. It is important to know the degree of agility and speed of the species in question. The importance of gaining as much knowledge as possible of the biology and physiology of any species to be restrained cannot be overstated.

Diminishing Senses

Reduction or elimination of an animal's visual communication with its environment is an important restraint technique (Fig. 1, Chapter 6). Stress to a parakeet can be reduced if it is placed in a darkened room before it is grasped for examination or medication. Obviously, it is impossible to blindfold mist wild animals until they are already in hand. One can frequently place animals in a darkened environment, however, and subsequent to capture, much stress can be relieved if the animal is blindfolded. A blindfolded animal my lie quietly for a long period of time while nonpainful manipulations are carried out. Sedation and anesthesia are required for painful procedures.

Sedated animals handled in sunlight should always be blindfolded to prevent damage to the retina by direct rays of the sun on an eye that cannot accommodate with normal rapidity.

Confinement

The acceptable degree of confinement may vary considerably, depending on the species and the situation. To the free-living adult wild animal, placement in a large, fenced area represents confinement and results in a certain degree of stress. Confinement stress can be progressively intensified by a gradual shift to smaller enclosures. In a zoo situation, this may be in an alleyway; for a domestic animal, this may be confinement in a stall or in a shed. The closest and most stressful confinement for an animal is to be placed into a special holding area such as a transfer cage, a special night-box or bedroom, a shipping crate, or one of the many different types of squeeze cages.

Squeeze cages are extremely valuable restraint tools for wild animals. It is important to recognize that no squeeze cage can be adapted for universal use. Animals vary in both anatomical conformation and physiological requirements, and the design of the squeeze cage must fulfill these in order to be safe and useful for carrying out various procedures.

Extension of the Handler's Arms

Ropes are excellent means of extending the working range of the arm. Snares are also used to capture and restrain animals in a variety of situation. A snare is an important tool, but if used carelessly, it can cause suffocation or unnecessary pain. Commercial snares are usually designed with swivels for more humane and effective manipulation. A quick-release snare that permits the animal to twist without being suffocated is shown in Home made snares can be constructed from either metal or plastic pipe; the flexible part can be made of rope or cable.

Nets are important tools for animal restraint. They come in all sizes and shapes, from those used to capture tiny insects to the very large cargo net used restrain a musk ox. A variety of sizes should be obtained so that a wide range of species can be manipulated. When a net is placed on animal, many procedures, such s medication, expels for laboratory work, can be carried out. The size of the mesh should correspond to the size of the animal to be netted. If the mesh of the net is too large, the animal may force its head through the mesh and strangle before it can be released.

It is important to know the characteristics of the materials from which a net is constructed. Nylon, cotton, and manila are all used, and each will with stand different degrees of stretch and wear. Carnivorous species are apt to chew at netting and may effect escape by chewing holes.

Physical Barriers

Physical barriers may be used to protect both handler and animal or to allow handlers closer proximity without alarming the animal.

Shields are important tools of restraint. They may consist of simple plywood sheets or may be equipped with handles on the back to be held by the manipulator. Shields can allow close approach to an animal and can be used between two transfer cages having swinging instead of guillotine doors. Plastic shields are useful in handling large, nonvenomous reptiles, some of the smaller mammals, and some birds.

It is obvious that this technique is suitable only for small- to medium-sized primates, carnivores, rodents and other small mammals not strong enough to push away the shield.

A blanket may be used to shield the animal from the handler. It will also protect the handler from legs, horns, or antlers. A small antelope can frequently be captured by allowing it to jump into a blanket. It can then be completely enclosed and held in the blanket. Small mattresses can be used similarly.

Animals recognize an opaque plastic sheet as a barrier, whereas they may not recognize a wire or wooden fence as such. Thus, animals can be directed into loading crates or into chutes with opaque, plastic sheeting in a manner heretofore impossible. Plastic sheeting has a its greatest application in the herding of hoofed animals.

Physical Force

Most manipulative procedures require use of the hands, and the wise restrainer takes every precaution to protect them. The restrainer must know where and how to grasp the animal in order to protect himself and to accomplish the restraint required. The greatest protection for the hands is detailed knowledge of the animal to be restrained.

The amount of force applied must be appropriate to the species. Handling a 50-gm parakeet is indeed different from holding a 12-kg macaque. Suffocation may result from the application of too great a pressure, and limbs or ribs may be fractured by applying too much force.

Many people use gloves, important tools of restraint. Glove material varies from thin cotton, used to handle small rodents, to heavy, double-layered, coarse leather, used to handle large primates. Leather welder's gloves are excellent for general use.

Restrainers should realize that wearing gloves diminishes tactile discrimination. The thicker and heavier the glove, the less the ability of the handler to determine how tightly he is grasping an animal or to feel the response of the animal. Because of this, many handlers refuse to use gloves.Carnivorous are usually able to bit through the thickest glove available, so a glove is no guarantee of protection from biting. In addition, gloves do not protect from crushing by powerful jaws. A tiger can crush the bones of the hand or a large macaw can fracture finger bones without breaking the skin of gloved hand.

Animals	**Equipment required**
Small carnivores	Nets, pole syringes, blow dart equipment, crates, squeeze cage
Hoofed stock	Projectile guns and darts, equipment, crates blow dart
Small mammals (e.g. primates)	Nets, surgical gloves, pole syringe, blow dart equipment, crates, squeeze cage
Reptiles	Nets, bags, plastic tubes, snake tong, snake hook
Amphibians and fish	Nets, gloves

Physical restraint should be used in preference to drugs if the animal can be handeled safely without exhaustion or injury to the animal.as a general rule the least restraint is best restraint. The following comments provide information on methods of restraint that are appropriate to the broad taxonomic groups listed :

1. **Large Mammals :** The choice of season for handling and restraining ungulates is an important consideration. For example, bison and elk tend to be less aggressive and most easily handled in winter, while large herbivores may be baited into traps and holding areas more readily when natural conditions are poorest.If possible, researchers should avoid working with large mammals during calving and breeding seasons. Pregnant animals should not be handled or restrained late in the gestation period. However, if the capture of newborn animals is part of the research it will be necessary to handle animals during spring. In some species of ungulates the mother will separate from her young when disturbed, leaving the young in hiding and accessible to capture.

   Wild ungulates may fail to recognize chain link or wire as a barrier; consequently, corral and run fences made from these materials should be draped with burlap or opaque plastic. Crossboards or "bellyrails" may also be used for this purpose.Body squeezes are commonly used to handle ungulates. A head squeeze is required to permit examination or marking of the head or front quarters of large mammals. Circular corrals and curved runways or chutes will improve ease of handling. A covered center insert in a corral provides a hiding place where animals can conceal themselves during the holding process. An animal should be able to see the rear of the animal preceding it, but must also be provided with a "pathway of escape" or "light at the end of the tunnel" during movement through solid sided chutes. As with cattle and sheep, a natural following tendency and group movement can facilitate handling. Within the processing corral, chutes for confining individual animals should be equipped with drop hatches or escape doors to provide additional access for working on the animal and to permit emergency exit of animals that become distressed.Immature bears can be hand held or controlled by nets or snares. Squeeze cages for large bears must be heavily constructed.When using muzzles to restrain canids, the investigator must not allow the animal to overheat, as the muzzle will not permit it to pant. Large members of this family should be handled only with special squeeze cages or by chemical restraint.Large felines are frequently handled using squeeze cages. If nets are used, the mesh should be small enough that the animal cannot poke head, paws, or claws through. When prolonged treatment of large felines is required, the immobilized animal should be strapped to the table to preclude injuries.

2. **Small Mammals :** Many small mammals can be adequately restrained by holding them firmly across the shoulders with a gloved hand so that they are secure, but cannot twist and bite the handler. Tiny species may be held by grasping the loose skin that runs down the neck and back. Conical cloth handling bags with a zippered opening at the narrow end, net bags, and wire cones are useful for large or more vigorous species, and when weighing or marking an animal.
3. **Birds :** When birds are restrained by hand, the hold must include the wings and legs in order to prevent damage or fractures to these appendages. When using a cloth bag, sack, or hood to restrain and settle a bird, care must be taken to prevent both hyperthermia and damage to plumage. Excessive loss or damage of feathers may lead to death from hypothermia after release. Birds that become immobile when suspended upside down can be inspected and weighed while secured by a leather thong placed around the legs. The capture and/or marking of some species of birds may alter their behavior and predispose them to death.
4. **Amphibians and Non-hazardous Reptiles :** Most amphibians and reptiles are relatively small and slow moving, and so they can be captured and restrained by hand or in a net. Small animals are easily injured if excessive force is used. Tail autotomy occurs in most lizards if they are restrained by the tail; although this is not a serious injury, it may affect the usability of the specimen.
5. **Hazardous Reptiles :** Venomous snakes, crocodilians and some of the large turtles are potentially dangerous, and thus require special methods of restraint. The particular method chosen will vary with the species and the purpose of the project. Adherence to the following general guidelines is recommended when working with hazardous reptiles:
   (a) procedures chosen should minimize the amount of handling time required, and reduce or eliminate contact between handler and reptile.
   (b) whenever possible, an anesthesic and/or physical restraint should be applied **before** actually handling the specimen.

(c) one should never work alone. A second person, knowledgeable of capture/handling techniques and emergency measures, should be present at all times.

(d) prior consultation with workers experienced in these species, in addition to reviewing the relevant is of particular importance here as much of the information on handling dangerous species is not published, but is simply passed from one investigator to another.

6. **Fish :** The most common method of capturing fish is by netting. Care should be taken to avoid abrasions, as removal or damage to mucous secretions and scales can lower the resistance of the animal to bacterial infection. Small fish can be captured using plastic bags

CHEMICAL RESTRAINT

The last decade has seen a phenomenal increase in the use of drugs for restraint and immobilization of wild animals. Commonly used drugs now permit manipulative procedures that were heretofore impossible to attempt. The lives of many animals have been spared by the judicious use of drugs that minimize stress and trauma.

An ideal drug should have a high therapeutic index (lethal dose/effective dose). Since the majority of chemical restraint agents are administered intramuscularly, the ideal drug also should not irritate the muscle. Some agents may sting or cause transient localized pain upon injection but cause no damage.

A short induction period is desirable. Movies and television have fostered the belief that darted animals fall immediately. Chemical restraint agents available at present require 10 to 20 minutes following intramuscular injection before immobilization is effected.

The ideal drug should also have an available antidote, which reverses its effects and prevents death from respiratory arrest or other problems that may arise during the course of immobilization.

Solution stability is important. Many restraint drugs are used in situation in which refrigeration is not available. The ideal drug should remain stable in solution for long periods of time at room temperature.

Finally, it is important that the effective dose of the drugs is low enough to allow its use in the small volume syringes necessary for dart injection.

Delivery Systems

The first challenge facing the person using a chemical agent for immobilization and restraint is to administer it to a site that allows absorption. Satisfactory techniques vary from species to species and from animal to animal, according to size, distance from the operator, ability to partially confine the animal, operator skill, and the effectiveness of the available equipment.

ORAL

Numerous unsuccessful experiences have taught the author the futility of depending upon oral medication to sedate wild animals. Aside from the lack of tolerance of such drugs by many species, the effectiveness of oral medication is often minimal, since many chemical restraint agents are either unabsorbed or destroyed in the digestive tract.

HAND-HELD SYRINGE

Intramuscular injections can be given very quickly with a syringe held in the hand. A large-gauge needle should be used to deliver the liquid quickly. The needle should tightened securely on the hub so that pressure buildup by rapid injection will not blow the needle from the syringe. A Luer-Lok connection is desirable.

PROJECTED SYRINGES OR DARTS

Modern chemical restraint requires equipment capable of projecting a syringe some distance and discharging the contents upon impact.

The first suitable weapons were developed by Jack Crockford and coworkers. Palmer Cap-Chur equipment is the standard used todway in the United States. Other companies in New Zealand and Africa are now competing for the world market.

There are three types of Palmer projectors (guns). The short-range projector (pistol) is a modified pellet gun powered by compressed carbon dioxide (CO_2). Its range is 15 m (16 yd). The long-range projector (rifle) is also powered compressed CO_2, and its range is 35 m (38 yd). The extralong-range projector is powered by percussion caps according to the distance from the target. The maximum range is 80 m (87 yd).

The blowgun is becoming popular with animal restrainers, since it offers certain advantages over other delivery systems. A major advantage of the blowgun is silent projection with less trauma on impact. It is adaptable for use on small animals, it can be easily sighted, and it has no mechanical parts requiring maintenance. The only disadvantages of the blowgun are its length and its short range. It is unwieldy in confined spaces, and the range is limited to approximately 15 m (16 yd).

Various homemade and commercial stick syringes extend the hand for administering drugs to dangerous animals. All require injection immediately upon insertion of the needle. A quick jab is necessary to effect administration, and the operator must maintain pressure against the animal until all the material has been injected or until the animal has jumped away.

Prerestraint Consideration

When preparing to chemically immobilize an animal, consider the following: (1) the species; (2) the physiological alarm status of the animal, including its age, sex, general health and state of lactation; (3) the physical condition; and (4) the emotional status of the animal. Deviations from the normal in a given species can have a profound influence on the outcome of any chemical restraint procedure.

No chemical restraint agent available at present is equally effective and safe for use wit all 45,000 vertebrate species. Experience has shown that most agents have strict species limitations.

The most important modifying factor to be considered in chemical restraint is the emotional status of the animal at the time of injection. Injecting a drug into an animal that is in a state of alarm (with high catecholamine and cortisol release) may produce effects opposite to those occurring in a normal, quiet animal. Emotions such as excitement, fear, and rage all produce the alarm response. Excitement may cause aimless running, which, in turn, produces acidosis and sensitizes the cardiac muscle to the effects of atecholamines. The end result may be an episode of ventricular fibrillation.

Adverse Effects of Restraint

Successful immobilization of a wild animal is an art. Many factors are involved. The operator must consider not only what equipment to use and the animal's condition but personal ability as well. If the operator is not skilled in the use of the chosen device, it will be difficult to utilize it to its greatest advantage.

Other causes for failure of immobilization procedures can be categorized into three major areas: (1) equipment failure, (2) operator fault, and (3) miscellaneous conditions over which the operator has little control.

MISCELLANEOUS CONDITIONS

Climatic conditions affect the functioning of equipment and the flight path of the dart. Wind can have a marked effect on trajectory. Sometimes because of wind it is necessary to select another time for injection. Warm weather increases gas efficiency, adding to the range of the Co_2 projector. Conversely, cold weather decreases the range.

IMMOBILIZING AGENTS

1. **Etorphine Hydrochloride :** It is a opium alkaloid (thebaine). It is particularly useful for immobilizing large ungulates such as Elephants, rhinos or hippopotamus. On set of anesthesia occurs in 10 to 20 minutes after i/m injection.

   Antidote : Diprenorphine at double the dose of etorphine. Recovery with in 4 to 10 minutes.

2. **Fentanyl and Droperidol :** It is a morphine derivative; Droperidol is a tranquilizer. The combination may induce a slight decrease in blood pressure. Used in a wide range of wild species, particularly carnivors, non-human primates and various small

mammals. Perticularly desirable for short procedures. Effect is seen with in 10 to 15 minutes and lasts for approx. 40 minutes.

Antidote : Naloxen hydrochloride 0.006mg./kg – i/v or i/m injection. Immediate eversal.

3. **Ketamine Hydrochloride :** A lerivative of phencyclidine Hcl Animal retains pharyngeal – leryngeal reflexes. This minimizes hazards of aspiration. Ketamine crocess placental barrier in all species. Anesthatic effects are noted in fetus also but is not known to induce abortion.

   It is particularly effective in wild carnivores, reptiles and birds but it is not suitable for most ungulates can be given orally or particularly Dose : 2 to 50 mg/kg – Tremendous variation Effect : within 50 to 10 min. Smooth recovery.

   Side effect and precaution : Causes tonic clonic convulsions in a small % of domestic and wild felids and other carnivores. (Taken care by giving Diazepam – 0.25 to 0.50 mg/kg) Antidote: Yohibin (0.125 mg/kg).

4. **Tiletamine hydrochloride and Zolazepam hydrochloride :** Tiletamine is a cyclohexanone dissociative agent related to ketamine hydrochloride. Zolazepam hydrochloride is nonphenothiazine pyrazolodia zepinone tranquilizer combition capitalizes on desivable characteristic of each while minimizing the side effect. Onset occur within 5-12 min after I/m ing.

   Antidote : No antidote.

5. **Xylazine :**

   Non-nacotic selative.

   Animals under the influence of xylazine appears to be sleeping. A populat immovilizing agent used signally or in combination with other drugs for a wide variety of spp can be given I/m or I/v. Immobliziation occurs within 15-30 min. by I/m or 3-5 min. after I/v inj. And lasts for 1-2 hours.

   Antidote : Dopram help in hestaning recovery yohimbin (0.125 mg/kg) can also be used.

6. **Acepromazine Meleatc :** Rarcly used singly but is must often used incombination with ctrophine or ketamine.

7. **Diazepam :** Effective anticonvulsant. It can be used as a preanesthetic medication to calm can excited animal.

   Dose : 0.5 to 1.5 mg/kg. onset within 1-2 min. with I/v inj.

BASIC PRINCIPLES

Use of chemical restraint is also increasing day by day for restraint and immobilisation wild animals in developed countries with advanced system and latest drugs. The advantage of chemical restraint compared to earlier technique of mechanical method is decreases the chances of physical damage and shock to animals, but requires past experience and safety margin know how of drugs used for the wild animals with specific antidotes.

i. **Route of Administration of drugs :** Oral route of medication is not effective because of difficulty in dosing the animal accurately due to uncertain drug intake and gut absorption. Intravenous route can only be used when animal is kept in squeeze cage. So, the most desirable route is intra muscular route.

ii. **Equipment used for Chemical Restraint :** There are many types of equipment, which can be used for intramuscular injection of drugs to free ranging animals, and zoo kept wild animals. The use of *(a) Stick Syringe, (b) Blow Pipe, (c) Powder Charged Rifles (Dist-Inject or Cap-chur gun):* In Dist-Inject or Cap chur guns, syringe is propelled by means of an explosive charge similar to 0.22 blank cartridge, which is loaded in a light weight 32 calibre (13mm) rifle. The distance covered is about 120 meters and the dose of drug is up to 20 ml. *(d). Blow Gun Rifle (Teleject):* Disadvantage of this gun is its small syringe capacity, low range and lightweight of dart due to which it is subjected to cross wind.

iii. **Estimation of body weight average :**

Table 1: Average body weight of wild animals of India.

S.No.	Animal	Male (Kgs)	Female (Kgs)
1.	Tiger	180-230	145
2.	Lion	100-120	90-100
3.	Leopard	68	50
4.	Snow leopard	50	40
5.	Elephant	2500-5000	2000-4500
6.	Rhinoceros	1000-2200	900-2000
7.	Sambhar	150-250	100-200
8.	Barasingha	125-200	100-150
9.	Nilgai	150-200	100-200

iv. **Selection of drugs :** A large variety of drugs are available to day in the world market, which can be used, for tranquillisation or anaesthesia of wild animals. The drug should have following criteria.

It should have high therapeutic index, suitable for intramuscular use, short induction time, effective antidote, good stability at room temperature, effective even with a low dose, subjected to restrictive legislature and readily be available at reasonable cost.

Some of the important drugs used in wild species are widely used are discussed bellow.

I. CARNIVORES :

a. **Canidae :** Ketamine hydrochloride or Xylazine. Acepromezine maleat minimises convulsions seen with ketamine.

b. **Ursidae :** Ketamine hydrochloride intramuscularly is drug of choice for immobilisation of small bears.

c. **Procyonidae and Mustalidae :** (Racoon, mink, otter, skunk, beaver, badger): Ketamine hydrochloride and Phencyclidine hydrochloride are commonly used drugs

d. **Viveridae :** (mangoes, civet): Ketamine hydrochloride and Xylazine

e. **Hyaenidae :** Ketamine hydrochloride is effective sedative.

f. **Felidae :** Ketamine and Xylazine is drug of choice (Fowler 1986). provides analgesia adequate for surgical operations like castration, laparotomy and tooth extraction, in lions, tigers, leopards and pumas. Inadequate analgesia contributed to cardiac arrhythmia.

II. **PRIMATES :** Ketamine hydrochloride is an excellent immobilising agent for all classes of primates.

III. **ELEPHANTS :**

Etorphine hydrochloride is valuable at the dose is 0.0022 mg/kg for the Asian elephant and 0.0017 mg/kg for the African elephant.

IV. **RHINOCEROS :** Etorphine hydrochloride is the drug of choice for immobilising rhinoceros. A sedative dose varies from 0.5 mg to 1 mg (total dose).

V. **CERVIDAE :** Etorphine, xylazine and tiletamine, zolazepam have been used in various species of cervids.

VI. **BIRDS :**

Psittaciformes & Passeri forms : (Parrot, macaw, cockatoos & perching birds): Ketamine hydrochloride 0.05 to0.2 mg/kg. Ketamine hydrochloride (25-50 mg) is the standard immobilising and anaesthetic agent for use in birds (Amand 1974).

VII. **REPTILES :** Ketamine hydrochloride is a drug of choice of anaesthesia for snakes, lizard and turtles. Small snakes, lizards and wimens can be deeply sedated by Ketamine hydrochloride with the dose rate of 44 mg/kg.

LAWS AND REGULATIONS

Investigators must obtain written permission from a competent authority (Forest officials) or collector of the appropriate jurisdiction depending upon the case before actual capture of the wild animal. In addition, Investigators must be familiar with the current list of threatened and endangered species and must comply with all rules and regulations pertaining to capture and handling of these animals.

1. Handling must be introduced slowly and carefully to avoid triggering, Try to avoid “fear” memory in the animal, if it is formed before handling, then rescue becomes difficult, particularly in captive animals.

2. Use food bait as a treats which are highly palatable.
3. Handling must be as quaint as possible to prevent a "fear" memory, If the animal begins to exhibit signs of agitation and stress - DISCONTINUE THE PROCEDURE.
4. The person handling must be a person the animal trusts.
5. Wild animals should be handled quickly and without sudden movements, utilizing the minimum number of personnel required for safe rescue operation.
6. Dark chamber for holding the captured animal to reduce stress. Excessive noise from loud equipment, vehicles, or talking should be minimized. In addition, the handlers should be aware of the negative responses wild animals.
7. Excessive struggling or stress during rescue can lead to hyperthermia and muscle damage (capture myopathy) in the animal, especially during warm or hot days. In some cases, the time of day will also be an important consideration; therefore it is advisable for rescuing critical cases handling efforts should be made during cooler periods of the day.
8. When restraining an animal by hand is required, then technique should be appropriate as per the species to be handled. The use of gloves may reduce the dexterity of the handler.
9. If muzzles or holding bags are being used as part of the restraint, the investigator must ensure that the animal is breathing must remained smooth.
10. When animals are confined in chutes or corrals, stress can be reduced by providing visual barriers, which will allow the animals to conceal themselves from handlers.
11. As different types of cages are available, cages should be used as per the need to restrain and handle wild animals, they should be properly designed and/ or padded in order to avoid injury to the animal.

COMMON PROBLEM ASSOCIATED WITH CAPTURE AND HANDLING OF WILD ANIMALS

Injury : The chances of injury to the handler as well as to the animals are a cornmon incidence during handling of animals. Sometime it leads to the death of animal as well as handler. This incidence can be minimizing by the way of scientific method of handling.

Stress : Wild animals are far more susceptible to stress and injury than domestic species, particularly during capture, handling, restraint and transportation. Even apparently, innocuous procedures such as blood collections and clinical examinations can be as dish'essing to a wild animal. In addition, many wildlife species are dangerous to the handler, and so human safety must be taken into consideration.

Capture Myopathy : Capture myopathy can be induced by a combination of many stressors (e.g. terror, chase, capture, restraint), and that it is associated with exhaustion of the normal physiological reserves that provide energy for escape. Capture myopathy is characterized by ataxia, paralysis, myoglobinuria, and acute muscle degeneration. General clinical pathology of capture myopathy include elevated blood serum enzymes (AST, CK, LDH), and blood urea nitrogen (BUN) levels. Initial clinical symptoms of the condition include increased respiratory and cardiac rates and elevated body temperature (hyperthermia) and death. Other early signs include lack of responsiveness to the environment, weakness, stiffness and paralysis of locomotory muscles.

CHAPTER 8

TRANSPORTATION

Not only at national but also at international levels, for the sake of conservation, education, research and enrichment of wildlife diversity, transfer of animal species are encouraged. To accomplish the translocation mission protocols are formulated for transfer between captive-to-captive, wild to captive (rarely vice-versa) and wild-to-wild facilities (reintroduction, supplementation). In these projects for movement of animals, ideal guidelines for quarantine, and medical and husbandry practices are implemented.

In future not far off, the transportation of wildlife may be in the form of transport of embryos or zygotes and their further implantation in surrogate mothers. This would be much easier, economical and will help us to remove all types of quarantine measures.

Guidelines on the basic welfare, technical requirements and practices to be adopted for ethical loading, transportation and unloading of animals are formulated. The species specific size of transportation crate/cage and carrier space requirements and other related instructions stipulated under the Acts, Rules and Regulations of the Governments and Transportation Authorities for the purposes are to be followed in letter and spirit.

Points to be Considered

(i) The registered Veterinary Officer is the competent officer under whom the operation of Handling and Transport of any kind of animals shall be conducted.

(ii) Animal feeding schedule, breeding history, medical history and quarantine health certificate shall be prepared in triplicate, a copy to be sent in advance about the schedule and details of the shipping, second copy to be handed over to the shipping authority along with other documents and waybills and the third copy for the office record.

(iii) The cage / crates will be thoroughly checked for fitness, disinfected well in advance before the animals are to be transported. The inner dimensions of the cages should be considered for all measurements.

(iv) Incompatible animals (even individuals of the same species) should not be crated together.

(v) Loading and unloading of animals into or out of the container shall only be undertaken during the daylight. After the delivery of the animals at destination, the containers must be cleaned thoroughly.

(vi) During ground transportation, most species of the region can withstand reasonable variations in ambient temperature but exposure to the wind or a

(vii) draft; extreme temperature (hot / chill) can be fatal.

(vii) Transportation may increase defecation. The floor of containers shall be leak-proof with sufficient absorbent material contained therein.

(viii) It is pertinent that deer stags with soft velvet antlers should never be selected.

(ix) Pregnant females and females in estrus should not be transported.

(x) For long distance transportation, journey should be performed during cooler periods of the day preferably in the morning or evening hours. Traveling during night is more convenient.

(xi) During journey, feeding should be avoided. But if the journey lasts more than 24 hours, some feeding and watering arrangements must be made.

(xii) Sedation of animal during transportation is only required if the animal is disturbed and is frequently striking, dashing or falling in crate or there is a risk of its breaking away.

(xiii) The crates/cages on the truck should be placed at least 1 to 2 feet apart for sufficient air circulation and ventilation and be securely fastened to avoid shaking or tilting and sliding forwards or backwards.

(xiv) Any morbidity/mortality during journey reflects about the management affairs on the part of shipping authority and if that is due to defect in crate/cage the exporting authority is held responsible.

Small cats, civets and procyonids : Crates for transport of wild cats *Felis chaus, Felis viverrina, Profelis teminmcki,* lynx *(Felis lynx),* caracal *(Felis caracal)* arid civets *(Viverricula Indica, Viverra zibetha, Paguma larvata)* by rail/road/inland water way/sea/air may be designed as recommended by ISI: 4746-1968 for the domesticated cats mentioned as below.

Table 1: Transportation of small cats specifications

Length (L)	Ax2	Ax2	= Tip of nose to root of tail (A)
Width (W)	A	A	= Width across the shoulder (D)
Height (H)	B+15	B+10	= Tip of ear to toe while standing (B)
Elbow height		-	= Toe to tip of elbow (C)

Large wild cats ; Suitable size of the transportation crates have been prescribed for tiger, lion, leopard and cheetah which are as follows:

(a) Tiger *(P.tigris)* & Lion *(P.leopersica)*

2.20 m (L) x 0.80 m (B) x 1.20 m (H)

(b) *Leopard (P.pardus)* & Cheetah *(A. jubatus)*

1.2 m (L) x 0.60 m (B) x 0.75 m (H)

Table 2: Size (in cm) of transportation crates for an adult bear species

Species	Animal size: Prater, 1971		IATA		Crate inside proposed Approximate dimensions		
	L (HB)	SH	L	B	L	B	H
Sloth bear *(M. ursinns)*	140 -170*	65-85	178	84	190	70	110
Black bear *(S. thib etanus)*	130-195	65-70	188	81	200+	90	no
Brown bear *(U. arctos)*	170-245	-	203	119	220+	90	130
Malayan bear *(H.malayanus)*	104-140	70	137	69	145+	50	90

Note : L(HB) = Head-body length, B = Breadth, SH = Shoulder height, H = Height, * = Aniaml standing height

Size and type of crate for transport of monkeys pregnant and nursing, by air ISI 3059, 1965 is also recommended Under Rule 40 (3 a, b) of Animal Transportation Rules, (1978)

(a) 460 mm x 460 mm to contain not more than ten monkeys weighing from 0.8 to 3.0 kilograms each or four monkeys weighing from 3.1 to 5.0 kilograms each; and

(b) 750 mm x 530 mm x 460 mm to contain not more than ten monkeys weighing from 1.8 to 3.0 kilograms each or eight monkeys weighing from 3.1 to 5.0 kilograms each.

Equids : Wild equids are very sensitive and they generally have a high value. Therefore, special care is required during all phases of their transportation to avoid the risk of damage. Like in horses, it is preferred to water the animals not less than 2 hours before loading for transportation. To ensure safety, the presence of keeper/attendant is generally necessary to supervise the behaviour of the animals and for intervening if needed. Keeper must have received adequate training especially to administer tranquilizers. Suggestive dimensions of transportation crates for equine species are mentioned in the table below:

Table 3: Size (in cm) of transportation crate for an adult wild Perissodactylid individual

Species	Animal size: Prater, 1971		IATA		Crate's inside dimensions Proposed (approx)		
	L (HB)	SH	L	B	L	B	H
Wild Ass *(E.h.Khur)*	200-210	110	216 to 130	127	225	80	128
Kiang *(E. h. kiang)*	>to above	156			260	95	180
Rhinoceros *(R. unicornis)*	305-380	170 to 186	304	170	400 (for Average sized animal	200	240

Asian elephant *(Elephas maximus)* : Livestock Importation Act. 1898, and Model Rules 1961 dictate that no export shall be accomplished without health fitness certification by the competent authorized Government Veterinary Officer or Veterinary Officer of the Department of Animal Husbandry, Ministry of Agriculture, Govt. of India.

Elephant should be properly fed and given water about 1 to 2 hours before it is taken for boarding for journey. If the journey is for more than 24 hours the required quantity of the fodder shall be carried for the journey. If the arranged fodder happens to be inadequate (due to some unavoidable reasons) it has to be arranged at the Journey break camping place locally. If an animal is reluctant to step on to ramp it should not be forced. It should be given some time so that the Mahout is able to persuade the animal.

Marine mammals : Sensory dependent animals must be allowed visual and olfactory contact and shall not be transported for more than 36 hours. An attendant or receiver shall accompany cetaceans, Sirenians, pennies and sea otters during transport. For the transportation, modalities contained in IATA regulations shall be adopted.

CHAPTER 9

HEALTH EXAMINATION AND RECORDING OF VITAL SIGNS

Clinical Examination is a comprehensive holistic examination of wild animals to know its physiological health status. Thus, neglect of one aspect of the clinical examination can render valueless and great deal of work on the other aspects can lead to an error in diagnosis. The examination of the affected animal represents only a part of the complete investigation.

It is comprises of 3 main aspects i.e. Anamnesis/History taking, examination of environment and examination of patient.

HISTORY TAKING (ANAMNESIS)

In wildlife medicine, history taking (is the most important of aspect of clinical examination. Successful history taking involves intangibles which may prove helpful to the clinician. The stakeholder/foresters/Zoo keepers or attendant must be handled with diplomacy and tact. The use of non-technical terms is essential since stakeholder/ foresters are likely to be confused by technical expressions or be reluctant to express themselves when confronted with terms they do not understand. Statements, particularly those concerned with time, should be tested for accuracy. Owners and more especially stakeholder and foresters, often attempt to disguise their neglect by condensing time or varying the chronology of events. For completeness and accuracy in history- taking the clinician should conform to a set routine. The system outlined below includes patient data, disease history and managemental history. History recording should be made in the following heads-

(i) Past diseases history : It include recording nature and duration of previous illness, morbidity, mortality, necropsy findings, case fatality rates, clinical signs observed, treatment, control measures adopted and results obtained.

(ii) Present disease history : It include recording of

(a) Duration of illness,

(b) Appetite is normal, decreased, increased or abnormal.

(c) Water intake is normal, decreased, increased or abnormal

(d) Faeces frequency, quantity, color, consistency or odour

(e) Urine frequency, quantity and color of urine

(f) Posture (it is denotes anatomical configuration of the animal when they remain in the stationary condition i.e. standing, sitting or lying down)

(g) Respiration,

(h) Physical activity,

(i) Gait (recorded while animal is in motion),

(j) Behavior of the animal-it is the responsiveness of an animal towards a stimuli may be considered as a criteria to judge the behavior of the animal. The responses may be normal (react and respond very promptly against stimuli), if response is decreased then the animal may show dullness, dummy response or go in coma state)

(k) Drugs used, doses given and duration maintained if animal is treated.

(iii) Managemental history : Depending upon the food they eat, wild animals are classified into carnivores, herbivores, omnivores, frugivores and insectivores. Depending upon the natural feeding habits, their nutritional requirement will vary. In addition the digestive system of various wild animals adopted to such diet. Therefore by providing the selected diet which should be proximate to the diet of the species like providing a mixed diet of many ingredients fortified with vitamins and minerals to minimize nutritional deficiency ,inbreeding policy and practice a planed and scientific breeding programmers can eliminate deleterious effects of inbreeding like congenital abnormalities, still birth, early mortality, abortion etc, enclosure (in terms of accurate space, ventilation, proper draining, situation and suitability of troughs, opportunity of exercise, protecting them from adverse climatic condition), transport practice and general handling of an animal, maintenance of proper sanitation and hygiene including disinfection and destruction of pathogens and vectors, inside and outside of the animal premises free from water logging and dry.

EXAMINATION OF ENVIRONMENT

To establish possible relationship between environmental factors and the incidence of disease an examination of the environment is a necessary part of any clinical investigation. It comprises of-

(i) Out door environment examination : It include topography and soil type, population density (overcrowding is the most common predisposing factors for many diseases), feed and water supplies should be fresh and clean, proper waste disposal etc.

(ii) Indoor environment examination : It include maintenance of adequate housing facilities and adequate ventilation, strictly avoiding overcrowding and maintaining proper sanitation and hygiene in and around animal enclosure.

EXAMINATION OF PATIENT

A complete clinical examination of an animal includes, in addition to history-taking and an examination of the environment.

(i) **Inspection : Distant/General Inspection** - The importance of a general inspection of the animal cannot be overemphasized. Apart from the general impression gained from observation at a distance, there are some signs that can be best assessed before the animal is disturbed like behavior and general appearance, voice, eating behavior, defecation, urination, posture, gait, body condition, conformation, skin etc. Distance of general inspection of various species of wild animals is variable. It varies from few feet to small size to many metres for large size.

(ii) **Close Inspection of body regions :** It include examination of head(corneal abnormalities, size of the eyeball, abnormal eyeball movement, nostrils, mouth part (buccal mucosa, tongue, submaxillary region, neck region, thoracic region (cardiac area, lung area), respiratory rate, rhythm, depth, type of respiration, chest symmetry, respiratory noises, abdomen region, external genitalia, mammary gland and limbs respectively.

(iii) **Close Examination :**

(a) **Systemic Examination :** General examination of digestive system, respiratory system, urinary system, central nervous system, liver and pancreas, Musculo-skeletalsystem, blood and blood forming system, lymphatic system etc.

PHYSICAL EXAMINATION

1. **Palpation :** Direct palpation with the fingers or indirect palpation with a probe is aimed at determining the shape and size, consistency, temperature and sensitivity of a lesion or organ. Types of palpation done in animals by

   (i) **Direct/ Immediate palpation:** It is done with finger tips or palm.

   (ii) **Indirect/ Mediate palpation :** It is done with a probe or catheter or any such instruments.

   Palpation findings is doughy when the structure pits on pressure as in edema, firm when the structure has the consistency of normal liver, hard when the consistency is bone like, fluctuating when structure is soft, elastic and undulating on pressure but does not retain of the fingers, emphysematous, when the structure is puffy and swollen, moves and crackles under pressure because of the presence of gas in the tissue.

2. **Percussion :** In percussion the body surface is struck, so as to set deep parts in vibration and cause them to emit audible sounds. Types of percussion done in animals are-

   (i) **Direct/ Immediate percussion :** It is made by striking a part directly with finger tips or plexor.

(ii) **Indirect/ Mediate percussion:** It is conducted by keeping the fingers of the left hand flat on the region and stricking it with flexed middle finger of the right hand or by keeping pleximeter on the region and stricking it with plexor.

The sounds may vary with the density of the part set in vibration like resonant sound emitted by organs containing air, tympanitic, a drum like sound emitted by an organ containing gas, dull sound emitted solid organs.

(3) **Auscultation :** It is direct listening to the sounds produced by organ movement is performed by placing the ear to the body surface over the organ.

(i) **Direct/ Immediate Auscultation:** Done by placing the ear on the body surface over the organ to be examined.

(ii) **Indirect/ Mediate auscultation:** Done with the help of stethoscope or phonandoscope.

RECORDING OF VITAL SIGNS

Vital signs are objective manifestation of animals which can be recorded by clinician/ physician. Value of vital signs reflect physiological reaction of the body and its organ and system. The commonly recorded vital signs in wild animals are temperature, respiration rate, heart rate, pulse rate, B.P., hydration status. All the vitals sings are recorded quietly and gently without disturbing the animals. The normal value of the vital signs are shown in Table 1 of this chapter.

Recording of Temperature : Temperature helps in establishing diagnosis of febrile diseases like shock, dehydration, hypothermia, hyperthermia etc. Temperature is commonly recorded through per rectum site in majority of wild animals but in some species some alternative site like *per vaginum*, *per os* etc. may be preferred. In case of failing to record internal temperature directly with animals, external body temperature is recorded from freshly voided feaces of this chapter.

Recording of Pulse Rate: Pulse is the expansion and elongation of arterial wall imparted by the column of the arterial blood due to contraction of the left ventricle. The importance of recording of pulse is that it helps to know the state of cardiovascular system and also assist in giving prognosis of diseases. Different site for recording pulse in different wild animals has been shown in table 2 of this chapter.

Common site of *i/m* and *s/c* injection in wild animals - For administration of different drugs, a suitable mode of administration is adopted to achieve a desired therapeutic responses for a particular diseases. The common site of *i/m* and *s/c* injection in wild animals has been shown in table 3 of this chapter.

Site of blood collection in different species: Blood is collected with an aseptic manner with an suitable anticoagulant or without anticoagulant for carrying out different haematological, biochemical tests which aid in diagnosis of diseases. The different sites in different wild animals has been shown in table 4 of this chapter.

Table 1: Value of Physiological Parameter/ vital Parameters in Wild Animals.

Sr.No	Name	Body temperature	Respiration rate per minute	Heat rate orpulse rateper minute
1	Tiger	37.8-39.9°C	12	40-50
2	Lion	37.06-38.6°C	10-12	40-50
3	Leopard (Panther)	38.0-39.0°C	12-16	42-45
4	Cheetah	38.0-38.5°C	10-16	43-45
5	Wolf	35.90-36.5 °c	10-15	28-35
6	Elephant	36.0-37.0°C	10-16	25-30
7	Rhinoceros	37.0-39.0 ° C	20-40	70-140
8	Giraffe	38.0-38.8°C	12-20	40-50
9	Kangaroo	37.0-39.0 ° C	20-40	70-140
10	Bear	37.5-38.3°C	15-20	60-90
11	Antlers	38.0-39.0°C	18-25	45-60
12	Yak	38.0-38.5°C	12-16	40-50
13	Alligator	80-90°F		
14	Chameleons	55-75°F		
15	Turtles	75-85°F		
16	Snakes	75-90°F		
17	Scorpians	18-32°C		
18	Hare	38-40°C	32-60	135-325
19	Avifauna	106-108°F		233 22

Table 2: Site of recording pulse in different specieis of wild animals.

S. No.	Name	Site of pulse recording
1.	Tiger & Lion	femoral / tail artery
2.	Rhinoceros	ear / tail artery
3.	Elephantear	artery
4.	Giraffe	facialj coccygeal artery
5.	Bear	femoral artery
6.	Antlers	femoral artery
7.	Yak	facial artery
8.	Cheetah	femoral/ tail artery
9.	Wolf	femoral artery
10.	Leopard (Panther)	femoral artery
11.	Nilgai	middle coccygeal artery

Table 3: Site of i/m and s/c injection in wild animals.

S.No.	Name	Subcutaneous	Intramuscular
1	Tortoise	corapecae	Plastern
2	Snake	ventral site of abdominal region abdominal inter between scales	abdominal inter scutes
3	Rhino	between skin fold	shoulder & rump
4	Boar	neck	hind quarter
5	Porcupine	ventral site of abdominal	neck
6	Elephant	between skin fold	hind quarter/ shoulder
7	Nilgai	neck	hind quarter/shoulder
8	Snow leopard	neck skin fold	rump
9	Tiger & musk deer	skin fold	rump and shoulder
10	Lizard and crocodile		Lateral aspect of the arm and forearm of the front limb
11	Turtle		Lateral aspect of front limb
12	Avifauna	Wing web,interscapular region & gron	Superficial pectoral muscles on either side of neck bone

Table 4: Site of collection of blood in wild animals.

S.No	Name	Site of collection
1.	Snake	Dorsal buccal vein
2.	Turtle	Jugular vein/Occipital sinus
3.	Lizard	Ventral tail vein
4.	Hare	Lateral saphenous vein
5	Avifauna	Jugular vein(small birds under 100 gnns) Cutaneous ulnavein/ Caudal tibial vein(Layer bird)
6.	Dear	Jugular vein
7.	Samber	Jugular vein
8.	Leopard	tarsal vein/Saphenous vein
9.	Lion	Tarsal vein
10.	Monkey	Jugular vein
11.	Nilgai	Jugular vein
12.	Tiger	Tarsal vein
13.	Elephant	Ear vein
14.	Snow Leopard	Tarsal vein/Saphenous vein
15.	Boar	Ear vein/ Anterior Vena cava

CHAPTER 10

WILD LIFE DISEASES: AN OVERVIEW

Wild/zoo animal medicine is a relatively recent development as compared to man and farm animal medicine. Certain species possess important innate or acquired peculiarities. For example, there are individual which may be eccentric, frivolous, frigid, affluent, destitude, timid, notorious, caddish etc. Some animals breed once or twice in a year, produce one or more off-springs, have little or intense maternal urge etc. There are species which remain in pairs or groups, arboreal, open land dwelling possessing distinct or indistinct territory, diurnal or nocturnal, etc.Based on their food preferences they are classified as herbivores, carnivores and omnivores. Characteristic anatomical features are present in some species viz., absence of gall bladder in Chital, lack of pleural cavity and lacrimal apparatus in elephant, incomplete ossification of hyoid bone in tiger and lion , etc. Knowledge of such peculiarities will help in understanding the physiology of the species.

In wild, the cause of mortality could be due to intrinsic and extrinsic factors and generally they are: starvation, diseases, parasite, predation, pollution, poisoning, accidents, etc. Health monitoring in such environment primarily rests upon the practice of population medicine. For diagnosis and adaption of effective control and preventive measures, epidemiological inquisitions to determine the specific (causal) factor responsible for decimating the population are necessary.

In captivity, for morbidity and mortality, usually responsible factors are in order of: management, malnutrition reproductive disorders, infectious and parasitic diseases, etc. They are more influenced by the prevailing conditions than population of the species *per se*. Although early detection of disease symptoms is an important principle in any medical programme, it is often difficult to determine whether a wild animals is ill. (Many animals in the wild appear to have the uncanny ability of masking serious illnesses even until the very end. This may be a type of defense mechanism so that predators are not easily alerted) Movement of vultures and scavengers in a particular area, however, may indicate the presence of animals carcass. Trekking should regularly be done to pools, brooklets and rivulets that transect the habitat. It is easier to monitor heath of the exhibits in captivity through many of these animals do not frequently exhibit overt clinical signs until near death. Daily rapport is, therefore, necessary between the veterinary staff and the keeper staff. Keepers can detect subtle abnormalities because they have more contact and familiarity with the individual animals looking after their eating habits, general attitude, body carriage etc. The ailing animal should be captured cautiously for detailed investigation. The whole fresh carcass 24 hours or less since death should be subjected to autopsy examination.

NUTRITIONAL AND METABOLIC PROBLEM

Malnutrition is defined as a state of ineffective or deficient procurement, ingestion, digestion, absorption and utilization of food. Starvation, by contrast, refers to a condition arising from prolonged deprivation of food. Malnutrition is frequently encountered in free ranging wildlife populations and individuals. Young animals are most severly affected in situations of limited feed because they have higher nutritional demands associated with growth, greater heat loss due to smaller body size, lack of fat reserves, and subordination in the social hierarchy resulting in decreased access to feed. Situations which can lead to malnutrition include poor or limited ranges (both winter and summer ranges) , severe environmental conditions increasing energy demands, and diseases or injuries decreasing the ability of an animal to procure and utilize food .It is easy to blame the deaths of many animals on a pathogen, but it is far less easy to determine that animal died from a deleterious food supply.Even if the food supply had remained constant the increasing population would have made food available per capita. The surplus individuals usually juveniles or subordinates living at or to near carrying capacity have two options either to stay and perish or to leave the area. If they move out, the odds will be greater and they will have to survive predators, diseases, accidents and even starvation.

Evaluation of the nutritional status of the live animals is very difficult, particularly when free living. In captivity the variables are under direct observation and can be measured in respect of their physical condition such as size, vigour and weight. High fertility, low mortality and increased population in wild (unless held down by hunting or predation) allude the soundness of the habitat. In period of stress, poor nutrition results in repression of growth, delayed sexual maturity, low conception rates and increased intrauterine mortality (malnourished animals maintained pregnancy for longer period of time).Reduced physical vigour in adult individuals is marked by increased prominence of bones (back bone, shoulder, pelvis, etc.) due to atropy of fat and muscle. Severely malnourished animals will also often have a listless appearance and rough hair coat.

Diagnosis of malnutrition after death is relatively easy. Postmortem examination of an emaciated animals should be thorough in order to rule out certain diseases which have malnutrition as a sign. The most obvious and important gross change is lack of fat in normal depots in the subcutaneous, cardiac, omental, kidney and bone marrow tissues. In a severely malnourished animal fatty tissue in these areas will take on a gelatinous appearance that refered to as serous atropy of fat. Bone marrow in the heavy long bones of the legs (femur,humerus) should be examined. Estimates of bone marrow fat of growing animals must be made with care because of the normal reddish colour and small amount of fat stored in this location in young animals. Besides fat indices, development and beam diameters of antlers are also affected by many nutritional factors. Minerals such as calcium and phosphorus have marked effect on antler growth, as also in cases of both energy and protein.

As propagation of rare and endangered species is one of the major objectives of the modern zoos, one should be enlightened with reproductive physiology, nutrition and behavior. Reproductive health can be a barometer of overall heath status of the species in facilities. Incestuous inbreeding in captivity in most species is unavoidable. Inbreeding though increases genetic uniformity within strain/breed by increasing homozygosity and herterozygosity, lowers the vitality, vigour and fecundity and increases susceptibility to disease. Its cumulative effects are reflected in having frequent reproductive problems such as abortions, still birth, metritis, infertility, sterility, congenitial abnormality and prenatal mortality. Inbreeding depression in many endangered wild mammals in captivity is not uncomman. Many of the reproductive problems akin to the livestock are imparted by infectious diseases (such as Blue tongue, MCD, BVD, Brucellosis, Listeriosis, Leptospirosis, Trichomoniasis, Vibriosis etc., in artiodactylids) and nutritional factors in primates and in Carnivores High perinatal mortalities are important factors bearing great impact particularly in endangered species conservation and propogation. Only limited records were available to screen and analyse postural abnormalities in various species of wild animals. There is need for exhaustive surveys to retrieve the data embodied in the veterinary medical records in all- large, medium and small-zoos.

REPRODUCTIVE DISORDERS

The common reproductive disorders encountered so far in wild animals are as under:

1. Abortion
2. Dystocia
3. Stillbirth
4. Endometritis/metritis/vaginitis
5. Infertility and sterility
6. Neoplasms
7. Congenital disorders
8. Miscellaneous

PARASITISM OF WILDLIFE

The perception of health of wild animals has changed over a period of time. The opinion has fluctuated in the past between the two extreme views namely, wildlife diseases are natural phenomena that need no active management and any disease or parasite is a vermin that should be erradicated. It is mostly agreed now that are balanced view of carefully monitoring wildlife health and intervening at appropriate time is needed.

Wild animals and birds suffer from various diseases in a free living state. Diseases and parasites can potentially play a major role in the population dynamics of wild animals and is one of the major factors in long term population viability. However, the precise role of these in wild populations has not been seriously studied.

Ectoparasites and endoparasites also play an important role in health status of wild animals and environmental health. They have a potential to affect the fitness and reproductive success of host individual. The effects of parasites on domestic animals are well studied. It is largely assumed that the same holds true for free ranging wild animals. But the ecology of parasites in the wild is likely to be much different. The role of parasites need particular attention in the present day environment of habitat encroachment, disturbance and fragmentation. There is a competition between domestic livestock for resources. Because of the ever increasing contact between wildlife and domestic animals, not only have new diseases evolved, but also the epidemiology have become complex. Host-parasite relationship need to be studied before formulating preventive measures.

The host-parasite relationship remains static as a subtle infection without causing a disease. In this situation, monitoring of health became still complicated and therefore be of concern.

The concept of health is not restricted to parasites and disease. Wild animals can suffer for example, from mineral deficiencies. Certain areas and certain soils may be deficient in some minerals. These aspects may be important in the population biology and health of animals but has not been investigated adequately.

Due to human interference, natural habitats of wild animals are changing. Human pressure on a protected area appears in a variety of forms; air and water pollution, construction of dams and roads, livestock grazing. Some types of pressure affect entire areas like grazing and others re concentrated as a specific spot like use of water holes by domestic animals. Visitor carrying capacity is an elusive concept and vehicular traffic of the tourist is likely to release toxic chemicals in the environment.

Condition of an animal responds to the changes in its habitat quality, which is governed by many interrelated component factors. The deterioration of condition could only be not due to chronic diseases and biological processes but could also be due to nutritional stress. Such nutritional deficiencies can have a noticeable effect on the body condition of wild animals, but has received scant attention.

Wildlife health evaluation is based upon the premise that animals in a population acts as indicators of the status of groups relationship with the environment.

Very little attention has been paid to the study of extremely varied nature of many parasitic species in regard to the host-parasite relationship, their pathogenic role and determination of the host ranges of their definitive hosts, besides assessing the role of some of these animals acting as reservoirs of infection, for domestic stock and also human beings.

The commonly protozoan parasites recorded in wildlife in our country. Trypanosomiasis/ Nagana/ Sleeping sicknesss/ chagas/ Surra, Leishmaniasis, Histomoniasis orenterohepatitis (blackad), Trichomoniasis, Giardia, Plasmodium, Amoebiasis, Haemoproteus, Leucocytozoonosis, Babesiosis, Coccidiosis, Toxoplasmosis, Besnoitiosis, Balanti diosis, Anaplasmosis, Theileriosis, Toxoplasmosis,.

In wild carnivores and primates.

Trematodes : Distome flukes, paragonimosis, paramphistomosis

Cestodes : Dipylid, taenid, mesocestoid, monieziosis, diphyllobothrid and echinococuss.

Nematodes : Ascarids, rhabditoid , strongyloids , trichuriosis, oesophagostomiasis, stephananuriosis, ancylostomatidosis, Trichinellosis, cappillariosis, oxyurosis, filaridosis.

COMMON PARASITIC DISEASES OF WILD GALLIFORMES / WILDBIRDS

I. Helminth parasites (Round worms)
   (a) Heterakiosis.
   (b) Capillariasis
   (c) Syngamiasis
   (d) Prosthogonimiasis

II. Tapeworms : Davainea proglottina, Raillientina cesticillus, Raillietina tetragona.

III. Protozoan parasites.
   (a) Histomoniasis
   (b) Coccidiosis
   (c) Trichomoniaisis.
   (d) Plasmodium

IV Ectoparasites
   Lice (Mallophaga)
   Bugs (Hemiptera)
   Fleas (Siphonaptera)
   Mites (Dermanyssus gallinae, Arachnida, ornithonyssus sylviarum, O. bursa.)

CHAPTER **11**

MAJOR INFECTIOUS DISEASES OF WILD ANIMALS

Understanding the **health** as "a state of well being: free from diseases, soundness of body and mind, in harmony with the environment" where as of **DISEASE** is a condition resulted from the collision between a pathological agent and a susceptible individual. Hippocrate defined disease as a disharmony with in the body, between the body and mind, and between man/animal and the environment. By putting the narrow definition in our common use, we neglected the population and preventive medicine in human, domestic and zoo animal medicine. Wildlife medicine or zoological medicine of captive and free living animals is a relatively new field in India and we must be sure that our application of disease principles is based upon the broad definition of disease.

GENERAL OBSERVATION OF ANIMALS

Each animal should be carefully observed every day. Notice should be taken not only of its physical state, but whether or not it is eating, drinking, defecating and urinating normally and an assessment made of its activity. Slight changes in activity may be the only outward sign of systemic disease. Careful examination of an animal's enclosure or den can also be rewarding. There are occasions when direct observation of animals would involve too many disturbances - immediately before, during and after giving birth for example. Rather than trust that all is well in these circumstances, consideration should be given to the use of microphones or even video cameras with which animals may be monitored remotely.

INDICATION OF ILLNESS

The sick animals behaves sluggish in their activity, it looks dull and depressed. It restricts the movements of ear, trunk, tail and legs. The animal becomes partial or complete anorexic (less interest in feeding). The abdominal pain can be expressed by grunting or groaning sound with restlessness, lying down and getting up, elephant can place the trunk in the mouth, biting the tip of the trunk or express the abnormal posture (Personnel observation). The urination, rumination, defecation and lacrymation are also demarcated (reduced or increased) in some condition.

DISEASES OF WILD ANIMALS

The diseases of captive/wild animals can be grouped broadly in to non-infectious one and infectious one. The major non-infectious diseases of captive animals are described as a common for all species where as infectious diseases has been described separately according to the major species wise.

INFECTIOUS DISEASES

Diseases, which are caused by infectious agents, like bacteria, virus, parasites or fungi are known as infectious one. The infectious diseases can be contagious in nature. All the contagious diseases are infectious one, but not all the infectious diseases are contagious in nature

The diseases of captive/wild animals can be grouped broadly in to non-infectious one and infectious one. Diseases, which are caused by infectious agents, like bacteria, virus, parasites or fungi are known as infectious one. The infectious diseases can be contagious in nature. All the contagious diseases are infectious one, but not all the infectious diseases are contagious in nature.

HEALTH MANAGEMENT OF FREE RANGING ANIMALS

Health management in captive wild animals is very different as compare to free ranging wild animals. Some major epizootics observed in free ranging wild animals in India,

Year	Disease	Animals Affected
1871	Rinderpest(RP)	Deer,wild buff,Gaurs,Boars,Blackbuck,Barasingha.
1935	FMD	Entire herd of Gaur wiped out in Hyderabad state.
1949	Anthrax	150 wild elephant in Assam.
1957	Anthrax	Killed many Gaurs, Deer and wild boars in Goalpara Dist. Of Assam.
1960	African Horse sickness	Cause heavy mortality among the wild Asses of little ran of kutch , Gajarat.
1970	H.S.	Killed 1500 Sambars in Sariska tiger reserve, Rajasthan.
1973	Kyasanaur forest diesease (KFD)	Killed 1808 Langurs and 315 Macaques in the Kyasanaur forest of Karnataka.
		Caused mortality among wild buff. Of Kaziranga NP, Assam.
1980	R.P.	Killed 12 Rhinos Kaziranga NP,Assam.
1980	H.S.	Affected wild buff. Kaziranga NP Assam.
1987	FMD	Killed three wild buff. Kaziranga NP Assam.
1999	HS	Vultures population decreasing all over India.
2005 & Cont.	Pesticides and unidentified Etiology.	Vultures,myena,house sparrows

SOME MAJOR EPIZOOTICS ABROAD

1994	Canine Distemper	Killed 30% of the 3000 Lions in Serengeti NP Tanzania.
1994-1999	Avian vacuolar, Myelino pathy.	66 Bald Eagles killed in Arkansas U.S.A.
1999	Anthrax	Killed 75% of the lesser Kudu population in mago NP , Ethopia.
1999	Pesticide	27,000 birds in illinois, USA.
1999	Botulinum	1000 fish eating birds (90% of local population) were found dead in lake Eire, USA.

The above are only a sampling of some of the major epizootics that have been reported. Many others have gone unreported. Though disease is a major decimating factor among India's wildlife, with some instances of species being extirpated locally, no systemic study has been instituted to find the impact of diseases on the country's wild animal population.

Past incidences of FMD, Rinderpest and Anthrax outbreaks in wild animals are believed to have been transmitted from the intermingling domestic livestock population.

Green house effect, acid rain, and ozone holes are regular phenomena now and they also affect wild animals.

Disease management in free-ranging wild animals is attempted not by treating individual sick animals but by manipulating those environmental factors that play a role in the transmission of the disease.

However, it has to be realized that this is a new science and that new techniques of disease management are being evolved everyday while old methods are being refined. Further, a technique that has proved successful in one place may not be feasible in other country or an other situation. For example, Western countries do not hesitate to carry out mass killing of a species to contain a disease; this is unthinkable in India.

The diseases of captive/wild animals can be grouped broadly in to non-infectious one and infectious one. Diseases, which are caused by infectious agents, like bacteria, virus, parasites or fungi are known as infectious one. The infectious diseases can be contagious in nature. All the contagious diseases are infectious one, but not all the infectious diseases are contagious in nature.

DISEASES OF UNGULATES

Most of the ungulates can suffer with bacterial, viral and parasitic diseases as that of domesticated cattle and horses. The most common infectious diseases includes

Bacterial diseases : Major reported bacterial diseases of ungulates include anthrax, pasteurellosis, tuberculosis, johnes disease, brucellosis and leptospirosis. The other diseases that are sporadic in nature include tetanus, salmonellosis, colibacillosis and clostridial infection.

Viral diseases : Major viral diseases of ungulates include foot and mouth disease, rabies and rinderpest.

Parasitic diseases : Haemoprotozoan diseases like trypanosomiasis and theilariasis have been documented in captive wild animals. Apart from these diseases internal parasites consists of cestodes, nematodes and trematodes, as well as external parasites like tick, louse, mite and flea infestation have been documented in wild and captive animals .

DISEASES OF ELEPHANTS

Elephants can suffer with bacterial, viral and parasitic diseases as that of cattle and horses. The most common infectious diseases of elephants includes.

Bacterial diseases : Major reported bacterial diseases of elephants include anthrax, tuberculosis, tetanus, pasteurellosis, colibacillosis and clostridial infection.

Viral diseases : Major viral diseases of elephant include foot and mouth disease, rabies, and encephalomyocarditis virus and elephant pox.

Parasitic diseases : Trypanosomiasis theilariasis and babesiasis have been reported. Apart from these diseases internal and external parasites consists of cestodes, nematodes and trematodes, tick, louse, mite and flea infestation have been documented.

DISEASES OF CANINES & FELINES

Bacteria, virus, parasites and other etiological agents cause illness in the wild large cats. Some of the important diseases and conditions reported in large cats are discussed here.

Bacterial diseases : Major diseases includes Anthrax, bacterial e*nteritis/ enterotoxaemia due to Salmonella spp, Shigella spp, Campylobacter spp., E.coli, Clostridium perfringens* etc.Localised and systemic bacterial infections like Staphylococcus *spp, Streptococcus spp, Pasteurella, Klebsiella*, *Yersinia pseudotuberculosis, Listeria monocytogenes* etc. Pasteurellosis frequently results from bite wounds. Bacterial pneumonia is common in non-domestic felids. Dental abscesses common following exposure of pulp cavities. apart from them the other diseases are systemic pseudotuberculosis, tuberculosis, listeriosis, leptospirosis, and botulism.

Viral Diseases : Amongst the common viral diseases of felids, feline infectious peritonitis feline panleukopenia, feline respiratory infections, canine distemper and rabies are commonly reported.

Fungal Diseases : *Microsporum canis* reported as cause of hair loss in young tigers.

Parasitic disease : Many external and internal parasites that can be seen in domestic cat and dog are also reported in captive large cats It includes as Toxocara spp., *Spirocerca lupi*, *Dirophylaria immitis*, Anchylostoma spp., Uncineria spp., Aeleurostrongylus spp., Physaloptera spp., Trichinella spp., Taenea spp., Echinococcus spp., *Paragonumus westermani,* Babasia spp., Eimeria spp., Hepatozoon spp., *Toxoplasma gondi,* Trypanosome spp.

DISEASES OF PRIMATES

The order primate comprises of 11 families and 60 genera. Indian monkeys are belongs to two subfamily of **cercopithecidae** family. The **cercopithecinae** (macaques) and **Colobinae** (langurs). In Asia, Macaques kind is the common omnivorous monkeys of which rhesus monkey being commonest found in India. Monkeys are also source of entertainment in zoos and as pet animals.

Bacterial diseases : Tuberculosis, pseudo tuberculosis, sylvatic plague, leptospirosis and other enteric disease like *Salmonella spp.*, Shigellosis and *E.coli* can produce digestive disturbances in primates.

Viral diseases : Herpes, infectious hepatitis virus, rabies, monkey pox, measles, Kyasanur Forest Disease (KFD) and Yellow Fever are some viral diseases of primates.

Parasitic diseases : Major parasitic diseases of primates includes zoonotic malaria which is mainly caused by *Plasmodium cynomolgi, P. Knowlesi* and *P.simium* and amoebiasis by *Entamoeba histolytica* and other endo parasitic condition such as ancylostomiasis, ascariasis, strongyloidosis, filarids, taenaisis, toxoplasmosis, coccidiosis have also been reported in primates.

DISEASES OF WILD EQUINES

The diseases are same as domestic horses, Rhinoceros comes under this group. Strangles, infectious equine anaemia, South African Horse Sickness, equine influenza, infectious equine encephalomyelitis, respiratory disease complex, equine abortion (*Salmonella abortusequi*), shigellosis, Glanders, ascariasis, strongyloidosis, strongylosis, babesiosis, surra, stomach botes (*Gastrophillus equi*), parafilariasis, mange tuberculosis, leptospirosis and rabies have been documented in wild equines.

DISEASES OF REPTILES

Gram negative bacterial infections are very common however Gram positive can also produces diseases of pulmonary system. Streptococci, *Acrosoma hydrophilia*, *Proteus rettgeri* and Pseudomonas spp. are responsible for respiratory problems in reptiles. Aspergillus, Beauveria spp. and Paecilomyces spp. causes mycotic gangrenous pneumonia in reptiles, where as *Entamoeba invadens* causes amoebic dysentery in snakes. Isopsora spp. has also been reported from cobra. Several round and tape worm infections have also been recorded from several reptiles apart from common ectoparasites like mite (Ophinonyssus spp.).

COMMON HAEMO-PROTOZOAN DISEASES

Like domestic animals, zoo and wild animals are vulnerable to a wide variety of ecto and endoparasites, and similar drugs are used for treatment. Care must be exercised in the choice of medications due to species-specific sensitivities to some drugs. Young animals and those stressed by shipment, disease or injury are the most likely to be adversely affected by parasites. At these times, commensal parasites, especially protozoa, can cause disease.

Important haemoprotozoan diseases reported in wild animals include babesiosis, theileriosis, plasmodiosis, leishmaniosis and trypanosomosis.

(a) Babesiosis

Babesiosis is a tick borne haemoprotozoan disease caused by *Babesia* sp. and characterized by pyrexia, haemolytic anaemia, haemoglobinuria, jaundice and death. Under stress (due to malnutrition, parturition, inclement weather) carriers may act as source of infection to newly susceptible host.

Ticks involved : *Babesia bigemina*, B. *bergalensis* and B. *gibsoni* Boophilus microplus, Boophilus annulatus, *Haemaphysalis punctata and Rhipicephalus spp.*

Susceptible wild animals : All types of deer and leopard, tiger, mongoose, Jackal, and primates

Mode of transmission :

i. Through tick bite
ii. Intrauterine
iii. Parentral injection of infected blood or organ emulsion.

Pathogenesis

Following infection, multiplication of protozoa occurs in peripheral vessel and there is intravascular haemolysis. The proteolytic enzymes are liberated from the infected erythrocytes and these enzymes interact with the components of blood and thus lead to increased erythrocytic fragitility, hypotensive shock and disseminated intravascular coagulation. The coating of erythrocyte by parasitic antigen neutralizes the normal surface and thereby favours auto agglutination.

Clinical Manifestations

The primary sign of disease is a haemolytic anemia. In some species, haemoglobinuria, haemoglobinemia and fever accompany the rapid red blood destruction. In some cases there is neurological sign due to cerebral form of the disease which is caused by sequestration of infected erythrocytes in the cerebral capillaries. Ataxia is produced due to anoxic condition in the cerebellum as sequestration blocks the capillaries.

Treatment

Diminazine aceturate @ 0.8 - 1.6 g/100 kg b.wt., i/m

Imidocarb 10% solution @ 0.45 ml/40 kg b.wt. i/m

Supportive Treatment

Analgesic, antipyretic for control of fever,

Glucose saline for dehydration.

Liver extract, Vitamin B complex, iron (orally or parentally) and in severe cases of anaemia whole blood transfusion

Control and Managemental Implication

Programme is to be taken to control tick population as far as possible with combined chemoprophylaxis and chemotherapy.

Vaccination using live vaccine, killed vaccine, inactivated cell culture vaccine and irradiated vaccine are being evaluated

(b) Theileriosis

Theileriosis is a tick-borne protozoan disease caused by *Theileria* spp. in wild animals

Ticks involved : *Rhipicephalus* spp.

Susceptible wild animals : White tailed deer, Mule deer, Sika deer, Fallow deer, Sa is deer for' *Theileria cervi* and bison, Bobat, Florida Panther, Tenas Cougar and wild buffalo for *Theileria annulata*

Mode of transmission :
i. Through tick bite
ii. Blood inoculation (*Theileria annulata*)

Pathogenesis

Gross lesions related to theleriosis include hyperplasia of lymphoid tissue, edema of the lungs and punched ulcers in the abomasal mucosa. Lesion usually associated with anaemia and icterus like erythrocyte haemolysis and circulatory failure. Obstruction of venous blood flow by infected and distended macrophages appears to be responsible for circulatory impairment.

Clinical Manifestations

Multiplication of piroplasms resulting in anaemia and icterus. In T. *annulata*, symptoms of acute septicaemia occurs, with laboured breathing occurring during the terminal phase. Death is due to asphyxia following edema of lungs. In T. *mutans* and T. *orientalis* infections, febrile periods are short followed by anaemia ,and icterus bubral forms with central nervous system signs are known for T. *parva*, T. *annulata* and T. *tauretragi.*Clinical signs with T. *cervi* in naturally infected white tailed deer are rare, and only mild to non apparent anaemia has been reported. Signs include fever, anaemia, pale mucus membrane, emaciation, weakness and general debilitation.

Diagnosis

1. Clinical manifestations
2. Microscopic demonstration of piroplasma in Geimsa stained blood or schizonts in smears of lymph node biopsy material.
3. Histopathological lesions
4. Indirect fluorescent antibody test (IF AT) are used as serological test to identify infected animals

Treatment and Control Management

1. Buparvaquone @ 2.5 mg/kg b.wt. i/m is effective.
2. Oxytetracycline @ 20 mg/kg b.wt. i/m is also used.

Control

1. Control of tick population
2. New approach/ efforts at immunization against bovine theileriosis infections using an infection and treatment method with sporozoite and treatment with oxytetracycline or buparvaquone are promising.
3. Use of low concentration of T. *parva* and T. *lawrenci* with oxytetracyclines.
4. Continued development of immunization method against tick infestations. Presently, antigens from saliva and midgut fractions of select ticks are used to immunize mammalian host against infestations.
5. Use of boluses that release compounds in order to give long term control against tick attachment.
6. Acaricidal treatments includes dips, sprays and quarantine

(c) Plasmodiosis (Malaria)

It is a zoonotic disease. The most significant of the blood-inhabiting haemosporidian parasites of mammals and birds are those of the genus *Plasmodium.* Clinical impact of the *Plasmodium* infection is relatively little known in wild mammals. Wild animals are also susceptible to P. *falciparum* (causing human malaria).

Susceptible wild animals : Monkeys, fly fox and bats only for *Plasmodium inui,* antellers for *Plasmodium semnopitheci,* Simian monkey for P. *Cynomolgi* and flying fox for P. *pteropi*

Clinical Manifestations

It includes anaemia, listlessness, loss of appetite and greenish diarrhoea.

Treatment

Sulphamonomethoxine sodium @ 1 g/litre of water for prophylaxis, continuous medication at the level of 1 g/20 litres drinking water is effective.

Prevention and Control

It is mandatory that each captive facility should have regular staff for daily disposal of the animal wastes, cleaning t11e left over food wastes of the visitors, decaying vegetative matters, roads, footpaths, water moats, toilets, drinking water points and release of insect eating fishes in the water pods/lakes at stipulated period, spray of insecticides, cutting the rainy seasonal weeds and vegetation around the exhibits, drainage of water from exhibits to ensure non-conducive environment for mosquito breeding

(d) Leishmaniosis

It is primarily, a disease of man and dog. *Leishmania donovani* is the causative agent of visceral leshmaniosis and *Leishmania tropica* is responsible for cutaneous form.

Susceptible wild animals : Desert rodents, gerbil, and ground squirrels are prone for both types

The third form mucocutaneous due to L. *brazilensis* occurs in S. America and rodent, is incriminated as the reservoir.

L. tropicalis has been reported in gerbils.

Mode of transmission : Through sand fly bite

Treatment

Pentavalent sodium stibogluconate

(e) Trypanosomosis

Trypanosomes occur in vertebrates, principally in their blood and tissue fluids as intercellular parasites. It is an flagellated (haemoflagellates) characterized by high rise of temperature, anaemia, wasting and cutaneous eruptions.

Susceptible wild animals:

Trypanosoma brucei	:	Dog, horse, pig, camel, cattle)
Trypanosoma cruzi	:	Non-human primates
Trypanosoma equinum	:	Dog, horse
Trypanosoma evansi	:	Tiger, chital, leopard, jaguar etc.
Trypanosoma simiae and *T. congolense*	:	Pig, camels
Trypanosoma vivax	:	Horse, camel, cattle, sheep and goat

Mode of Transmission

1. Flies (Tabanus, Stomoxys, Musca spp.)
2. Blood inoculation
3. Ingestion of infected carcass in dogs. IV. Blood sucking bugs (reduviid) in *Trypanosoma cruzi*.

Clinical Manifestations

There is intermittent fever and anaemia. Affected animals appear dull and sleepy, show staggering gait, eye staring and wide open, circling movements, nervous excitement, profuse salivation, shivering of the body followed by coma and death.

Chitals : Pyrexia, anaemia, irritability, progressive muscular weakness, paresis of hind limbs, refusal to take food.

Tigers, jaguars, leopard : Dyspnoea and convulsions in the terminal stage

Necropsy Findings

Anaemia, congestion of lungs, liver, spleen, kidneys, brain, subcutaneous haemorrhages.

Treatment

Diminazine aceturate @ 0.8 - 1.6 g/100 kg b.wt., i/m

Quinapyramine (sulfate and chloride in the ratio of 3:2) @ 0.025 ml/kg b.wt. (after adding 15 ml sterile water for injection to 2.5 gm vial).

Prevention and Control

1. All possible precautions should be taken to keep away the flies (*Tabanus*, *Stomoxys* and *Musca* sp.).
2. Meat should be given only after 6-8 hours of slaughter.
3. To prevent food borne infection, zoo inmates are to be ensured that the procured food is safely stored away from arthropod vermin, stray free -ranging animals such as rats, mongoose, squirrels, stray cat etc.
4. Each carnivore should be fed in isolation in the fly and vermin proof feeding cell.
5. Disinfection of premises once a month.
6. Regular cleaning of unwanted bushes and vegetation for every three surroundings.
7. Avoid breeding of flies especially in the rainy season.

CHAPTER 12

MAJOR NON INFECTIOUS DISEASES OF WILD ANIMALS

In free living wild animals the territorial behaviour,anthropological impact, stress due to many reasons and natural calamities predispose the animal for several non infectious diseases and disorders.The majority of them are grouped either as systemic diseases and or based on etiological factors such as physical,chemical, deficiencies,poisoning or stress induced one. Few of them are mentioned here.

SYSTEMIC DISEASES

a. **Digestive system:** Major disease problems of digestive system include tooth problem, tusk problem, choke, tympany, indigestion, constipation, intestinal obstruction, colic and enteritis.

b. **Respiratory system:** Major respiratory problems are rhinitis, cough, pharyngitis, pneumonia; tracheitis and bronchitis are common problems.

c. **Nervous system:** Heat (Sun) stroke due to high ambient temperature, poor ventilation, prolonged improper transportation and over crowding in small space can predispose to the condition leading to stress.

d. **Musculo skeletal system**: Wound, dislocation, fractures, sprains, degenerative joint disorders, rickets, capture myopathy and physical injuries (Cooper, 1968) are common problems in captive animals. Trunk injuries like crushing, laceration, penetration of foreign bodies and paralysis have been documented in elephants (Fowler, 1986).

e. **Foot Injuries:** Over grown sole, overworn sole, cracked sole and heel, over grown nails, split nails, ingrown nails, wound, abscess, laminitis on account of prolonged work, over feeding with rich food, interdigital fibroma, uneven wear and tear of hooves have been reported. The removal of causal agents for such disorders, proper care and management of feet can mitigate the problem.

f. **Poisoning:** Accidental ingestion of toxic materials like plastic can cause choke, plants or plant material can lead to toxicity, licking of paints leads to lead poisoning, pre flowering stage of sorghum feeding or ornamental plants of enclosures (Ratigan, 1921) can cause cyanide poisoning in captive animals. Apart from this spray of insecticides, disinfectant or pesticides can also causes toxicity and mortality in captive animals of zoo.

g. **Metabolic and nutritional disorders:** Over use or deficient elements in food can lead to metabolic and deficiency disease in captive wild animals .

PHYSICAL INJURY

Trauma : This is produced by sudden violent force that results in a crushing as separation of the tissues. E.g. Collision between wild species and motor vehicles, trains or aircraft result into injury or death.

Electrocution: Sources to free living animals are lightening and electrocution from the power lines.

Burning : Burns caused by heat, chemical frictions. Forest fires are an important cause of death from burning and suffocation in free living population.

Drowning : Considerable animals are drawn but this factor is not play role as major except for rodents.

CAPTURE MYOPATHY

It is also known as over straining disease, stress myopathy, polymyopathy, white muscle disease, transport myopathy, spastic paresis, muscle necrosis and leg rodents. This syndrome is characterized clinically by depression, muscular stiffness, incordination, paralysis, metabolic acidosis, myoglobinuria and death.

Aetiology : It is not fully understood but following are possible reason

(a) Fear and anxiety

(b) Nutritional deficiencies.

Signs : Includes increased respiratory rates and cardiac rates and elevated temperature. Other signs are unsteady gait, incordination, stiffness, tremors, torticollis, weakness, lameness and difficulty in swallowing may occur also. Depression and partial or total muscular paralysis are frequent findings.

Treatment and Prevention

Injecting of preparations containing selenium and Vit E, Vit B12, Calcium borogluconate, antibiotics, detoxicants, corticosteroids and antihistamines,

AMYLOIDOSIS

Also known as amyloid degeneration. It is a disease resulting from deposition in tissue of a homogenous, acidophilic proteinceous infiltrate called amyloid. It tends to occur in vertebrates in association with aging, chronic stress or diseases associated with considerable chronic tissue destruction.

PLANT POISONS

Factors

Climate and soil: The ecology of an area is altered by mechanical, climate or other means. Plants normally found in limited amount may become abundance and therefore troublesome.

Adaptation, learned behaviour, species resistance: Wild animal appear to demonstrate some ability to differentiate between to demonstrate some ability to differentiate between toxic and non-toxic plants or to limit consumption of the toxic variety to nonharmful amounts. This characteristic is generally thought to occur through a learning process in animals living in a geographic area where certain toxin plants are present

Anatomical and Physiological differences: Difference between monogastric and ruminants. Difference between young, adult and old.

Signs

Signs vary with type of poison ingested. In general the sudden inset of illness without a visible cause should provide sufficient reason for including possible toxicity. In absence of fever and with acute disorder of digestive tract or central nervous system leading to loss weight or prostration, plant poisoning may be suspected. Other clinical signs are distress and repeat attempts to defecate, stomach or intestinal irritation and an increase or variability in the heart rate. Most of above sign are followed by weakness, labour breathing, collapse and coma with death following within hours.

Treatment: Removal of unabsorbed poison: By gastric lavage by mineral oil, saline laxative.Elimination of absorbed poison: Provide ample amount of water, fluid therapy and diuretics.Antidote: If specific antidote is available. Supportive therapy: e.g. Antihistaminic, corticosteroids, warm condition to animal body.

VITAMIN DEFICIENCY DISEASES

1. Vitamin A Deficiency: Keratinisation of the third eyelid,Pustular pharyngitis and oesophagitis, Nightblindness and Nutritional roup.
2. Vitamin B1 Deficiency : Cause inflammation neuron (Polyneuritis).
3. Vitamin B2 Deficiency : Cause ventral bending of toe in birds. (Curl toe paralysis)
4. Vitamin B3 Deficiency : Black tongue: produce black discolouration of the tongue and oral mucous. Hock disorder similar to perosis.

5. Vitamin B5 Deficiency : Generalized epidermal desquamation. Crusty exfoliation of the eyelids and commissure of the beak.
6. Vitamin B6 Deficiency : Excitability, convulsions, aimless running, paddling of the legs in the air, twisted neck. Polyneuritis and complete exhaustion fallowing convulsions.
7. Choline Deficiency : Disrupted fat metabolism and perosis.
8. Biotin Deficiency : Exfoliation dermatitis similar to that found in panthothenic acid deficiency. Lesions are found on dorsal surface of feet.
9. Folic acid deficiency : Cause hock disorder, poor feather development and pigmentation and achromocytic anaemia with fragility in red blood cells.
10. Vitamin D Deficiency : Rickets in young and osteomalacia in adult
11. Vitamin E deficiency : It result to nutritional encephalomalacia (Crazy chick disease), nutritional myopathy and white muscle disease
12. Vit. K deficiency : Delayed clotting with resultant haemorrhages from minor trauma.

MINERAL DEFICIENCY DISEASES

1. Calcium and Phosphorus Deficiency : It causes rickets in young and osteomalacia in adult. (along with Vit D). In layer it cause production of shell less egg and decrease hatchability.
2. Manganese Deficiency : It is characterised by a swelling and a flattening of the lateral condoyle of the hock joint, allowing the gastronemius tendon to slip laterally out of articulation. In also result to chondrodystropy (Parrot beak) in embryo chick.
3. Sodium and Chloride Deficiency : It result to dehydration, diarrhoea in animals and develop cannibalism among birds.
4. Iodine Deficiency : It cause enlargement of thyroid gland i.e. goitre.
5. Iron and copper deficiency : Produce anaemia that is characterized by lower haemoglobin.
6. Zinc Deficiency : In birds cause poor feathering, enlarge hock, slipped tendon and scaling of skin and feet (parakeratosis)

CHAPTER 13

RESCUE AND CRITICAL CARE MANAGEMENT

The present matter will enlighten and emphases that wildlife health management is the process of using science to identify appropriate management principles and the practices for safeguarding gene pool of wildlife, in which veterinary science plays a vital role for critical care.

REASONS FOR RESCUE

The wild animals are being rescued for several reasons. The rescued animals are always remaining in a state of excitement and feel uneasiness in the presence of man. This affects the health and mortalities of such animals. The animals, which are being rescued, may be juvenile, very young, sub adult or adult or old in age. Some time the animals that are

socially rejected,

neglected by their own group,

trapped in trapping,

being injured by vehicle or trauma,

fall in the wall or in deep water or in mud,

healthy or sick one

Attending a rescue case, the operation of strategy varies and its depends upon the situation under which the captured animals are being rescued.

Apart from the psychological changes the newly rescued animals also suffer from ecological, physiological, climatic, nutritional and geographical changes that take place by transferring them from their original place to other unknown situation. Utmost care is primarily required during transportation.

During rescue operation the stress conditions play a greater role in health status of animal. As during capture, crating, transportation, lag period for settlement in captivity the animal always remain under stress. The social structure of a species may influence the degree of stress of an individual. It has been observed that the most of the animals trapped from the free-living wild areas are probably the subordinates and that is why more than 50% of rescued animals fail to survive. The animal's susceptibility to stress usually diminishes as it gets accustomed to the new surroundings.

The rescued animals should be attempted as emergencies and should be taken care with utmost all care. The animal should be kept into suitable enclosure where all kind of good hygiene and sanitation measures have been ensured. Suitable diet and clean water supply is also a part of better management. The period of settling down for the rescued animal is crucial and such animals should not be disturbed frequently. The physical examination, collection of biological materials and the critical care measures should be synchronised in such a manner that animal may not get much stress.

The rescued animals should be kept away from the visitors and unknown attendant. The specific person with mothering interest should be deputed for managing such animals. After settlement the rescued animals should be placed for some specific period of quarantine. Quarantine helps the animal to adjust to the new environment, built up strength and offset the effects of trapping, crating and transportation.

Once the initial signs of physical injuries, electrocution, burning, drowning, hypo or hyperthermic condition or sicknesses are detected, diagnosis and treatment should be attempted first. The sub-clinical stage of infection should be ruled out by through clinical examination and by special diagnostic aids. The specific treatment should be provided along with needed prophylactic measures that should be ensured for the safe guard of rescued animal.

The basic optimum needs such as type of diet, housing, space requirement, fresh air, ventilation, sunshine etc. which vary from species to species should be considered during management. Knowing the biological need of rescued animal the enrichment of housing can be planned out (climbing, perching, swimming etc.). The quality and quantity of food selected should be nutritive, hygienic and palatable to the animal. The required amount of nutritive parameter and calories can be manage by procuring available materials (if mother milk is not available then use of skim milk powder, replacing formulae, other animal's milk, corn milk etc.) from nearby resources. Hand rearing of neonates should be planned based on the experience and referring the available literature on the species. The success rate can be increased if above steps are being considered seriously to reduce the failure rates.

The trained personnel staffs with enough experience of rescue, full flagged rescue materials/equipment, quick and safe transport facilities, hospital's critical care management and dedicated team members are the basic need for up keeping the high success rate of orphaned wild animal's rescue operation. Scientific approach in providing every animal congenial environment, better upkeep and health care to ensure them good quality of life and longevity.

CRITICAL CARE OF WILDLIFE

Majority of free living or captive wild animal met with accidental injury and traumatized, poisoning or drowning. Some time they may go in the shock for several other reasons. Here is some of the basic critical care management and priority based material required to keep ready for attending emergency.

TRAUMA AND SHOCK

It is desirable to make overall evaluation of the traumatized patients. All body tissues may be affected by trauma, so examine the WHOLE animal apart from the obvious injuries, such as, fractures of limbs. After identification of all injuries they should be prioritized based on their ability to threaten the animal's life.

Supplying oxygen, ensuring a patent air way and maintaining the B.P. and circulation are the keys to effective resuscitation. Oxygen delivery to the tissues is determined by pulmonary function/cardiac output and the oxygen carrying capacity of the blood. In shock the oxygen supply is of critical importance.

Oxygen Supply : Materials required: Masks, oxygen tents and intranasal oxygen catheters

Indicators of effective oxygen delivery are : mucous membrane colour, body temperature and respiratory rate.

Intravenous catheterization : It is the pipeline to life Place in i.v. catheter to take initial blood sample for determining the P.C.V. and to administer fluids and medication.

It is desirable to have or to keep the P.C.V. greater than 20%.

Options of Fluids in hypovolemia/SHOCK :

1. Give 0.9% sodium chloride at 60-100 ml per kg per hour for approximately one hour or
2. 3% sodium chloride 20 ml per kg @ 1 ml/kg/minute Or
3. 7% sodium chloride at 4-8 ml per kg @ 1 ml/kg/minute, or
4. 6% Dextran 70 at a dose of 10-20 ml/kg/day, @ 1ml/kg/minute

If the P.C.V. falls to less than 20%, give whole blood @ 20-30 ml/kg. Monitor the effect by color, blood pressure and P.C.V. ***Plasma can be given in hypoproteinemia at the dose of 10-20 ml per kg.*** If B.P. is still weak, in spite of aggressive fluid therapy, consider the use of constant rate infusion of: Dobutamine (Dobutrex: Eli Lilly) @ 2-15 mg/kg/minute or use of Dopamine (Dopinga: Inga) @ 3-5 mg/kg/minute.

Skull/Head Injuries : If the animal is unconscious, then brain damage must be suspected. In brain concussion: i.v. mannitol 1-2 g/kg of 5-10% solution at 4 ml per minute as a single dose is indicated to reduce oedema. Vigorous i.v. fluid therapy may worsen brain edema. Corticosteroid is not as useful as mannitol.

Eye Injuries : Assessment of severe eye injuries is important factor in determining the rest of the treatment because few clients would want a blind animal. Early treatment can often save what looks to be a hopelessly damaged eye.

Fractures of Mandible/Maxilla : Surgical intervention

if Presenceo of Nystagmus :

Horizontal is a sign of peripheral vestibular disease

Vertical nystagmus is a sign of central vestibular disease

Upward nystagmus indicate brain stem lesion.

Downward nystagmus indicate cerebellar lesion.

Respiration and Thoracic Injury : Wild animal met with mobile accident and with a limb injury may also have some degree of thoracic trauma. Observe the breathing .A chest radiograph can identify possible thoracic injuries (mild pulmonary contusion, pneumothorax, fractured ribs, and diaphragmatic hernia).

Snake Bite : Major clinical presentation in envenomation have signs occur within 30-60 minutes. Local pain, swelling / ecchymosed dark sera-sanguineous fluid may ooze from the punctured wound with fanged mark if visible. Hypotension, tachycardia, pulmonary edema, salivation, shock after several hours. Hemolytic anemia or hemoglobinurea. Acute renal failure and respiratory distress.

Management :

1. Anti-snake venom, polyvalent I/v 10-20 ml I/v or intra arterial.
2. Broad spectrum antibiotic
3. Aspirin or NSAID.
4. Prevent anaphylactic reaction, use adrenaline if required.
5. Intravenous fluid therapy.
6. Neostigmine (myostigmine) 20-40 mg/kg, I/m or I/v.
7. Corticosteroid to prevent shock.
8. Tetanus antitoxin.

EMERGENCY KITS AND DESIRABLE DRUGS

Priority materials : *m*eans of providing oxygen, provision of positive-pressure ventilation, assortment of endotracheal tubes, sterile peripheral and venous jugular catheters. surgical instruments and towels, surgical lights, stomach tubes and stomach pumps, electrocardiogram, anesthetic machine, refractometer, defibrillator, suction apparatus and catheter, chemistry strips for blood glucose and urea nitrogen, ophthalmoscope, face masks, coagulation testing method, blood pressure and radiographic equipment, laryngoscope and assortment of blades, circulating warm water blanket / Infra red lamp, Administration sets- regular and pediatric / blood filtration sets, Chest drain cannulas, Urinary catheters, Closed urine collection bags, Tape, cotton, gauze and other bandaging materials, Fleece pads for cage padding, sand bags for positioning.

Desirable : Tracheostomy tube, Harleco total CO2 apparatus or oxford titrator, flame photometer, osmometer, fluid infusion pump, and nebuliser.

Fluids : Blood (DEA-4), Dextrans (high & low molecular weight), 5% Dextrose, 2.5% Dextrose and 0.45% saline, 50% dextrose in water, Lactated Ringer's solution,Mannitol, Potassium chloride, 0.9% saline, Sodium Bicarbonate, sterile water for injection, Heparin,

DRUGS : Specific anaesthetic of choice, epinephrine, sodium bicarbonate, atropine, blood pressurestimulant, (Mephentermine, dopamine, dobutamine, ephedrine), Vasoconstriction (Phenylephrine, Norephinephrine), Cardio tonics (Calcium, Digitalis, Isoproterenol, Blood pressure stimulant, Glucagon), Antiarrhythmic agents (Lidocaine, Procaineamide, Quinidine, Beta-receptor blocking agent, Tranquilizers, Sedatives (Meperidine, Acepromazine), Anaesthetics for cardiovascular debilitated patients, (Oxymorphone and Diazepam, Ketamine and diazepam, Analeptics (Narcotic reversing agent (naloxone), Doxapram, Glucocorticoids (Dexamethasone, Prednisolone succinate), Diuretics (Furosemide, Mannitol, dextrose, Dopamine), Antipyretics (Dipyrone, Antiprostaglandins), Antiemitics (Thorazine, Prochloperazine), Antitoxins for specific cause (Activated charcoal, Pralidoxieme (for organphosphates, Kaolin-pectate, Vitamin K), Broad-spectrumantibiotics, Antihistamine (diphenhydramine), Aminophylline,.Vasodilator (Nitroprusside, Nitroglycerin, Phentolamin, Clonidine, Prazosin), Neuromuscular blocking agent and reversing agents.

CHAPTER 14

POST MORTEM EXAMINATION OF WILD ANIMALS

Post-mortem or necropsy examination is one of the most important aspect of diagnosis of diseases in wild animals as in domestic animals. The examination is conducted to identify cause of death in a given herd/flock. The history, signs, clinical examination of affected animals, laboratory analysis and gross and histopathology are the usual methods of disease diagnosis in most animals. however, diagnosis of diseases is very difficult in free ranging animals because mostly these animals are simply found dead without any previous clinical signs or partly eaten by scavengers. The pathologist seldom possesses a history of illness that preceded death. The animals in wild are not usually observed until they are found dead. Frequently they do not even exhibit symptoms of illness until they are too weak to walk, fly or crawl and sudden death may occur in the absence of visible gross or microscopic lesions. Necropsy problems are further complicated by the fact that most examiners are not likely to be familiar with normal habits and anatomy of wild animals. We do not have exact information regarding anatomy, physiology, behaviour and diseases of wild animals. The status of various diseases of Indian wild animals is poorly known.

Since the loss of a wild animal may have major economic or legal implications, it is the duty of pathologist/examiner to exhaust every known means of establishing the cause of death. Special efforts will sometimes yield at least a minimal history, and particular attention should be paid to the external examination. Frequently, helpful clues can be obtained from the quarters in which the animal has been kept or in the area it has inhabited. Therefore, postmortem examination of all dead captive and free ranging animals may provide valuable information about prevailing disease conditions in Indian wild animals. It may be of immense help in formulating effective control measures.

Obviously, it is impractical - indeed, impossible to outline a necropsy procedure to *all* wild animals. Not only would complete knowledge of comparative anatomy be required to the author, but similar knowledge would have to be possessed by anyone expected to follow the prescribed technique.

The basic principles of post mortem examination are same as in case of domestic animals. Post mortem examination should be conducted immediately after death of wild animal to avoid postmortem changes eg. putrefaction and autolysis. Similarly, if animal is suspected died of anthrax, necropy should not be conducted. We have domestic parallels for most of wild animals including birds, that are excellent guides in necropsy procedures. Wild ruminants such as deer, antelopes, cattle, camels, llamas, sheep and goats are necropsied in the same manner as domestic cattle, sheep and goats. The domestic horse will serve as a pattern for the examination of wild horses, asses, zebras, and certain other ungulates such as elephants, hippopotamuses, rhinoceroses and tapirs. Wild swine are the equivalent of domestic swine. Wild birds of domestic chickens, turkeys, ducks and geese. All wild carnivores- wolves, coyotes, bears, lions, tigers, sea lions and the like are necropsied like domestic cats and dogs. Therefore, similar techniques can be followed by the pathologist while conducting the necropsy examination.

However, there are certain species of wild animals for whom domestic parallels are not available like reptiles, fish, marsupiales, rodents, edentates, armadillo and pangolins. Pathologist under such conditions may follow a common sense based on general knowledge and basic principles of post mortem examination. For exotic species of wild animals, one has to see the literature. Wild animals belonging to different groups, posses anatomical peculiarities. For example, all reptiles (snakes) except crocodile, have three chambered heart, lung in snakes is single or double and elongated, deposition of fat in body tissue is normal in hibernating animals and trachea in certain storks and cranes is very long and tortuous, single, double and semidivided. The gall bladder is absent in all deer except in musk deer; moose deer has got only three compartments of fore stomachs. The elephant lacks a pleural cavity, in the sense that lungs are diffusely adherent to the chest wall and it may give a false impression that animal has diffuse pleural adhesions. The neotropical birds known as screamers have diffuse subcutaneous air spaces that crepitate on palpation- a condition that is normal and not produced by gas gangrene. Fish eating birds, such as penguins have very dark, reddish black breast muscles.

REQUIREMENTS FOR POST-MORTEM EXAMINATION

1. Disinfectant/ detergent solution to wet the body of animal / bird.
2. Post-mortem knives of different sizes, electric saw/bone cutter.
3. Post-mortem gloves (disposable).
4. Scissors, forceps, bone forceps and scalpels of varied size.
5. A table and adequate light arrangement for small animals and birds.
6. Provision for sifting carcase of heavy animals.
7. History sheet for recording the findings.
8. Sterile syringes, needles, vials and petri dishes for collecting blood samples and tissue specimens.
9. Specimen bottles with 10% formalin to collect tissues for histopathology.

NECROPSY PRECAUTIONS

1. Good quality rubber gloves should be used and care should be taken that neither the pathologist nor assistant puncture the skin of their hands or inhale dust or aerosols from tissues or feces.
2. If there is reason to suspect that animal/bird to be examined are infected with disease that may be contagious to humans (tuberculosis, ornithosis, erysipelas, equine encephalomyelitis), extra precautions are advisable.
3. The carcase and necropsy table surface should be wet thoroughly with a disinfectant.
4. All laboratory staff who may come in contact with carcases, tissues or cultures should be apprised of their possible infectious nature and precautions to be taken.
5. Examining specimens on metal treys that will fit into autoclave and facilitates quick sterilization of carcases after necropsy.
6. Submit post-mortem report in standard proforma (Annexure -I).

Before opening the carcase in all wild animals including birds, it should be examined externally for any ectoparasite, injury, swelling, abnormal growth and damage/ changes in external hair coat or plumage in birds. It will be advisable to ask back history related to death/ mortality. Record of animal in case it is from zoo should be seen to know about vaccination and previous history of disease. Try to collect information regarding food habits, anatomical features and clinical history of wild animal under question.

Wild ruminants : Carcase is kept on surface that back faces ground. Mid line incision is given on ventral surface from anterior end to posterior end. Body cavities are opened to examine different visceral organs.

Wild cat and other carnivores : As in case of dogs and cats, put carcase on surface that back faces ground. Incision is given on ventral surface in mid line. Open the body cavities and examine internal organs.

Primates : Carcase should be placed in such way that back faces ground. The incision should be given on ventral surface (mid line) starting from mandibular symphysis to pubic symphysis. Body cavities are opened and examined. Care should be taken to avoid cutting abdominal organs. Note the position and colour of organs in abdominal cavity and presence and absence of exudate, transudate, parasites and adhesions. If blood, pus or any other fluid is present, its quantity, consistency and nature should be measured and noted.

Thereafter, individual organs belonging to different systems are examined carefully and post-mortem changes are recorded.

Reptiles : Mouth is examined carefully for any lesion in buccal cavity, particularly necrotic stomatitis or mouth rot which is very common in reptiles. Before opening the carcase, it should be examined externally for any lesions or ecto-parasites which are very common. Skin and scales are observed for presence of ulcerationa or other lesions, particularly on ventral surface. Following external examination, snake is placed on back and mid line incision is made from anterior cervical region to cloaca to expose visceral organs. The skin is than reflected over sides and usual examination is made. Starting from anterior cervical extremity and proceeding caudally, examine the organs *in situ*. Then remove the organs serially and examine for the pathological lesions.

Turtles : The carcase is placed on back on a concave surface. Longitudinal incisions are made on each side of junctures of plaston (ventral shell plate) and carapace (dorsal shell) with the help of electric saw. Then soft tissues at the plaston are incised which is then removed to expose all internal organs.

Rodents : Members of this order are examined by method used for carnivora. The examiner may expect to find in some rodents an alimentary tract considerably unlike that of the carnivores, since the differing food habits of rodents account for differences in anatomical structures.

Edentates : The edentates or toothless mammals such as sloths, armadillos and anteaters, vary widely both anatomically and in size. Because of their dermal covering, which in case of the armadillo, consists of bony plates embedded in skin, it is easier to place them in dorsal recumbency and make a ventral midline incision through the softer tissue, leaving the dermis over the sides and back intact. The technique for primates is then followed.

Pangolins and aardvarks are similar to the edentates and should be examined in the same manner. The entire group, with exception of the sloths, are meat eaters. The alimentary tract therefore most nearly approximates that of carnivora.

Birds : Small birds such as humming birds and canaries should be examined as soon as possible after death. Otherwise they either decompose rapidly from lack of refrigeration or dehydrate as a result of refrigeration. Necropsy technique for wild birds is similar to that employed for domestic birds. The bird is laid on its back and each leg in turn drawn outward away from the body while the skin is incised between leg and abdomen. Each leg is then grasped firmly in the area of femur and bent forward, downward and outward, until the head of femur is broken free of the acetabulum attachment so that the leg will lie on the table.

The skin is cut between the two previous incisions at a point midway between keel and vent. The cut edge is then forcibly reflected forward, cutting as necessary, until the entire ventral aspect of body including neck is exposed. The necropsy knife is used to cut through the abdominal wall transversely midway between keel and vent and then through breast muscles on each side. Bone shears are used to cut the rib cage and then the coracoid and clavicle on both sides. With some care, this can be done without severing the large blood vessels. The sternum and attached structures can now be removed from the body and laid aside. The organs are now in full view and may be removed as they are examined. If a blood sample has not been previously collected and the bird was killed just prior to necropsy, a sample can be collected by heart puncture at any stage when heart is exposed.

After necessary cultures have been made, intestine may be laid out on table top or news paper for examination of any inflammation, exudate, parasite, foreign bodies, tumour or abscess. Various nerves, bone structures and marrow condition and joints can now be examined. The sciatic nerve can be examined by dissecting away musculature on medial side of thigh. Similarily sciatic plexus is examined by removing kidney tissue with blunt dissection. Nerves of brachial plexus are easily found on either side near thoracic inlet and should be examined for any enlargement.

Joint exudate, if present can be removed after first plucking the feathers and shearing the overlying skin with a hot iron, which may be incised with a sterile scalpel and exudate removed with a sterile inoculation loop or swab. Exudate from paranasal sinuses can be removed in similar manner.

Post mortem examination, if conducted sincerely keeping all precautions in mind and considering all available informations may be of great help in unearthing the cause of death in wild animals as in domestic animals. It is the experience and knowledge of the pathologist which will decide the specimen to be collected. Laboratory investigations on these samples will further strengthen the diagnosis.

CHAPTER 15

FORENSIC AND VETEROLEGAL ASPECTS OF CRIME INVESTIGATION

Several factors are responsible for causing variable morbidity and mortality in wild animals. The ecological imbalance, natural calamities, anthropological factors, accidents, poaching, malicious poisoning, nutritional disorders and epidemics of infectious diseases are some of the examples where the wild animals are victims. The managers of protected areas in such circumstances approaches the near by field veterinary officers to deal with the issues and for conducting post mortem examination and report. Some times the veterolegal cases, involvement of endangered species and legal issues related with WL protection act 1972 involves the veterinary officers as an expert witness and in such situation the veterinarian has to be specific in some of the aspects as a forensic investigator. Based on experiences here are some of the important points, which could help the protected area manager and veterinarian to deal with the disease diagnosis from dead wild animals.

The approaches for such incidences should be prompt and with all efforts to get maximum clue and evidence to confirm the cause of death or event. The investigation officer should report and get maximum photographic evidence with all angles to support and explain the story even in the court.

The following points can be looked in to

The scene of the crime

The condition of animal

Circumstantial evidences

Collection of evidential materials

Preservation and dispatch of the materials for court evidence and for laboratory investigation.

Report and photographical records with signature.

EVIDENCE WHERE TO LOOK IT?

1. Scene of crime
2. Victims
3. Surroundings
4. Disposal site

1. **Scene of Crime :** The route can be inspected from a distance by several techniques. Close observation of every aspects of scene of crime.

   What to Look?

   Mark : Finger print, pug marks, teeth mark, tyre marks, oil and other items

   Fluids : Blood, urine, faeces, vomitus, remnants of food, empty utensils, chemicals, seeds, gun powder, flash bone, skin etc.

   Fibers : Hair, textiles, feathers, fur, artificial fiber.

   Fire arms : cartridge, case, bullets, marks, burning marks, shell, gun powder, box

   Body fluids : Blood : Pool of the blood, blood stain, smear of blood, blood on soil

   Collection Methods : Blood, hairs, fibers, food specimen

2. **Victims :** Observe the position of victim and external condition of the victim. The natural orifices and the skin and hair coat is a good indicator of several evidence to rule out. The cadaver/carcass condition can give the clue to many situation and diseases. The duration and probable time of death can also be judge from the condition of carcass based on several physiological and pathological changes took place in the body.

   Duration and Time of Death :

   Early changes :

   Change in eye lusters, corneal reflex, Change in skin: elasticity, pale ness, Cooling of body: after 12 hrs. Hypostasis: PM lividity: Discoloration of skin, 6-8 hrs.

   Moderate changes :

   Primary changes : Whole muscle relaxed, flexed eyes, 1-3 hrs. **Rigor mortis (RM)**: death stiff ness anterior to posterior stiffness passes out in the same sequence, RM sets in one to two hrs. After death, well developed with in 12 hrs. remains for next 12 hrs. It disappears from the body and decomposition of body starts.

   Secondary changes : Soft and flaccid muscle, autolysis, putrifaction, foul smell, colour changes: greenish discoloration of abdominal area, skin, distension of abdomen,1-2 days. soft eye ball, cornea white milky, eye collapse later on, prominent superficial vein.

   After 35 hrs : foul smell: anal sphincter relax, leakage of faecal material.1

   After 48 hrs : loose hairs, easy to pluck, PM blisters on skin, bulged eye ball, protruded tongue, frothy blood mixed fluid from mouth and nostrils

   After 48 hrs : up to 2-5 days whole hair falls, down, sunken body and abdomen may burst, flies ,beetle lays eggs, maggots, liquification of internal organs.

   5-10 days : Severe degree of putrifaction-Colliquative putrifaction, whole muscle become decomposed, separated from bone, stick to bone with black mass and bone exposes from the several places.

Late Changes :

More than 10 days : it is difficult to estimate the duration of death as there is masking of all injuries and features except fracture of bone.

Flies : House fly, green or blue bottle e fly eggs turn in to maggots with in 8-24 hrs during hot weather, maggots crawls in the body, destroy tissues, pupa in 4-5 days, adult in 3-5 days, whole cycle takes places with in 11-14 days. Eggs –maggots, pupa-larvae should be collected for investigation. In star pupae (Pale brown, maroon and deep chocolate colour).

See the external appearance of carcass and grade them, check hair, skin and coat for ectoparsites, injuries and any mark on the body also.

Injuries/Wound : Caused by: arrow, gun, run over of vehicle, snaring, trapping and blunt njuries.Contusion/abrasionIncised, stab, puncture, lacerated, chopping, fire arm wound, Burns, Electrocution.

Stages of wound healing : May vary with dampness and complication of wound.

Abrasion: oozing: 3 hrs.

Red scab: one day

Brown scab: 2 days

Black scab: 3-5 days

Cicatrisation: 7 days

Scab fall down: 7-10 days

Categories the injuries :

a. **TYPE:** SIMPLE,MULTIPLE

b. **SITE:** ON WHICH BODY PART?

c. **SIZE:** MEASUEMENT, LENGTH, WIDTH, DEPTH etc.

d. **SHAPE**: ROUND,OVAL,CIRCULAR, ELLIPTICAL

e. **PLAIN:** HORIZONTAL,VERTICLE,OBLIQUE,PARALLEL

f. **ENDS:** SHARP,BLUNT,POINTED,CIRCULAR,

g. **EDGES** : PLAIN,SHARP

h. **CONDITION:** BLEEDING,CLOT,WET,DRY,INFECTED,HEALING

i. **MARGIN:** PLAIN,SERRATED,ZIG ZAC

j. **DEPTH:** MM/CM/UP TO CAVITY,CAGE,BONES

k. **DIRECTION:** Examine the direction of wound features

Examine the carcass for the following

CUT THROAT

DECAPITATION

DISMEMBERMENT

DE SKINING

BURN:

Weapons:

Object	Type of weapons	Injury type
Blunt	Lathi, vehicle	Bruises, contusion, abrasion, crushed lacerated wound
Sharp	Weapons, axes, arrow, farsa, knife,	Incised wound, stab, perforating, penetrating wound
Trapping/snaring	Rope, wire, cloth, plastic	Ligature, friction cut deep mark
Gun shot	Gun, rifle, pistol	Fire arm injuries
Electrocusion	Electric shock, live wire,	Charring-singing mark

Difference Between Ante Mortem and Post Mortem Injuries:

Ante Mortem	Post Mortem
Ecchymosis	absent
Hemorrhage jet like	In drops
Swollen edges of wound	absent
Red everted edges with gapping	absent

Drowning :

Ante Mortem	Post Mortem
Tenacious leather froath, blood tinged,	absent
Lungs –oematous, voluminous, froth fluid blood	In drops
Water weed, mud in respi.tract	absent
Diatoms	absent

Electrocution : Charing, Zigzac, Singing Marks

BURNS :

BLISTERS over the skin, filled with fluid blood tinges

Line of red inflamed area

Demarcation of healthy skin

Thick and cherry coloured blood, -CO investigation

Internal Changes : Observe the opened skin, fascia, omentum ,muscle, different organs of the different system, specifically digestive see oral cavity for any object inside, stomach content, intestine, liver, kidney),respiratory system(trachea, lung).Collect the content in respective preservatives, take pictures of the abnormalities, check for the bullet, injuries inside the cavity or organs, pathognomonic lesions if any.

3. **Surroundings :** The surrounding of victim animals should thoroughly investigated for direct and indirect evidence for suspection of crime.
4. **Disposal Site :** The disposal site should be plan out with view to dispose the animal so that zonatic trade of diseases can be avoided and left over part of dead animal should not be taken by SCAVENGER and Vermins.

IMPORTANT FEATURES

The recording of each matter, manner, mechanism and cause of death with detailed past history is an important clue to judge the matter for the forensic aspects.

It is better to use audio visual aids with good quality of photographs or video with close up views which supports the circumstantial evidences.

Recognition and description of arrow wound, gun shot wound, anatomical position and location and velocity of caliber guns (low, high) supports the cause of death in wild animals.

Complete full flagged field kit for post mortem and sample collection is an essential tool.

Use of disposable gloves and mask, sterile bottles and protocol of post mortem helps to go the veterinarian in systemic diagnosis of diseases.

Massive blunt trauma caused by mobile accident can be looked for fractures, bruising or internal hemorrhage, which should be closely examined with patience.

Head, neck and limbs covered with hair should be thoroughly examined for the presence of marking of traps and snares.

Trauma caused by predators and scavengers must be differentiated from other cause of trauma.

Electrocution, snake bite and drawing should be confirmed with circumstantial evidences.

The poisoning or poaching with bait should also be ruled out by several field study of the area and near by surroundings.

Forensic laboratory investigation of carbamate, OP compounds, strychnine, anticoagulant rodenticides, thallium and cyanide held for confirmation of the diagnosis.

The hairs from the stomach of poisoned animals, knives, pickups, clothings, traps and other evidences should be closely observed and collected in separate bags with marking.

The postmortem changes are often influenced by several factors like ambient temperature, health of animals, degree of muscular activities before death, mode of death and medium of death etc. should be marked to rule out there time of death.

The development of rigor mortis from head, neck and lateral spread to back and limbs, cooling of body after a few hours of death (8-12 hrs.) also depends on several factors.

The putrefaction of carcasses with distension of abdomen with gases, blood stained fluids from nostrils and mouth, liquification of eye balls, foul smell, bursting of abdomen, thorax and liquefaction of body should be recorded .

Samples To Be Collected :

Laboratory For Analysis	Sample	Preservative	Collection Tool	Remark
Histopathology Lab.	Heart, Lungs, Kidney, Brain, Spleen, Liver, Intestine etc.	10% Formaline	Glass/Plastic bottle of Broad lid.	Sample 1 .. x 2.. (May be different depends on animal to animal formalin should be 10 times of sample.
Poison/ Chemical	Fat, food found in intestine, Parts of Guts	Ice in thermos or ice box 70% Alcohol or10% Formaline	Glass/Plastic bottle	Leak proof bottle, Avoid touch on the body
Blood	Whole blood (With anti-coagulant or EDTA	Ice box or thermos	Small bottle of Glass/Plastic	Send as soon as possible.
Species Identification Parasites	Hairs, Faecal, Ecto-parasite, Blood smear, Endo-parasite	Clean and Air dried, 5% Formaline Air dried	Cellophane bags Glass/ Plastic bottle Glass Slides	Seal properly Send with proforma enclosed

Field Visit Kit

Chemicals:

EDTA vials:	5ml X 4 nos
Plastic Bottles:	200 ml X 6 nos
Dropper:	5 nos.
Glass slides:	1 box
Coverslip (Round & Square):	1 set
Benzidine Reagent:	200 ml X 1
Pottasium iodide:	200 ml X 1
Salt (Nacl):	500 ml X 1
Glycerine:	50 ml X 1
Filter Paper:	25 nos.

Glasswares:

Hand gloves:	2 pairs
Hand gloves (Disposable):	5 pairs
Brush:	1
Metal measuring scale:	1
Swiss knife:	1
Forceps (Large toothed):	1
Forceps (Small/ Plain):	1
Knife:	1
B.P. Handle and Handle:	2
Specule/ Spoon:	1
Scissor:	1
Cotton swab:	1

Miscellaneous:

Camera: 1

Binocular (20 X 25 UCF): 1

Measuring steel tape:	1
Torch:	1
Hand lamp:	1
Packing Paper:	1
Needle/ harced:	1
Watch glass/ Paper slip:	1

Stationary:

Compass needle: 50 mm X 1

Sticky lable

Writing paper

Permanent marker

Pen

Pencil/ Sharpner/ Eraser

Wax

Cotton cloth

Nylon Apron

Book let

CHAPTER 16

ZOONOTIC DISEASES

The sanctuaries, zoological gardens and parks are the areas to encounter a side variety of zoonotic diseases. Generally animals come to the zoos from wide varieties of geographic areas of the world after their capture, and they often spend weeks or months in close contact with the native people, which allow them to pick up human infections endemic to the locality. Added to this the "STRESS" of capture, captivity, drastic diet changes, transport and varying climate sones through which the animal passes, sets the stage for these animals to shed pathogenic organisms communicable to human. In addition to these factors, the frequency of shedding of the contageons is also associated with over crowding of the animals.

Unlike the other places (animal and pet markets, infected sanctuaries etc.) zoos do not report a high incidence of zoonotic diseases. The possible reasons are as follows:

Zoo animals are usually not imported in great quantity in a single shipment.

Most zoo animals are procured through local animal dealers or other zoos

Rather than directly from the country of origin.

Because of high cost of species, zoo will not accept or pay for animals not in good condition of health.

Most zoos carefully check up animals after their arrival. This procedure includes Quarantine, a through veterinary examination, tuberculin testing, serological screening for disease, checkup for ecto endoparasite, acclimatization, vaccinations and careful diet before importes are added to the zoo collections.

DEFINITION OF ZOONOSES

Zoonoses are diseases the agents of which are transmitted between vertebrate animals and man.

It is the interaction of agent, host and the environment they share that determines whether or not transmission of the agent will be successful, leading to infection and ultimately, occurrence of disease. Carrier host individuals infected without overt signs of diseases are important in the persistence of many zoonotic agents. Vertebrate animals are the reservoirs of zoonoses. The agents may be transmitted either directly or indirectly by fomites or vectors.

It is estimated that there are over 175 infections and diseases of animals that are transmissible to humans under certain conditions. There are diseases common to man and animals in which they both generally acquire the infection from the same source. In certain cases, Animals contribute in varying degrees to the distribution and actual transmission of infections. Lastly there are certain diseases in which humans are the primary hosts and animals are only infected when they are in contact with them.

Wild animals can transmit disease to humans in the following way:-

By Direct Skin Penetration

Anthrax

Leptospirosis

Meliodosis

Glanders

Tularemia

By Animal Bite

Rat bite fever

Herpes B.encephalomyelitis

Rabies

By Arthropod Vectors

Lyme disease - Tick

Plague - Flea

Relapsing fever - Tick

Tularemia - Tick and Biting flies

Rocky Mountain spotted fever - Tick

Scrub typhus - Mite

Yellow fever - Mosquito

Encephalitis - Mosquito

Rift valley fever - Mosquito

PREVENTIVE STEPS IN WILDLIFE ZONOSES

The capture of wild animals for trade or laboratories or for use as pets should be Discouraged.

Quarantine procedure should be implemented for such animals.

Pets caught in wild or belonging to exotic species are dangerous because they may carry diseases that are zoonotic.

Wild animals should not be kept as pets.

Avoid infected animal tissues and premises contaminated by animal urine, blood or tissues.

People should wear protective clothing such as apron, rubber gloves, face mask when handling infected materials.

Swimming in fresh water ponds and streams likely to be contaminated by urine of wild animals should be discouraged.

Avoid contact with flies and ticks in enzootic areas during the seasonal incidence of biting arthropods.

CONTROL OF ZOONOTIC DISEASES IN WILD ANIMALS

The control of wildlife zoonoses must be approached in a multidisciplinary measures, because of the issues regarding the animal reservoir and mode of transmission of the contagion, the measures that must be taken across human and veterinary medicine, sanitary engineers and in some cases the entomologist and wildlife zoologist.

The Measure of Control Includes

Eradication of infected animal reservoir/population reduction. There are many techniques of wildlife population reduction and selection of proper one depends on local conditions and regulations. Acceptable methods used are:

1. Hormonal inhibitors
2. Automatic vaccination devices to vaccinate selected wild animals.
3. Oral bait vaccination.
4. Protection of animals before they become infected
5. Improved sanitary Measures and vaccination.
6. Eradication of biting arthropods : Use of insect repellents, insecticide

A LIST OF ZOONOSES (SOME DISEASES NATURALLY TRANSMITTED BETWEEN VETRTEBRATE ANIMALS AND MAN)

The following list of zoonoses is not comprehensive, although all the known major ones have been included. The diseases listed have been confined to those in which the animal link in the chain of infection to man is considered to be importance, although not always essential.

VIRAL DISEASES

Sr. No.	Disease	Causative Organism	Animals Principally involved
Arthropod-borne virus infections			
1	Colorado tick fever	Colorado tick fever virus	Rodents
2.	Eastern equine encephalitis	Eastern equine virus	Birds, Equines
3.	Encephalomyocarditis	Encephalomyocarditis virus	Rodents
4.	Japanese B encephalitis	Japanese B virus	Birds, horse, swine and other mammals
5.	Murry Valley encephalitis	Murry valley virus	Birds
6.	Rift valley fever	Rift valley fever virus	Birds
7.	St. Louis encephalitis	St. Louis virus	Birds
8.	Tick-borne spring-summer group (including louping ill, Russian spring-summer encephalitis, Omsk haemorrrhagic fever, Kyassanur forest disease)	Russian spring-summer-louping-ill group of viruses	Goats, sheep, birds, wild mammals
9.	Venezuelan equine encephalitis	Venezuelan equine virus	Equines, rodents
10.	Wesselsborn fever	Wesselsborn virus	Sheep
11.	Western equine encephalitis	Western equine virus	Birds, Equine
12.	West Nile fever	West Nile fever virus	Birds
13.	Yellow fever (Jungle)	Yellow fever virus	Monkeys
14.	Aujeszky's disease (Pseudo rabies)	Aujeszky's virus	Ruminants, swine, dogs
15.	B virus disease	B virus	Monkeys
16.	Cat-scratch disease	Cat-scratch disease vir.	Cats
17.	Cow pox (Milker nodule)	Cow pox or Vaccinia virus	Cattle
18.	Equine infectious anaemia	Equine infectious anaemia virus	Equines
19.	Foot-and-mouth disease	Foot-and-mouth disease virus	Ruminants, swine
20.	Influenza	Influenza virus type A	Swine, horses
21.	Lymphocytic choriomeningitis	Lymphocytic choriomengitis virus	Mice, dogs, Monkeys
22.	Newcastle disease	Newcastle disease virus	Chickens, birds
23.	Ovine pustular dermatitis (Contagious ecthyma)	Ovine pustular dermatitis virus	Sheep, goats
24.	Psittacosis (Ornithosis)	Psittacosis virus	Psittacines, Poultry, Pigeons
25.	Rabies	Rabies virus	Dogs, cat, wolves, foxes,Jackals, Bats, other wild animals
26.	Sendai virus disease	Sendai virus	Swine, rodents
27.	Vesicular stomatitis	Vesicular stomatitis virus	Equines, cattle, swine

RICKETTSIAL DISEASES

Sr. No.	Disease	Causative Organism	Animals Principally involved
1.	Murine (endemic) typhus	Rickettsia typhi (mooseri)	Rats
2.	Northen Queensland tick typhus	Rickettsia australis	Bandicoots, rodents
3.	Q fever	Coxiella burnetii	Cattle, sheep, goats, wild and domestic birds and mammals
4.	Rickettsial pox	Rickettsia akari	Mice
5.	Scrub typhus (Tsutsugamushi)	Rickettsia tsutsugamushi	Rodents
6.	Spotted fever (including Rocky Mountain, Brazilian and Colombia spotted fever)	Rickettsia rickettsii	Dogs, rodents and other animals

BACTERIAL DISEASES

Sr. No.	Disease	Causative Organism	Animals Principally involved
1.	Anthrax	Bacillus anthracis	Ruminants, equines, swine
2.	Brucellosis	Brucella abortus, Br.suis, Br. melitensis	cattle, swine, goat, sheep, hares
3.	Bacterial food poisoning and intoxications	Salmonella spp., Staphylococcal Enterotoxin, Clostridium welchi, and others	Ruminants, swine, poultry, rodents
4.	Colibacillosis	Escherichia spp., Arizona group of Enterobacteriaceae	Poultry, swine, dogs
5.	Erysipeloid	Erysipelothrix rhusiophathiae	Swine, poultry, fish
6.	Glanders	Actinobacillus mallei	Equines
7.	Leptospirosis	Leptospira spp.	Rodents, dogs, swine, cattle
8.	Listeriosis	Listeria monocytogenes	Rodents, sheep, cattle, swine
9.	Melioidosis	Pseudomonas pseudomallei	Rodents, sheep, goats, equines, swine
10.	Plague	Pasteurella pestis	Rodents
11.	Pseudotuberculosis	Pasteurella pseudotuberculosis	Rodents, cats, fowls
12.	Rat-bite fever	Spirillum muinus, Streptobacillus moniliformis	Rodents
13.	Relapsing fever (endemic)	Borelia spp.	Rodents
14.	Salmonellosis	Salmonella spp.	Mammals, birds,

[Table Contd...

Contd. Table]

Sr. No.	Disease	Causative Organism	Animals Principally involved
			poultry
15.	Staphylococcosis	Staphylococcus spp.	Mammals
16.	Tuberculosis	Mycobacterium tuberculosis var. bovis var. hominis var. avium	Cattle, goats, swine, cats, wild animals
		Var. hominis	Dogs, swine, cattle, wild animals
		Var. avium	Poultry, swine, cattle, wild animals
17.	Tularemia	Pasteurella tularensis	Rabbits, hares, sheep, wild rodents
18.	Vibriosis	Vibrio foetus	Cattle, sheep

FUNGAL DISEASES

Sr. No.	Disease	Causative Organism	Animals Principally involved
1.	Ring worm (Favus)	Microsporum spp.	Dogs, cats, horses, wild animals
		Trichophyton spp.	Horses, cattle, poultry, small animas

Protozoal Diseases:

Sr. No.	Disease	Causative Organism	Animals Principally involved
1.	Balantidiasis	Balantidium coli	Swine
2.	Leishmaniasis:		
	Espundia (American leishmaniasis)	Leishmania brazaliensis	Dogs, cats, rodents
	Kala-azar	Leishmania donovani	Dogs, cats, rodents
	Oriental sore	Leishmania tropica	Dogs, cats, rodents
3.	Toxoplasmosis	Toxoplasma gondii	Mammals, birds
4.	Trypanosomiasis:		
	African sleeping sickness	Trypanosoma gambiense	Wild and domestic ruminants
	Chagas' disease	Trypanosoma rhodesiense	Wild games
		Trypanosoma cruzi	Cats, dogs, rodents

TREMATODE DISEASES

Sr. No.	Disease	Causative Organism	Animals Principally involved
1.	Amphistomiasis (gastrodiscoidiasis)	Gastrodiscoides hominis	Swine
2.	Bilharziasis	Schistosoma japanicum (occasionally other species)	Ruminants, swine, dogs, cats
3.	Clonorchiasis	Clonorchis sinensis	Dogs, cats, swine, wild mammals, fish
4.	Dicrocoeliasis	Dicrocoelium dendriticum	Ruminants, equines
5.	Echinostomiasis	Echinostoma ilocanum (occasionally other species)	Cats, dogs, rodents
6.	Fascioliasis	Fasciola hepatica,Fasciola gigentica	Ruminants
7.	Faciolopsiasis	Fasciolopsis buski	Swine, dogs
8.	Heterophyiasis	Heterophyes heterophyes	Cats, dogs, fish
9.	Metagonimiasis	Metagonimus yokogawal	Cats, dogs, fish
10.	Opisthorchiasis	Opisthorchis felineus (occasionally other species)	Cats, dogs, wildlife, fish
11.	Paragonimiasis	Paragonimus westermani	Cats, dogs, wildlife
12.	Swimmer's itch	Schisstosoma spp.	Birds, rodents

CESTODE DISEASES

Sr. No.	Disease	Causative Organism	Animals Principally involved
1.	Diphyllobothriasis	Diphyllobothrium latum	Fish, Carnivores
2.	Diphylidiasis	Diphylidium caninum	Dogs, cats
3.	Hydatidosis	Echinococcus granulosus (Larval cyst stage) and occasionally E. multilocularis	Dogs, ruminants, swine, foxes, rodents
4.	Hymenolepiasis	Hymenolepsis nana	Rats, mice
5.	Sparganosis	Sparganum mansonoides (other species)	Cats, mice and other mammals
6.	Taeniasis and cysticercosis	Taenia saginata, Taeniat solium	Cattle Swine

NEMATODE DISEASES

Sr. No.	Disease	Causative Organism	Animals Principally involved
1.	Ancylostomiasis and cutaneous larva migrans ("Creeping eruption")	Ancylostoma brazilliense (occasionally other species)	Dogs, cats
2.	Strongyloidiasis	Strongloides stercoralis	Dogs
3.	Toxocariasis (Visceral larva migration)	Toxacara canisToxacara cati	DogsCats
4.	Trichinosis	Trichinella spiralis	Swine, rodents, wild carnivores
5.	Trichostrongyliasis	Trichostronglyus colubriformis (Occasionally other species)	Ruminants

ARTHROPOD AND INSECT INFESTATIONS

Sr. No.	Disease	Causative Organism	Animals Principally involved
1.	Acariasis	Dermanyssus spp. Sarcoptes spp.,Trombicula spp., etc.	Poultry, Domestic animals, wildlife
2.	Bug-bites	Cimex spp., Triatoma spp., etc.	Chickens, birds, small mammals
3.	Flea-bites	Xenopsylla, Ctenocephallus, Ceratophyllus, Tunga, etc.	Rats, dogs, cats, swine, birds,
4.	Myasis	Oestrus, Hypoderma, Gasterophilus, Cochliomyia, etc.	Ruminants, equines
5.	Tick-bites	Ixodes, Dermacentor, Rhipicephalus, Haemaphysalis, Amblyomma, Argas, etc.	Dogs, Cattle

CHAPTER 17

DISEASE CONTROL TECHNIQUES IN FREE-RANGING WILD ANIMAL

AN OUTLINE FOR DISEASE CONTROL OPERATIONS

1. Planning

A. Identify Needs

(i) Source of additional personnel to help during disease outbreaks. They can be:

a. State Animal Husbandry Department.

b. Nearest Veterinary Colleges.

c. Wildlife Institute of India, Dehra Dun.

d. Indian Wildlife Conservative Health Centres.

(ii) Special needs

a. Burning of infected materials.

b. Ability to attract and hold animals on a specific sites by food eater and other means.

c. Capturing animals for collection of samples.

d. Isolation and treatment of sick animals in case of endangered species.

(iii) Availability of equipments for disease control operations.

B. Record of biological informations.

(i) Daily and seasonal animal movements.

(ii) Migration pattern and population size in case of any endangered species.

(iii) History of diseases.

2. Initial Response

A. Identify problems

(i) Arrange post-mortem examination by specialist.

(i) Obtain diagnosis by submitting tissue material to the established diagnostic laboratory as soon as possible.

(iii) Conduct field investigation to determine extent of problem.

B. Establish control of area

(i) Close affected area.

(ii) Identify special work for disease control activities.

a. Carcass disposal sites.

b. Laboratory investigation area.

(iii) Dispose the carcass with guidance of specialist.

C. Comunications

(i) Notify appropriate agency of die-off.

3. Disease Control

A. Response

(i) Disease control actions dictated by the type of disease, species involved and other circumstances.

a. Bringing personnel, equipment and supplies on sites.

b. Organise work with briefing to the workers regarding the nature of the work.

c. Carcass pick up and disposal.

d. Monitoring case of mortality.

e. Decontamination of personnel and equipment.

B. Management

(i) Disease management activities.

a. Controlled movement if required relocation of the animals.

b. Habitat manipulation to prevent, attract or maintain animal use or use of an area.

(ii) Decontamination of infected environment

a. Chemical treatment to land and water.

b. Vegetation and water removal to allow air and sunlight to destroy micro organism.

c. Controlled burning of vegetation and dispose mechanical structures.

4. Surveillance

(a) Monitoring : After disease control operations have ended, the area should be kept under surveillance for 10 to 30 days.

(b) Investigations : Investigations should be made to know the disease exposure pattern, predisposing factors and environmental reservoirs of the disease.

5. Analysis : Disease control operations need to critically analyzed, for future operation, if required.

EQUIPMENTS AND SUPPLIES USED IN DISEASE CONTROL OPERATION

A. Carcass Collection

i. Large heavy duty plastic bags.

B. Carcass Disposal

i. Equipment for digging trenches or pits.

ii. Incineration

a. Fuel for burning carcasses.

C. Sanitation Procedures

1. Decontamination of environment and structures

i. Chemicals such as Chlorine bleach, Environ etc.

ii. Hand carried spray units

2. Protection of personnel and prevention of mechanical movement of disease agent to other locations by people and equipment.

i. Rubber gloves, foot coverings.

ii. Spray units and chemical disinfectants.

iii. Plastic bags for transportation of field cloths.

D. Surveillance and Observation

i. Monitoring animal population and environmental conditions.

ii. Binoculars and spotting scope.

iii. Maps for tracking the progress of events and animal populations associated with die off.

E. Animal Population and Habitat Manipulation

i. Denying animal use of a specific area.

a. Monitorized means of hazing animal population.

ii. Conservation and maintenance of animal in a specific area.

a. Grain and other source of food.

b. Water

c. 'Area closed' signs to prevent temporary refuge area.

F. Animal Sampling and Monitoring

i. Animal Capture

a. Nest and capture equipments

b. Animal marking.

MAJOR APPROACHES FOR CONTROL IN FREE LIVING ANIMALS

1. **Ring Vaccination/ Barrier Vaccination of Domestic Livestock :** Domestic animals living around protected areas often transmit diseases to wild animals. To prevent this, domestic livestock should be immunized at regular intervals against infectious disease like FMD, HS., Anthrax , etc. Section 3-A of Wildlife (Protection) Act, 1972, specifically suggested immunization of all livestock living within five kilometers of a Protected areas.
2. **Immunization of Wild Animals :** A number of programes have been carried out in Western countries to immunize wild carnivores against rabies by use of Oral Rabies Vaccine (ORV). In South Africa, roan antelopes were immunized during anthrax epizootic by shooting vaccine-loaded darts from helicopters.
3. **Eliminating the Causal Agent :** This method is more easily possible when the causal agent is of inanimate nature. The case of lead poisoning in birds is best for illustrating this method. Hunting of water-fowl is a very popular sport in the USA. Pellets made of lead are used to shoot the ducks. Pellets that fail to kill birds fall back to earth and remain in the bed of the water body where the shooting is generally takes place. Water fowls, during the feeding process, ingest lead pellets that then poison the birds, ultimately killing them. Lead pellets have replaced in the USA by non-toxic steel pellets.
4. **Mass Treatment of Wild Animals :** Bighorn Sheep found in the USA often fall victim to pneumonia caused by lungworms. De-worming drugs, mixed in apple-pulp, which the sheep like, are left at strategic places in the jungle for the animals to eat. This control method proved quite effective and mortality in the bighorn came down.
5. **Control of Insect Vectors :** This method is extensively used in control of human diseases and has been applied to a limited extent in wildlife as well. Control of mosquito population to check the transmission on malaria is a good example of this method.
6. **Animal Dispersal Technique :** Waterfowls are susceptible to the botulinum toxin released by the bacterial species, *Clostidium botulinum,* Type C. Certain condition lead to the build up of botulinum toxins in water bodies and ducks feeding in such lakes fall prey to the toxin. Dispersal techniques have been used quite successfully to move waterfowls away from toxin-contaminated water bodies, or from areas of disease outbreak. Dispersal technique involves the use of firecrackers, smoke-generating machines, signal flares, gunshot sound; human vice amplified over a public address sound system, beating of drums, lighting of fires, etc. The objective of such pyrotechnics is to frighten the birds/animal away from the contaminated site.
7. **Selective Removal of Diseased Animals :** Certain practical difficulties prevent this technique from being extensively used in wildlife The method is similar to the test and Slaughter programme practiced in domestic animals husbundry, i.e., sacrificing an animal when tested positive for an infectious disease. This method was however

used to eradicate brucellosis disease from wood bison in Elk Island National Park, Canada.

8. **Reduction of Population Density :** Many infectious diseases are dependent on the density of the host population for its spread-higher the density more favourable is the condition for the disease to take epizootic form. Reduction in population is generally achieved by de-population is killing. Foot and Mouth disease has been eradicated from the deer in California, USA, by depopulation. More than 20,000 deer were shot dead in a one-year period for this purpose.Free-ranging wild animals are as susceptible to diseases as any other living. Disease has been the cause of local extirpation of a number of specie in India and abroad. Pollution and other man-made condition have aggravated the situation to such an extent that veterinary intervention is now necessary in all the cases. Disease management techniques have been developed to meet the special requirement of free-ranging animal. In India, the practice of tackling disease problems in free-ranging have so far been that of the "fire-fighter" approach, i.e., taking steps only when problems makes their appearances. The need of the hour is to carry out immediately survey programs in all protected areas to understand the magnitude of the diseases problem in wild animals.

CHAPTER 18

FIELD APPROACH FOR DISEASE DIAGNOSIS FROM DEAD ANIMALS

Several factors are responsible for causing variable morbidity and mortality in wild animals. The ecological imbalance, natural calamities, anthropological factors, accidents, poaching, malicious poisoning , nutritional disorders and epidemics of infectious diseases are some of the examples where the wild animals are victims.The managers of protected areas in such circumstancies approches the near by field veterinary officers to deal with the issues and for conducting post mortem examination and report.some times the veterolegal cases, involvement of endangered species and legal issues related with WL protection act 1972 involves the veterinary officers as an expert witness and in such situation the veterinarian has to be specific in some of the aspects as a forensic investigator. Based on experiences here are some of the important points, which could help the veterinarian to deal with the disease diagnosis from dead wild animals.

The recording of each matter, manner, mechanism and cause of death with detailed past history is an imporant clue to judge the matter for the forensic aspects.

It is better to use audio visual aids with good quality of photographs or vedio with close up views which supports the circumstancial evidences.

Recognition and description of arrow wound, gun shot wound, anatmical position and location and velocity of caliber guns (low, high) supports the cause of death in wild animals.

Complete full flagged field kit for post mortem and sample collection is an essential tool.

Use of disposable gloves and mask, sterile bottles and protocol of post mortem helps to go the veterinarian in systemic diagnosis of diseases.

Massive blunt truma caused by mobile accident can be looked for fractures,brusing or internal haemorrahge, which should be closely examined with patience.

Head, neck and limbs covered with hair should be thoroughly examined for the presence of marking of traps and snares.

Truma caused by predatrs and scanvengers must be differntialted from other cause of truma.

Electrocution, snake bite and drawing should be confirmed with circumstancial evidences.

The poisoning or poaching with bait should also be ruled out by several field study of the area and near by surroundings.

Forensic laboratory investigation of carbamate,OP compounds,strychnine,anticoagulant rodenticides,thallium and cyanide hepd for confirmation of the diagnosis.

The hairs from the stomach of poisoned animals, knives,pickups,clothings ,traps and other evidences should be closely observed and collected in separate bags with marking.

The postmortemchanges are often influenced by several factors like ambient temperature,health of animals,degree of muscuklar activities before death, mode of death and medium of death etc. should be marked to rule out ther time of death.

The development of rigor mortis from head, neck and latral spread to back and limbs, cooling of body after a few hours of death(8-12 hrs.) also depends on several facotrs.

The putrifaction of carcasses with distension of abdomen with gases, blood stained fluids from nostrils and mouth,liquification of eye balss, foul smell, bursting of abdomen, thorax and liquification of body should be recorded.

CHAPTER 19

EX SITU CAPTIVE MANAGEMENT IN ZOOS

Ex-situ conservation means literally, "off-site conservation". It is the process of protecting an endangered species (plant/animal) by removing it from an unsafe or threatened habitat and placing it or part of it under the care of humans. While ex-situ conservation is comprised of some of the oldest and best known conservation methods known to man, it also involves newer, sometimes controversial laboratory methods.The major approaches are :

i. Zoo
ii. Frozen zoos/semen bank

Central Zoo Authority and Captive Animal Health Management

When the wildlife (protection) Act of 1972 was ammended in 1991 a new chapter, viz, chapter IV A,was added to the Act. This particular ammendment paved the way for the establishment of Central Zoo Authority whose purpose, besides other functions, was to specify the minimum standards for housing, up keep, and veterinary care of the animals kept in a zoo. The CZA became operational in February, 1992, and one of its first task was to evolve a set of norms for the recognition of zoos. These norms came to be known as Recognition of zoo Rules, 1992.The zoos which are classsified by CZA as per below criteria

Classification of Zoo

Categories of zoo	Large	Medium	Small	Mini
Area of the zoo in hectares	More than 75 hectares	50-75 hectares	20-50 hectares	Less than 20 hectares
Numbers of animals exhibited	More than 750	500-750	200-499	200
Animals variety exhibited	More than 75 numbers	50-75 numbers	20-49 numbers	20 numbers
Number of endangered species exhibited	More than 15	10-15	5-9	Less than 5
Annual attendance of visitors per year	More than 7.5 lakhs	5-7.5 lakhs	2-5 lakhs	Less than 2 lakhs

The Central Zoo Authority shall grant recognition with due regard to the interests of protection and conservation of wild life, and such standards, norms and other matters as are specified below:

GENERAL GUIDELINES AND NORMS

1. The primary objective of operating any zoo shall be the conservation wildlife and no zoo shall take up any activity that is inconsistent with the objective.
2. No zoo shall acquire any animal in violation of the Act or rules made thereunder.
3. No zoo shall allow any animal to be subjected to the cruelties as defined under the Prevention of Cruelty to Animals Act, 1960 (59 of 19fi0) or permit any activity that exposes the animals to unnecessary pain, stress or provocation, including use of animals for performing purposes.
4. No zoo shall use any animal, other than the elephant in plains and yak in hilly areas for riding purposes or draughting any vehicle.
5. No zoo shall keep any animal chained or tethered unless doing so is essential for its own well being.
6. No zoo shall exhibit any animal that is seriously sick, injured or infirm.
7. Each zoo shall be closed to visitors at least once a week.
8. Each zoo shall be encompassed by a perimeter wall at least two metres high from the ground level. The existing zoos in the nature of safaries and deer parks will continue to have chain link fence of appropriate design and dimensions.
9. The zoo operators shall provide a clean and healthy environment in the zoo by planting trees, creating green belts and providing lawns and flower beds etc.
10. The built up area in any zoo shall not exceed twenty five per cent of the total area of the zoo. The built up area includes administrative buildings, stores, hospitals, restaurants, kiosks and visitor rest sheds etc: animal houses and 'pucca' roads.
11. No zoo shall have the residential complexes for the staff within the main campus of the zoo. Such complex, if any, shall be separated from the main campus of the zoo by a boundary wall with a minimum height of two metres from the ground level.

    Administrative and Staff Pattern :

12. Every zoo shall have one full-time officer incharge of the zoo. The said officer shall be delegated adequate administrative and financial powers as may be necessary for proper upkeep and care of zoo animals.
13. Every large and medium zoo shall have at least one full-time curator having the sole responsibility of looking after the upkeep of animals and maintenance of animal enclosures.
14. Each large zoo shall have at least two full-time veterinarians and medium and small zoo shall have at least one full-time veterinarian. The mini zoo may at least have

arrangement with any outside veterinarian for visiting the zoo every day to look after the animals.

Animal Enclosures – Design, Dimensions and other Essential Features :

15. All animal enclosures in a zoo shall be so designed as to fully ensure the safety of animals, caretakers and the visitors. Stand of barriers and adequate warning signs shall be provided for keeping the visitors at a safe distance from the animals.
16. All animal enclosures in a zoo shall be so designed as to meet the full biological requirements of the animal housed therein. The enclosures shall be of such size as to ensure that the animals get space for their free movement and exercise and the animals within herds and groups are not unduly dominated by individuals. The zoo operators shall take adequate safeguards to avoid the animals being unnaturally provoked for the benefit of viewing by public and excessive stress being caused by visibility of the animals in the adjoining enclosures.
17. The zoo operators shall endeavour to simulate the conditions of the natural habitat of the animal in the enclosures as closely as possible. Planting of appropriate species of trees for providing shade and constructing shelters which would merge in the overall environment of the enclosures, shall also be provided. Wherever it is technically feasible, only moats shall be provided as enclosure barriers.
18. The enclosures housing the endangered mammalian species, mentioned in appendix I to these rules, shall have feeding and retiring cubicles/cell of minimum dimensions given in the said appendix. Each cubicle/cell shall have resting, feeding, drinking water and exercising facilities, according to the biological needs of the species. Proper ventilation and lighting for the comfort and well being of animals shall be provided in each cell/cubicle/enclosure.
19. Proper arrangement of drainage of excess of water and arrangements for removal of excreta and residual water from each cell/cubicle/enclosure shall be made.
20. Designing of any new enclosure for endangered species shall be finalized in consultation with the Central Zoo Authority.

Hygiene, Feeding and Upkeep :

21. Every zoo shall ensure timely supply of wholesome and unadulterated food in sufficient quantity to each animal according to the requirement of the individual animals, so that no animal remains undernourished.
22. Every Zoo shall provide for a proper waste disposal system for treating both the solid and liquid wastes generated in the zoos.
23. All left over food items, animal excreta and rubbish shall be removed from each enclosure regularly and disposed of in a manner congenial to the general cleanliness of the zoo.
24. The zoo operators shall make available round the clock supply of potable water for drinking purposes in each cell/enclosure/cubicle.

25. Periodic application of disinfectants in each enclosure shall be made according to the directions of the authorised veterinary officer of the zoo.

Animal Care, Health and Treatment :

26. The animals shall be handled only by the staff having experience and training in handling the individual animals. Every care shall be taken to avoid discomfort, behavorial stress or physical harm to any animal.
27. The condition and health of all animals in the zoo shall be checked every day by the person in-charge of their care. If any animal is found sick, injured, or unduly stressed the matter shall be reported to the veterinary officer for providing treatment expeditiously.
28. Routine examination including parasite checks shall be carried out regularly and preventive medicines including vaccination be administered at such intervals as may be decided by the authorised veterinary officers.
29. The zoo operators shall arrange for medical check-ups of the staff responsible for upkeep of animals at least once in every six months to ensure that they do not have infections of such diseases that can infect the zoo animals.
30. Each zoo shall maintain animal history sheets and treatment cards in respect of each animal of endangered species, identified by the Central Zoo Authority.

Veterinary Facilities :

31. Every large and medium zoo shall have fullfledged veterinary facilities including a properly equipped veterinary hospital, basic diagnostic facilities and comprehensive range of drugs. Each veterinary hospital shall have isolation and quarantine wards for newly arriving animals and sick animals. These wards should be so located as to minimise the chances of infections spreading to other animals of the zoo.
32. Each veterinary hospital shall have facilities for restraining and handling sick animals including tranquilizing equipments and syringe projector. The hospital shall also have a reference library on animal health care and upkeep.
33. The small and mini zoos, where full-fledged veterinary hospital is not available, shall have at least a treatment room in the premises of the zoo where routine examination of animals can be undertaken and immediate treatment can be provided.
34. Every zoo shall have a post-mortem room. Any animal that dies in a zoo shall be subjected to a detailed post-mortem and the findings recorded and maintained for a period of at least six years.
35. Each zoo shall have a graveyard where the carcasses of dead animals can be buried without affecting the hygiene and the cleanliness of the zoo. The large and medium zoos shall have an inscinerator for disposal of the carcasses and other refuse material.

Breeding of Animals :

36. Every zoo shall formulate a programme for captive breeding of only such animals as are approved by the Central Zoo Authority for that zoo. They shall abide by the guidelines and directives of the Central Zoo Authority in this regard.

37. Every zoo shall keep the animals in viable, social groups. No animal will be kept without a mate for a period exceeding one year unless there is a legitimate reason for doing so or if the animal has already passed its prime and is of no use for breeding purposes. In the event of a zoo failing to find a mate for any single animal within this period, the animal shall be shifted to some other place according to the directions of the Central Zoo Authority.
38. No zoo shall be allowed to acquire a single animal of any variety except when doing so is Education and Research essential either for finding a mate for the single animal housed in the said zoo or for exchange of blood in a captive breeding group.
39. Every zoo shall take up regular exchange programmes of animals so as to prevent the traits or ill effects of inbreeding. To achieve this objective each zoo shall maintain a stud book in respect of every endangered species.
40. To safeguard against uncontrolled growth in the population of prolifically breeding animals, every zoo shall implement appropriate population control measures like separation of sexes, sterilization, vasectomy, tubectomy and implanting of pallets etc.
41. No zoo shall permit hybridization either between different species of animals or different races of the same species of animals.

Maintenance of Records and Submission of Inventory to the Central Zoo Authority :

42. Every zoo shall keep a record of the birth acquisitions, sales, disposals and deaths of all animals. The inventory of the animals housed in each zoo as on 31 st March of every year shall be submitted to the Central Zoo Authority by 30th April of the same year.
43. Every zoo shall also submit a brief summary of the death of animals in the zoo for every financial year, along with the reasons of death identified on the basis of post-mortem reports and other diagnostic tests, by 30th April of the following year.
44. Every zoo shall publish an annual report of the activities of the zoo in respect of each financial year. The copy of the said annual report shall be made available to the Central Zoo Authority, within two months, after the end of the financial year. The report shall also be made available to the general public at a reasonable cost.

Education and Research :

45. Every enclosure in a zoo shall bear a sign board displayng scientific information regarding the animals exhibited in it.
46. Every zoo shall publish leaflets, brochures and guidebooks and' make the same available to the visitors, either free of cost or at a reasonable price.
47. Every large and medium zoo shall make arrangements for recording, in writing, the detailed observations about the biological behaviour, population dynamics and veterinary care of the animals exhibited as per directions of the Central Zoo Authority so that a detailed database could be developed. The database shall be exchanged with other zoos as well as the Central Zoo Authority.

Visitor Facilities :

48. The zoo operators shall provide adequate civic facilities like toilets, visitor sheds, and drinking water points at convenient places in the zoo for visitors.
49. First-aid equipments including anti-venom shall be readily available in the premises of the zoo.
50. Arrangements shall be made to provide access to the zoo to disabled visitors including those in the wheel chair.

Development and Planning :

51. Each zoo shall prepare a long-term master plan for its development. The zoo shall also prepare a management plan, giving details of the proposal and activities of development for next six years. The copies of the said plans shall be sent to the Central Zoo Authority.

OBJECTIVES OF ZOO

The main objective of the zoos shall be to complement and strengthen the national efforts in conservation of the rich biodiversity of the country, particularly the wild fauna. This objective can be achieved through the following protocol:

1. Supporting the conservation of endangered species by giving species, which have no chance of survival in wild, a last chance of survival through coordinated breeding under *ex-situ* conditions and raise stocks for rehabilitating them in wild as and when it is appropriate and desirable.
2. To inspire amongst zoo visitors empathy for wild animals, an understanding and awareness about the need for conservation of natural resources and for maintaining the ecological balance.
3. Providing opportunities for scientific studies useful for conservation in general and creation of data base for sharing between the agencies involved in *in-situ* and *ex-situ* conservation.
4. Besides the aforesaid objectives, the zoos shall continue to function as rescue centres for orphaned wild animals, subject to the availability of appropriate housing and upkeep infrastructure. Where appropriate housing and upkeep is not available, State Governments and the Central Government would ascertain setting up rescue facilities in off - the display areas of the zoo, subject to the availability of land.

STRATEGY FOR ACHIEVING THE OBJECTIVES

1. General Policy About Zoos

(i) Since zoos require a significant amount of resources in the form of land, water, energy and money, no new zoo shall be set up unless a sustained supply of resources including finance and technical support are guaranteed.

(ii) Zoos shall prepare a long-term masterplan for development to ensure optimum utilisation of the land, water, energy and finance.

(iii) Every Zoo shall maintain a healthy, hygienic and natural environment in the zoo, so that the visitors get an adequate opportunity to experience' a natural environment.

(iv) Zoos shall give priority to endangered species in their collection and breeding plans. The order of preference for selection of species shall be (in descending order) locality, region, country and other areas.

(v) Zoos shall regulate the number of animals of various species in their collection in such a way that each animal serves the objectives of the zoo. For achieving this objective, a detailed management plan of every species in the zoo shall be prepared.

(vi) Every zoo shall endeavour to avoid keeping single animals of non-viable sex ratios of any species. They shall cooperate in pooling such animals into genetically, demographically and socially viable groups at zoos identified for the purpose.

(vii) Zoos shall avoid keeping surplus animals of prolifically breeding species and if required, appropriate population control measures shall be adopted.

2. Acquisition of Animals

(ii) Except for obtaining founder animals for approved breeding programme and infusion of new blood into inbred groups, no zoo shall collect animals from the wild.

(ii) Zoos shall not enter into any transaction involving violation of the law and provisions of international conventions on wildlife conservation.

(iii) Zoos shall not enter into any transaction in respect of their surplus animals with any commercial establishment. Even the animal products should not be utilised for commercial purposes. The trophies of the animals could, however, be used for educational or scientific purposes.

3. Animal Housing

(i) Every animal in a zoo shall be provided housing, upkeep and health care that can ensure a quality of life and longevity to enable the zoo population sustain itself through procreation.

(ii) The enclosure for all the species displayed or kept in a zoo shall be of such size that all animals get adequate space for free movement and exercise and no animal is unduly dominated or harassed by any other animal.

(iii) Each animal enclosure in a zoo shall have appropriate shelters, perches, withdrawal areas, wallow, pools, drinking water points and such other facilities which can provide the animals a chance to display the wide range of their natural behaviour as well as protect them from extremes of climate.

4. Upkeep of Animal Collections

(i) Zoos shall provide diet to each species, which is similar to its feed in nature. Where for unavoidable reasons any ingredients have to be substituted, due care will be taken to ensure that the substitute fulfils the nutritional requirement of the species.

(ii) For well being of the animals, round the clock supply of potable drinking water shall be made available to all animals kept in the zoo.

(iii) With the objective of avoiding human imprinting and domestication of animals, zoos shall prevent physical handling of animals by the staff to the extent possible.

(iv) Zoos shall not allow any animal to be provoked or tortured for the purpose of extracting any performance or tricks for the benefit of the visitors or for any other reason.

5. Health Care

(i) Zoos shall ensure availability of the highest standards of veterinary care to all the animals in their collection.

(ii) Adequate measures shall be taken by every zoo for implementing wildlife health and quarantine rules and regulations. Appropriate vaccination programmes shall also be taken up for safeguarding against infectious diseases. Timely action to isolate infected animals from the zoo population shall also be taken to avoid further spread of disease.

6. Research and Training

(i) The zoos shall encourage research on the biology, behaviour, nutrition and veterinary aspects of animals in their collection. They shall also endeavour for creation of expertise on zoo architecture and landscape designing, cooperation of recognised institutions already working in relevant fields in this regard shall be taken.

(ii) Zoos shall endeavour for transfer of technical skills available in the field for zoo personnel. The Central Government, Central Zoo Authority and State Governments shall provide due support to zoos in these efforts. Assistance of Wildlife Institute of India (WII), Indian Veterinary Research Institute (IVRI) and other institutions within India and abroad, having appropriate expertise shall be taken in this regard.

(iii) Zoos shall also endeavour for dissemination of information on scientific aspects of management through publication of periodicals, journals, newsletters and special bulletins. Help of non-governmental organisations (NGOs) and government institutions shall, also be availed in such efforts. The Central Zoo Authority shall provide technical and financial support to the Indian Zoo Directors Association (IZDA) and other institutions in this regard.

7. Breeding Programme for Species

(i) Before taking up breeding programmes of any species, zoos shall clearly identify the objectives up. The targeted numbers for the programme would be decided keeping in view the identified objectives.

(ii) All zoos shall cooperate in successful implementation of identified breeding programmes by way of loaning, pooling or exchanging animals for the programme and help creation of socially, genetically and demographically viable groups even at the cost of reducing the number of animals or number of species displayed in individual zoos.

(iii) Breeding programme shall be taken up by zoos after collection of adequate data like biology, behaviour and other demographic factors affecting the programme, including the minimum number of founder animals and the quantum of housing facilities available.

(iv) Programmes for breeding of zoo animals for re-introduction in the wild shall be taken up after getting approval of the State Government, the Central Zoo Authority and the Central Government as the case may be.

(v) Zoos shall give priority in their breeding programmes to endangered species representing the zoo-geographic zones in which they are located.

(vi) For carrying out breeding programmes in a scientific and planned manner the zoo shall every individual animal involved in the programme in an appropriate manner and maintains appropriate records.

(vii) Zoos shall take utmost precaution to prevent inbreeding. They shall avoid artificial selection, traits and make no explicit or implicit attempts to interbreed various genera, species and sub-species

(viii) Special efforts shall be made to avoid human imprinting of the stocks raised for reintroduction purposes by providing off exhibit breeding facilities.

8. Education and out reach Activity

(i) Each zoo should have a well drawn-up for educating the visitors as well as others in the community. Zoos shall keep a close liaison with other *ex-situ* facilities in this regard.

(ii) The central theme of the zoo education programme being the linkage between the survival of various species and protection of their natural habitat, enclosures, which allow the animals to display natural behaviour, are crucial to zoo education. Zoo shall, therefore, display animals in such enclosure only where the animals do not suffer physiological and psychological restraint.

(iii) Attractive and effective signage method interactive displays to explain activities of various species to visitors, published education material and audio-visual devices are proven methods for home the conservations message. A formal educational programme should also be perused for strengthening the education message.

(iv) Besides signage, the zoos shall also have guided tours, talks by knowledgeable persons, and audio-visual shows for effectively communicating message of conservation to the visitors.

(v) The help of universities, colleges and the governmental organisation shall be taken to educate the students about the benefits of supporting nature conservation programmes.

9. Extension Activities

(i) To provide the urban population with a window to nature and to serve as green lungs for the polluting environment, zoos shall extend their expertise and help to State Governments and local authorities to create nature parks extending over extensive areas near big cities.

10. Amenities to Visitors

(i) Zoos shall provide basic civic amenities to the visitors like toilets, drinking water points, shelters and first-aid facilities. Ramps shall also be provided for the benefit of visitors in wheel chairs for approach to animal enclosure and other civic amenities.

(ii) Zoos shall not provide any infrastructure for recreation/entertainment of visitors that is inconsistent with the stated objective of zoos.

FUTURE PLANNING FOR THE ZOOS

India is having 578 wildlife protected areas that include 489 wildlife sanctuaries and 89 National Parks (imbibing 27 tiger Reserves). During the last 3 decades of 20th Century some of the foreign countries have made frog-leaping advances in captive wildlife conservation by contributing substantially in developing potent anaesthetizing and chemical restraint drugs and their remote delivery equipments, instituting animal quarantine and transportation guidelines, health care and disease diagnostic techniques, assisted reproductive technologies (as most powerful, potential tools enhancing endangered species conservation), early pregnancy diagnosis (by hormone analysis through urine, feces and saliva), electronic numbering of the animals right at birth, remote tissue sampling, sonography (to differentiate the normal coelomic structures and to diagnose pathological conditions), population control, oral immunization, etc.

Efforts are continued to formulate guidelines on the other vital issues of wildlife welfare, to facilitate them to maintain socially, genetically and demographically viable in captivity. This will enable us to propagate the species effectively and provide self-sustaining population for the purposes of display and augmentation of their *in-situ* populations as well as re-introduction into devoid habitations. For species housing and biological and physiological enrichments, technical experts of animal ethology, physiology, husbandry, and bioengineering involved. While constructing exhibits for animal species following basics are being looked critically.

The open moat system, availability of water in metro cities has to be recognized for exhibits housing of the species.

The soil substrate, humidity and photoperiods of the area conducive to species natural activities and landscaping to provide natural camouflage look and face to face seeing visitors and animals to each other simultaneously.

Now top of the kraals coupled with exhibits are designed with provision of wire mesh cover to ensure protection against harmful effects of free ranging creatures.

Display yards/enclosures to have perfect drainage system to drain out animal wastes and rain water.

The species which are ecto-thermic have temperature dependent immune system. Sub-optimum environment conditions will lead to cause depression to their immune function. High humidity may act as predisposing factor in causing *Pasteurella* and mycotic respiratory infections and ecto-parasite infestations in zoo animals.

For handling and restrain of large carnivores in hospital or in their exhibits, provision for the squeeze cages has been existing but it has been observed that sqeeze cages in the hospital for bears, orangutan, chimpanzees are not available and the squeeze cages available for tiger/lions used to be utilized.

The International Health code formulated for quarantine and translocation recommended by O.I.E (Office International Des Epizootics) Paris are being look into by the Ministry of Agriculture, Government of India.

There is need to develop our own National standards on the aspects relevance to endemic wildlife species.

Transportation crates sizes for long distance journey recommended by IATA need to be reviewed in the light of animals sitting and standing comforts besides the length and height.

In the Zoological Parks, animal-swapping programme is a regular feature that cannot be accomplished without the Wildlife Veterinary Officer because it is a purely technical job. Therefore, the team member of veterinary officer staff should have training in animal restraint, crating, loading and transportation techniques.

To conduct the disease diagnosis and provide health management support, the nodal centers should be provided enough staff and fund.

CHAPTER 20

HOUSING OF CAPTIVE ANIMALS

WHY HOUSING IS IMPORTANT?

One should not think that wild animals do not require any proper housing, but this is a misconception. Housing is very important in zoos as wild animals are always under stress due to captive condition. Further, captive condition may also cause inflicted injuries/traumatic injuries due to limited space particularly during the breeding season. Due to presence of continuous stress animal may suffer from chronic infectious diseases, parasitic diseases, hoof problems etc. Faulty housing also may lead to faulty management.

From the point of view of the zoo staff, important advantages of natural habitat design would be in permitting animal's greater freedom of movement, with minimal hoof disorders, leading to their better health. Animals in natural simulated housing help in displaying normal behavioural patterns. Social relationships between animals are enhanced leading to natural sexual behaviour and mating resulting in healthy reproduction lives. Therefore, housing should have it

Facilities for indoor public viewing.

The animals may have access to outside enclosures or cages from the Zoo buildings.

The building should be comfortable for the animals with enough space to lie down or rear the young ones.

It should have easy accessibility for catching animals and for bringing in transport containers,

Proper ventilation and drainage are extremely essential.

Objectives

Housing or shelter and the enclosures are provided to the wild animals and birds in zoo should have the following objectives:

1. To protect from adverse environmental conditions such as severe summer or extreme cold, rain, snow fall etc.
2. To facilitate the breeding, feeding, watering and treatments (administration of drugs and wound dressing) captivity.

3. To protect the wild animals of one category from the others, (e.g. deers from tigers or lions).
4. To protect the wild animals from the stress due to the constant flow of visitor in the zoo.
5. Enclosure barrier are especially meant to protect the animals from teasing and mischievous behavior of visiting public.

Principles of Housing

Construction of the housing should be economic but it should provide maximum comfort to the captive animals. Overcrowding within the enclosure must be avoided. All the animal houses and enclosures should be equipped with resting platforms, bedding, boxes, open to sky raised platforms, etc as per the need of the individual species (Annexure I & II). The floor space and height of the roof must be adequate and as per the norms depending on species, size and behaviour of the animals. The outdoor area should have soft earthen floor while the indoor flooring should be made up of cement and concrete. For the animals which are having the burrowing habits, e.g. rabbits, mongoose, etc the floor must be cement and concrete. The floor should be smooth surfaced without any projected area. For the animals like rhinoceros or hippopotamus, there should be land as well as water ponds as per the biological need of these species. Roof is not mandatory. Generally natural shade is used. However, artificial roof like asbestos sheet, polythene sheet, tiles can also be used. Adequate provision of ventilation, air circulation and maximum exposure to sunlight, particularly in night kraals should be ensured. During construction it should be kept in mind that it should have proper drainage facility and should be easy to clean. Night cubicles are used for feeding of Carnivores. Hay racks/trough for roughage feeding and feeding trough for feeding of concentrate is used for herbivores. Bears like to feed from elevated place. There should be provision of water trough in different animal housing. While designing the animal house safety of animals, care takers and visitors must be considered. Separate enclosures should be provided to the animals during advanced pregnancy and also to the lactating mother with its recently born young ones. There should also be separate ward for the sick animals, in proximity with veterinary dispensary within premises of the zoo. There must be provision of post-mortem room nearby with the facilities of incarcination of dead animals. Around the enclosure of the zoo there should be plantation of large flowering plants to provide shade which may add to the aesthetic value of the zoo.

Housing depends upon the following factors:

1. Type of zoo
2. Landscape
3. Type of species
4. Behaviour of wild animals
5. Ecology

Types of Zoo

The guidelines for housing the wild animals under following categories should be obtained from CZA. Additionally if Zoological Garden (a zoo with a horticultural base either situated within a formal horticulture garden or a zoo with preponderance of formal horticultural features) or Zoological Park (a zoo designed informally like a woodland and park with group of trees, open spaces and water bodies and spacious animal enclosure with simulated habitat conditions) are to be established, then housing become furthermore easy. In case of Safaris (specialized zoos where the captive animals are housed in any large naturalistic enclosures) the area should be surrounded by a suitable peripheral chain-link fence or wall. A buffer zone of about 5 meter width is provided around the fence area. Large carnivores are kept here. Minimum height of 5 meter or 4 meter high non-scalable fence or wall or a 2.5 meter high chain-link fence preferably with overhangs is used as boundary. In case of Deer Park, night shelter or kraals should be constructed for the deer. It requires more free space with special design. Additionally require more mobile cubicles for emergency condition.

Rescue Centre is nothing but a mini zoo with specialized health care unit. Accidental/injured/orphan/rescued/sick animals are kept here. From rescue centre recovered animal can be given to other zoo on demand basis. Therefore, there should be proper distance between the two buildings to avoid any spread of any infection.

Landscape

The topography should also be kept in mind, while constructing a zoo. If the zoo is designed solely for the entertainment, then the growth potential of the population and the accessibility to the area should also be considered. The ideal site for a zoo should be an undulating or sloping terrain which would facilitate landscaping. It should be away from railway or air route. Susceptibility of the site to erosion, both wind and water, apart from the drainage condition, should be examined before construction.

Within the environmental type, critical components are equally important for different species.

Topography	:	flat, rolling, hilly, cliffs.
Rock formations	:	scattered, continuous, Igneous, metamorphic, sedimentary.
Water features	:	marsh, pond, stream, waterfall.
Vegetation	:	ground cover, shrubs, trees, vines.
Artifacts	:	lianas, termite, mounds, buttress trees

Behaviour

Behavioural aspects of animals are very important during construction of zoo. Based on the behavioural attributes of animals, the enclosures should be designed in near to their natural habitat. For some animals, height of the enclosure is more important than the floor space. Psychological space is more valuable than physical space. E.g. leopard like to rest on tree or crocodile like to wallow in water. Housing infrastructures should be enriched for proper behavioral expression of the animals.

Type of species : The habitat type to be represented is completely dependent on the species in question and there are roughly four environment types in which habitats occur. These are

1. Aquatic : Riverine, pond, marsh, estuarine (underwater, overwater).
2. Terrestrial : Jungle/ woodland/ scrubland, meadow, grass land, desert grazing, climbing)
3. Arboreal : jungle, woodland (upper canopy / lower canopy / trunk dwelling)
4. Flying : Jungle, woodland, scrubland, marsh

Ecological

Here, most of the animals originate in one type of habitat like forest, deserts, alpine etc. They are specialized collections which include members of several orders; e.g. the Arizona-Sonora desert museum, Tuscon, Arizona. Again ecology of animals depends on the environment.

ENCLOSURE BARRIERS

Enclosure barriers are the special features for the wild animals housing in zoo which are not fount necessary for normal domestic and pet animals. Each and every enclosure in zoo is provided with enclosure barrier for extra protection of wild animals. The enclosure barrier not only restricts the entry of visitors to certain permissible limit but also ensure against the escaping of the animals by jumping or by any other means.

Barriers help to keep the animals in and the public out. An animal may feel secured in a well-designed cage, especially when it finds the human being at a close distance. Further, fence or barrier is essential not only to prevent the public from teasing the animals, but also to prevent them from getting injured by coming in contact with the wild animals.

Various types of barriers are important like :

Wire mesh

Chain link mesh

Weld mesh bars,

Glass

High tensile wire

Rubber walls

Moat

In th.e modern zoos, moats are very frequently used. Moats should enclose the animals securely during all seasons and should be safe for the animals. The animals should be in a position to climb back to the enclosure from the moat; at the same time, it should not find its way out of the enclosure by crossing through the moat. The designing should not disturb public viewing.

Dry moat : Usually 'U' shaped dry moats are used for dangerous carnivores like tiger and lion; however, leopards are kept in a closed enclosure. Elephant, bears and also rhinos can also be kept ill enclosures with a *'V'* shaped moat.

In case a dry moat is used, it should be deeper than the wet moat. For elephants the moat bottom should be soft with loose gravel or sand and the width should be more. For heavy animals, it is advisable to have a ramp from the back of the moat into the enclosure, which can be used during emergency. In general, dry moats should not be used for non-aquatic animals.

While displaying carnivores, it is essential that the side towards the public should be more in height when compared to the inner one; this will prevent the animal psychologically as well as physically from leaping. The safety margin for width should be at least 10 % more than the leaping capacity of the animals. The edge of the moat, usually, becomes the physical as well as the psychological boundary for the enclosed animal. From a distance, the predator as well the prey animals may appear together, but in reality they remain separated by dry moats. Shallow ditches with a fence are ideal for separating hoofed animals.

Wet moat : Water moats should be functional so that the animals can use them for swimming, usually; tigers prefer to swim in wet moats. On the public side, the moat should be concealed from view by raising low hedges or plants which can spread over the area. The pathways of public can also be designed like concealed moats, so that visitors between two enclosures are not seen by others standing on the far end of those enclosures. Wet moats should only be used for non-aquatic animals. Wet moat should be less deep than dry moat, if used.

OTHER FACILITIES

1. There should be separate area for storage of food for the wild animals within the zoo premises.
2. Availability of wholesome water should also be ensured.
3. There should be a full-fledged veterinary unit with basic diagnostic facilities.
4. Isolation and quarantine wards should be there to take care of newly arriving animals and sick animals.
5. To prevent the spread of infection within zoo, installation of incinerator is essential.

6. Garden facility cum lawns should be created at the entrance of the zoo for the relaxation to the public.
7. Zoo should include restaurants and visitor rest sheds in its built-up area to avoid supply of food from outside area.
8. Adequate parking space should be provided outside the zoo for avoiding inconvenience to the visitors and to maintain cleanliness within the zoo.

Table 1: Minimum prescribed size for feeding/retiring cubicle for important mammalian species of captive animals

Name of the species	Minimum size outdoor enclosure (per pair)	Minimum area extra per of additional animal (sq. m)
	Family - Felidae	
Tiger and Lions	1000	250
Panther	500	60
Clouded leopard	400	40
Snow Leopard	450	50
	Family - Rhinoccrotidae	
One- horned Indian Rhinoceros	2000	375
	Family - Cervidae	
Brow antlered deer	1500	125
Hangul	1500	125
Swamp deer	1500	125
	Family- Bovidae	
Wild buffalo	1500	200
Indian bison	1500	200
Bharal/ Goralf Sheep	350	75
	Family-Equidae	
Wild Ass	1500	200
	Family-Ursidae	
All types of Indian bears	1000	100
	Family - Canidae.	
Jackal, Wolf and Wild dog	400	50
	Family - Procyonidae	
Red Panda	300	30
	Family - Ceropithecidae	
Monkeys and Langurs	500	20

Note:

1. No dimensions for outdoor enclosure have been prescribed for Chinkara and Chowsingha because of the problem of infighting injuries. 2. The designs of enclosures for Schedule I species, not covered by this Appendix, should be finalised only after approval of the CZA.

Table 2: Minimum prescribed size for outdoor open enclosure for important mammalian species of captive animals

Name of species	Size of feeding cubicle/night shelter			Name of species	Size of feeding cubicle/night shelter		
	Length	Breadth	Height		Length	Breadth	Height
Family- Felidae				**Family- Equidae**			
Tiger and Lions	2.75	1.8	3	Wild Ass	4	2	2.5
Panther	2	1.5	2	**Family-Ursidae**			
Clouded leopard & Snow Leopard	2	1.5	2	All types of Indian bears	2.5	1.8	2
Small cats	1.8	1.5	1.5	**Family- Canidae**			
Family- Elephantidae				Jackal, Wolf and Wild dog	2	1.5	1.5
Elephant	8	6	5.5	**Family- Vivviridae**			
Family- Rhinocerotidae				Palm civet	2	1	1
One-horned Indian Rhinoceros	5	3	2.5	Large Indian civet and binturong	2	1.5	1
Family- Cervidae				**Family- Mustellidae**			
Brow antlered deer	3	2	2.5	Otters all types	2.5	1.5	1
Hangul	3	2	2.5	Ratel/hogbadger	2.5	1.5	1
Swamp deer	3	2	2.5	Martens	2	1.5	1
Musk deer	2.5	1.5	2	**Family- Procyonidae**			
Mouse deer	1.5	1	1.5	Red Panda	3	1.5	1
Family- Bovidae				**Family- Lorisidae**			
Nilgiri tahr	2.5	1.5	2	Snow loris and slender loris	1	1	1.5
Chinkara	2.5	15	2	**Family- Cercopithecidae**			
Four horned antelope	2.5	1.5	2	Monkeys and Langurs	2	1	1.5
Wild buffalo	3	1.5	2				
Indian bison	3	2	2.5				
Yak	4	2	2.5				
Bharal/Goral/ wild sheep and markhor	2.5	1.5	2				

CHAPTER 21

HYGIENE AND SANITATION

Diseases of wild animals should be of concern to any one who comes in to contact with wild life, either in free living or in captivity. Many wildlife diseases have public health significance. Some wildlife population may be "**reservoir**" of infection for diseases that may be transmissible to man or his domestic animals; while others can serve as "**carrier**" for various diseases, providing a mechanism for dissemination of diseases from one animal to another.

As compare to free living wild animals the captive ones may contract more with wide variety of transmissible diseases. In general, wild animals are also susceptible to a wide variety of infectious and parasitic organisms capable of causing disease. Many of these organisms are quite specific and of significance to only one or a few species of related wild animals. Others, however, may be quite generalized and be capable of infecting a wide range of wild and domestic animals or even humans.

HYGIENICAL APPROACH

Personnel hygiene, sanitation and judicious use of disinfectant in the animal inhabited workplace are of important, as the contamination with the organic matter is a real concern. The preventive medicine programs for zoo personnel as recommended by American Association of Zoo Veterinarians suggests following.

1. Pre employment Physical examination: - Adequate base line data of all zoo employee should be maintained by regular and periodical collection of serum, faecal and of other required biological samples. Regular testing for tuberculosis by skin test and employee's immunization history should be reviewed to assure current protection. Annual health status for immunization and tuberculosis should be reviewed.
2. Prophylactic rabies vaccine in certain areas where rabies is endemic should be employed.
3. High-risk personnel like pregnant employees, person with cancer, long -term steroid therapy, recurrent diarrhoea and respiratory problems should have a limited contact with nonhuman primates.
4. Appropriate references on the medical handling of certain expected problems, animal bites, snakebite therapy and antirabies prophylaxis should be available to attending employees and to physicians.

Sanitation

The contamination occurring in the zoos is largely derived from the animals and visitors entering it. The contaminated meat diets of carnivores, vermins, birds and insects also play a vital role for dissemination of the contaminants in the captive wild animals. Keeping of the contamination to a minimum level and to supplement this with strict hygiene following steps are required to be taken.

Buildings

1. All buildings should be vermin proof. The surrounding area should be well maintained so that there is no risk to the captive animals from vermins.
2. There should be justifiable distance of wild animals from the visitors
3. Buildings should have easily cleaned walls and floor surfaces.
4. Maintenance should be of high standard.
5. There should be of seperate isolation space for keeping a suspected, under observed animals.

Cleansing

Cleanliness in the zoo is utmost at prime important. The use of most effective agents like water, soapy materials or enzyme based foam cleaning agents which are economic, noncorrosive, having sanitizing substance with biodegradability property should be tried. Periodical clearing of bedding materials if so should be replaced. Equipment and untensils should be effectively sterilized with the use of hot water and detergent solution. Develop an effective programme of sanitation with

(a) Identifying needs and defects.
(b) Provide the detailed cleaning instruction for all areas and equipment.
(c) Set up a working programme.
(d) Ensure that all personnel receive proper training inhygiene, environment and personal.
(e) Evaluate efficiency

Disinfection

It is important to establish sound principles of correct sanitation and disinfection in the animal exhibited work place. Due to the nature of the job, organic contamination is a real concern. The major steps for cleaning process include first, removal of soil by hosing, sweeping or by other means. Disinfectants would be of little use if organic debris id left on the surface to be treated. Urine scale should be taken when rinsing residual acids off the surface or it will interfere in the detergent wash.

The second step is cleaning the surface with a good detergent or with a disinfectants/ detergent combination made for that job. Be sure to rinse all the detergent off before applying the disinfectant if this method is to be used, because more often than not, residual detergents will inhibits the disinfection due to incompatibility.

The next step is the application of a disinfectant, sanitizer or steriliser. There are several types of chemicals –many specific in their action and function. Each has their own advantages and dis advantages when evaluating their safety, economic value, stability, kill time and rate, effectiveness and proper application concentrations. Be sure that the chemical is applied according to the instruction on the lable.

There are several conditions before the use of chemicals one should follow such as a thorough cleaning and removal of organic debris is essential as many of disinfectants are inactivated by or are used up on the residual organic debris. The use of hot water when mixing solutions increases the penetrations of the solutions. Because of the potential corrosiveness of certain chemicals, attention should be given to the materials to be disinfected. Again refer to the lable to evaluate their use on solid, porous, rubber coated, epoxy coated, glass or plastic surfaces. Application time is important—the longer the germicidal process is continued, the greater is its effectiveness. Be sure the target microorganisms are covered in the product testing stage of the chemical. Donot mix chemicals unless stated, as many times the corrosiveness is increased or noxious fumes develop. Follow, if possible the Environmental Protection Agency and Food and Drug Administration instructions for the use of certain chemicals.

MAJOR DISINFECTANTS

Chlorine compounds: Are bactericidal, pH influences on chlorine antimicrobial activity. An increase pH substantially decreases biocidal activity. Raising the temperature while holding pH constant reduces the kill time essentially making more efficient use of the chlorine mixture. Organic matter reduces the capacity of bactericidal activity. Adding iodine or bromine can greatly enhancethe mixture's effectiveness. Various types of bacteria, viruses, fungi and algae exhibits different resistance to chlorine compounds.

A mixture of 120-ml bleach per gallon hot water is best to kill a wide spectrum of microbes if allowed to stand for five-minute or more.

HALOGEN COMPOUNDS

Iodine and bromine works well on living and dead organic matter. Hard water impairs these compounds. Many metal surfaces are oxidised by iodine and can be altered. It also stains some plastics and rubbers too. Iodine is an excellent, prompt, effective microbiocides with a broad range of action on most important health –related microorganisms like bacteria, micobacteria, viruses, fungi, protozoan cysts and bacilli spores.

PHENOLIC COMPOUNDS

Cresols, (surface disinfectant), bio-phenols (Scrub soaps) and resoricols (topical antiseptics) are used as bactericidal. It inactivates the bacterial cell wall enzyme systems. At higher concentrations, this compound acts as protoplasmic poison by precipitation of cell proteins.

ALDEHYDE COMPOUNDS

Acid gluteraldehyde and buffered alkaline gluteraldehyde provides good sterilient properties. They can be used on plastics, rubber and on aluminum area. They are irritants and are sensitive to organic debris.

ALCOHOLS

It possess many desirable characters of good disinfectant Ethyl alcohol, ispropanol are good examples of bactericidal compound by denaturing the cell proteins and interfering with its metabolism. A concentration of 70 % is most effective for general use. Organic debris should be removed before its application.they are stainless, inexpensive, evaporate quickly and easily available.

QUARTERANY AMMONIUM COMPOUNDS

It can be use on plastic or rubber surface. It attacks on cell wall and cytoplasmic membrane causing a reversal of cell charge. Mixture of alcohol's and detergents are commonly applied for full spectrum of its action.

CHLORHEXIDINE

For scrubbing of hands acts on vegetative gram-positive and gram- negative organisms. It is available in an alcohol base with added skin emollients to reduce skin damage due to repeated washing of hands.

OTHER DISINFECTANTS

Nitrogen compounds, silver, mercurial and other heavy metals can have disinfectant properties but are seldom in routine use on account of more toxicity.

Employees

Apart from cleanliness, high standard of personal hygiene and responsibility towards cleanliness in the employees themselves is also desirable. Periodic medical examinations, hygiene training to achieve a high standard of personal cleanliness and clean workman like job should be oriented. Use may be made of suitable posters, supplemented by lectures, films, and suggestions as well as by group discussion etc. should be framed out. If possible each employee should be given a hygiene booklet along with important occupational hazards.

Visitors

Effective and good education to the visitors depending upon the objectives of Zoo should be planned out. Avoidance of vandalism, feeding and teasing of animals should be taken care with the help of proper audio-visual tools and with the use of interpretation, which will cover the zoonotic diseases, way of transmission and steps to prevent it.

HINTS

Don't approach or handle any wild animals that appear sick or are acting in an abnormal manner without proper care.

Always wear rubber gloves and other protective wears while skinning or examining any wild animal.

Discard any animal intended for consumption that has spots or lesions on the liver, spleen or lung.

Practice cleanliness when working with animals. Do not eat or drink while handling or skinning animals and wash hands thoroughly when you have finished.

Vaccination of staff where appropriate and prompt and effective treatment when required should be carried out.

Close liaison with medical experts should be developed and the detailed information of the possibility that the worker may have contracted the diseases from wild animals or from insect bite should be supplied.

CHAPTER 22

COLLECTION, PRESERVATION AND DISPATCH OF BIOMATERIALS

Investigation of disease problem whether involving a single animal or a group the first and most important part is to carry out an accurate clinical examination. The major aspects includes which physical examination of the animal, the history, and the environment. Inadequate examination of any of these may lead to error. Every disease has got its own clinical syndrome and a physician can come to conclusion by his past experience and depends upon recognition of a syndrome which is identical with one seen on an earlier occasion. He may be more certain of diagnosis, which is supported by laboratory findings.

In recent years there has been an enormous increase in the number of laboratory techniques for quick diagnosis of animal diseases and every veterinary clinician should be capable of making use of it. Veterinarian should also be able to choose and select suitable material submitted to laboratory for examination. Thus blood smears, blood, blood serum, pus, sputum, discharge, exudate, urine, milk etc. should be collected when there is every likelihood of the etiological agent being present, whose isolation and identification is to be made or when antibodies are at peak and sero-diagnosis is to be undertaken. The veterinarians who are engaged in diagnostic work in regional laboratory should also be well conversant with common laboratory techniques involved in the diagnosis of infectious diseases.

A. Pre Laboratory Procedure at Field Level :

1. **Recording and Submitting Specimen History Data :** Veterinarian should collect the following information and record it in proforma specified for specific diagnostic laboratory and dispatch it to the respective laboratory. The details of species affected, age, sex, number of sick/number of dead animals, major clinical signs and population at risk should be considered to determine what species, and in what numbers, are in the vicinity of the die-off. This information can provide clues about the transmissibility of disease, and it may be useful during control efforts. The information regarding the specific features of problem areas should also be collected in some special circumstances which include any available precise location data, such as global positioning information (GPS) or data that will facilitate entering of specific locations into geographical information system

(GIS) databases. Describe the problem area in terms as topography, soil, vegetation, climate, water conditions, and animal and human use.

2. **Specimen Collection and Preservation :** Specimens are used to provide supporting information leading to the diagnosis of a cause of disease or death. A specimen may be an intact carcass, tissues removed from carcasses, parasites, ingested food, feces, blood, serum, pus, milk or environmental samples. The specimen should be as fresh and undamaged as possible.

   i. **Choosing a Specimen :** An entire, fresh carcass is the best specimen to submit to the laboratory for diagnosis if the animal/bird is of small size. This allows the diagnostician to assess all of the organ systems and to use appropriate organs for different diagnostic tests. Obtain the best specimens possible for necropsy; decomposed or scavenged carcasses are usually of limited diagnostic value. A combination of sick animals, animals that were euthanized after clinical signs were observed and recorded, and some of the freshest available carcasses compose an ideal specimen collection.

   ii. **Collection Protocol :** The primary consideration when collecting carcasses or tissues for diagnosis should be personal safety. Some diseases are transmissible to humans, and every carcass should be treated as a potential health hazard.

   Wear disposable rubber or plastic gloves, coveralls, and rubber boots.

   If gloves are not available, inverted plastic bags may be used.

   Before leaving an area where carcasses are being collected, double-bag used gloves and coveralls, and disinfect boots and the outside of plastic bags with a commercial disinfectant or a 5 percent solution of household chlorine bleach.

   If possible, do not dissect carcasses in the field without first consulting disease specialists about methods of dissecting and preserving tissues or parasites or both.

   The basic supplies and equipment that should be included in a field kit for specimen collection will vary with the species being sampled and the types of analyses that will be conducted.

   Keep a small kit packed in a day pack for read use. (see appendix-)

   Specimen identification should be written directly on the bag with an indelible marker.

   If lesions are noted, collect separate tissue samples for microscopic examination, microbiology, toxicology, and other analyses.

   iii. **Sample Collection :** The samples should be specific for laboratory confirmation of specific diseases. The laboratory biological sample should be collected with all due precaution and aseptically if possible. The details of biological sample is specified in appendix-I.

iv. **Labeling Specimens :** Proper labeling, maintaining label readability, and preventing label separation from specimens are as critical as proper specimen selection and preservation. The label should be as close to the specimen as possible. Use soft lead pencil or waterproof ink on these tags; do not use ballpoint pen, nonpermanent ink, or hard lead pencil. The most durable tag is made of soft metal, such as copper or aluminum, and can be inscribed with ballpoint pen, pencil, or another instrument that leaves an impression on the tag.

   Information on the tag should include the name, address, and telephone number of the submitter, collection site, species; whether the animal was found dead or was euthanized(indicate method); and a brief summary of any clinical signs. Place each tagged carcass in a separate plastic bag and seal the bag.

v. **Specimen Preservation :** Chill or freeze all specimens, depending on how long it will take to ship to a diagnostic laboratory. Freezing reduces the diagnostic usefulness of carcasses and tissues, but if specimens must be held for 2 or more days, freezing the specimens as soon as possible after collecting them minimizes their decomposition. Formalin-fixed tissues should not be frozen.

vi. **Specimen Shipment :** Procedures for shipping specimens vary with different disease diagnostic laboratories. Therefore, it is important to contact the receiving laboratory and obtain specific shipping instructions. This will facilitate processing of specimens when they reach the laboratory and assure that the quality of specimens is not compromised. There are five important considerations for proper specimen shipment: (l) prevent cross-contamination from specimen to specimen, (2) prevent decomposition of the specimen, (3) prevent leakage of fluids, (4) preserve individual specimen identity, and (5) properly label the package.

vii. **Preventing Breakage and Leakage :** Isolate individual specimens from one another by enclosing them in separate packages such as plastic bags. Protect specimens from direct contact with any coolant used (e.g., wet ice or dry ice), and contain all materials within the package so that leakage to the outside of the shipment container is prevented if breakage occurs (e.g., blood tubes) or materials thaw (wet ice and frozen carcasses) due to transit delays.

viii. **Specimen Shipment:** Basic specimen shipment supplies at the bottom are more likely to break during transit than those with straight sides. Fill the space between the outside of the cooler and the cardboard box with newspaper or other packing material to avoid cooler breakage. If coolers are not available, cut sheets of insulation to fit the inside of cardboard boxes. Cardboard boxes are not needed when hard plastic or metal insulated chests are used for specimen shipment, but boxes can be used to protect those containers from damage and to provide a surface for attaching labels and addresses to the shipment.

ix. **Cooling and Refrigeration :** Chemical ice packs are preferable to wet ice because their packaging prevents them from leaking when they thaw. Ice cubes or block ice may be used if leakage can be prevented. The lids of these containers should be taped closed to prevent them from being jarred open during transit. Use dry ice to keep materials frozen, but do not use it to ship specimens that should remain chilled because it will freeze them. Also, the carbon dioxide given off by dry ice can destroy some disease agents; this is of concern when tissues, rather than whole carcasses, are being shipped.

x. **Preparing Specimens for Shipment :**

1. Specimens should be shipped on the same day or within 24 hours.
2. Double-bag carcasses and place them in a cooler lined with a plastic bag.
3. When using chemical ice packs, intersperse them among specimens; place within the container other types of coolants in locations that will provide maximum cooling for all contents or, if dry ice is used, will keep everything frozen.

xi. **Specimen Shipment :**

1. Individual carcasses should be double bagged to prevent leakage of fluids and cross-contamination of specimens.
2. Cooler with newspaper to prevent materials from moving during transit.
3. The insulating properties of newspaper will also help maintain cool temperatures within the package, and its absorbent qualities will help prevent fluid leakage outside of the box or container.
4. Close the plastic bag lining the cooler and seal the lid with strapping tape. Tape the specimen data sheet and history, contained in an envelope within a waterproof plastic bag, to the top of the cooler
5. Enclose the cooler in a cardboard box and secure the contents with strapping tape.

xii. **Completing the packaging process :**

(a) Tape specimen data sheet and history, contained in an envelope within a waterproof plastic bag, to top of cooler.

(b) Place cooler in cardboard box, secure box with several bands of strapping tape, and secure another copy of the specimen data sheet to the outside of the box. If the specimens were placed inside a cooler, then use crumpled newspaper or other packing material to fill all spaces between the cooler and the box.

(c) Write with marker pen as "DIAGNOSTIC SPECIMENS (Dairy)".

xiii. **Dispatch of Materials :** For the purpose of dispatch of material suitable containers should be ready at hand. For many types of materials, bottles with metallic screw cap and rubber lining are suitable. Plastic bottles, which are autoclavable, may also be chosen. All the materials used for collection of specimens should be sterilized in hot air oven or by autoclaving. Materials for cultural examination and virological be sent in sterile containers without any preservatives.

B. **LABORATORY PROCEDURES :** After receiving the sample in the laboratory, the container should be opened with all due care and as per the requirement. Keep the biological materials either in freeze or at room temperature and proceed further as per the request for conducting the specific diagnostic tests or laboratory procedure. The routine examination of faecal, blood, urine, milk and other samples should be labeled properly and results should be written in provided proforma. Here are some of the common lists of procedures for samples.

1. **Examination of Faeces :** There are two methods for the examination of faecal samples.
   (i) Gross examination and
   (ii) Microscopic examination
   (i) **Gross examination :** The faecal samples are examined for the following information: Consistency (hard, normal, and loose), colour, composition, presence of adult parasites or other artifacts, blood etc.
   (ii) **Microscopic examination :** The microscopic examination of faeces is done by three methods as
      (a) Direct Smear Method(putting a drop of faecal material along with water ,mix it and put cover glass on slide and examine under microscope).
      (b) Qualitative Concentration Method (Floatation Method): Saturated sugar solution (sheather's solution), saturated sodium nitrate solution, saturated sodium chloride solution, magnesium sulphate (41%) and zinc sulphate (33%) solutions are used for floatation technique. Mostly for flukes like Fasciolia spp., Amphistomes spp.
      (c) Quantitative concentration method : This includes Mac Master Technique and Field Method

2. **Examination for blood :** Mostly the blood can be collected from the following sites such as cephalic vein, recurrent Tarsal vein, jugular vein, ear vein and anterior venacava. About 5-10 ml of blood can be collected for routine haematobiochemical analysis. The anticoagulants can be used or vacutainers can also be used for blood collection. Commonly used anticoagulants are potassium and ammonium oxalate (HP solution), lithium oxalate, heparin, E.D.T.A. (Disodium Salt of Ethylene Diamine tetra Acetic Acid), sodium citrate and sodium fluoride.

3. **Examination of Urine:**

   (a) **Physical Examination of Urine:** This includes colour, transparency, smell, specific gravity, foam and pH.

   (b) **Chemical Examination of Urine:** Glucose, ketone bodies, blood, protein etc. are the major abnormal constituents which can be detected from the urine using specific tests.

   (c) **Microscopic Examination of Urine**: for detection of casts, crystals, parasites, RBCs. and pus cells it is used.

4. **Examination of Milk :**

   (i) Collect the milk sample aseptically in different labeled vials. Use preservatives like boric acid, thymol etc. and store it. (a) Physical examination of milk can be done like change in colour, odour, viscosity, presence of blood, pus, cells for detection of sub clinical mastitis. (b) Several chemical tests like California mastitis test, bromo thymol blue test, white side test etc are commonly used for detection of mastitis.

   (ii) Determination of leukocyte in milk can be done by somatic cell count.

5. **Examination of Skin Scrapings :** Examination of skin scraping for presence of mange mites and other parasites, fungi can be performed by examination of sample clinically and then by direct smear method, sedimentation method or by sugar floatation method. The skin scrapings for fungal infection can also be examined grossly and with the use of Wood's ultra violet lamp, microscopic examination or by culture method on SDA media.

6. **Examination of Rumen Fluid :** The collected rumen fluid can be tested for gross and microscopic examination. The pH, colour, odour, turbidity, sedimentation time, cellulose digestion test, MBRT are the common tests can be performed from rumen fluid.

7. **Examination of Cerebrospinal Fluid :** Collection of sample can be done from lumbo sacral area or from cisterna magna site. Examine the fluid for colour, turbidity, coagulation reaction and for protein (Pandy's test) and carry out tests for total cell count and differential cell count.

8. **Examination of Exudates and Transudates :** Collection of the sample from the affected area then go for specific test with physical, chemical or microscopic (Cell count) as well as for some samples can be examined for bacteriological parameters.

9. **Serology :** Several serological tests are used to know the level of antigen or anti bodies. Some of them are as

   (i) **Agglutination Test :** Tube Agglutination Test (TAT) and Plate Agglutination Test (PAT) are commonly used for detection of brucellosis and avian salmonellosis.

(ii) Precipitin Test (Gel-Diffusion) : like Ascolis test for Anthrax is used.

(iii) Complement Fixation Test:

(iv) Haemagglutination Test:

(v) Enzyme Linked Immunosorbent Assay :

Assay for antibody : which includes Indirect Method, Competitive Method or Anti-IgM Method

Assay for antigen: Competitive Method, Double Antibody Sandwich Method, Inhibition Method

Application of ELISA :

(a) Diagnosis of infectious diseases (Bacterial, Viral, Rickettsial, etc.)

(b) Immunodiagnosis of parasitic diseases like malaria, trypanosomiasis, schistosomiasis, toxocariasis, and toxoplasmosis.

(c) Used in endocrinological area, for the protein hormone such as chorionic gonodotropin, luteinizing hormone, follicles stimulating hormone and in steroid assay of estrogen, progestrone, testosterone and cortisol.

(d) For the detection of oncofetal proteins.

10. **Allergic Tests :** Such tests are used for detection of chronic infections likes TB, Brucellosis, glanders, JD. etc. The above tests can be performed by [i]. Single Intradermal Test ,[ii].Double Intradermal Test ,[iii]. Stormont's Test [iv]. Intra dermo palpebral method or by [v]. Comparative Test for the specific disease.

11. **Advanced Diagnostic Techniques :** Compton metabolic profile test for metabolic disorders, Polymerase chain reaction(PCR) for certain bacterial and viral diseases, computer aided diagnosis are some of the advanced laboratory techniques which can be used for diagnosis of diseases of dairy animals.

METHODS AND MATERIALS FOR COLLECTION AND THEIR DISPATCH TO LABORATORY

Smears : Smears are required to be prepared from blood, exudates, other body fluids, discharges from wounds, uterine or other discharges. The blood smear should preferably be collected at the height of the temp before administering, antibacterial, antibabesial, and antitrypanosomal drugs. In suspected trypanosomal, filarial infection blood smears should be of thick quality. Smears are dried in air and fixed with Methyl alcohol for 2-3 minutes. Slides with smears are packed with paper with matchstick in between two adjacent slides to keep the smears apart from each other.

Blood : Blood from ailing animals is collected from peripheral circulation and from dead animals from right auricle or ventricle. Blood should be collected with sterile syringe and needle whenever any septicaemic disease condition or viraemia in viral disease occurs. Blood should be collected in screw capped bottles having anticoagulants and transferred to freezer and dispatched over ice in thermos.

Serum : For collection of serum, blood is collected from convalescent or recovered animals with sterile precautions and allowed to clot in sterile test tube or Mc-cartney bottle. After clot when serum is separated it is transferred with the help of sterile Pasteur pipette into a screw-capped tube or bottle. For preservation of serum sample carbolic acid and Merthiolate are added so that final concentration of these two in the serum sample is 0.5% solution and 1: 10,000 respectively.

Cotton Swabs : For collection of scanty amount of body fluids, secretion, effusion or transudates etc. cotton swabs may be used. This cotton swab is then inserted into the test tube and the mouth of the tube is plugged with absorbent cotton and sterilized in hot air oven.

Pus, Sputum, Throat and Uterine Discharges, Exudates, Transudates and other Body Fluids

They are collected with sterile pipettes, Pasteur pipettes or with the help of syringes with sterile precautions and contamination, when discharges and fluids are scanty they are collected with the help of sterile cotton swabs. After collection the fluids are transferred in screw capped bottles and swabs into test tubes. They are dispatched over ice in thermos.

Milk : Wash the udder with some antiseptic solutions, wipe out with towel and dry, discard few strippings and collect 15-20 ml. from each quarter. Separate samples should be collected from each quarter. Dispatch in screw capped bottles over ice in thermos.

Urine : Collection with sterile catheter should be emphasized, or after cleaning the external genitalia with mild antiseptic and after drying.

Skin Scrapings and Hair : From dermatomycotic cases and Mange, deep scrapings from recently developed lesions mainly from periphery of the lesions should be collected. From dermatomycotic infections few hairs from affected lesions should also be pulled out and put in the specimen. The specimen may be sent in paper.

Faecal Samples : For diagnosis of helminth infection faeces should be collected from the animal itself and not from ground. Atleast 10 gm. of faecal sample should be sent after adding 10% formalin as preservative.

CHAPTER 23

NUTRITION OF ZOO ANIMALS

Nutrition is an integral component which is required for life, growth, production, reproduction and for longevity of individual. The psychological and physiological needs of species that have evolved under diverse environmental circumstances, to occupy specialized niches, with the ability to make diet choices according to seasonal, physiological, environmental or individual requirements The free living wild animal food preference and selection is quite variable compared to captive ones. Mostly it depends on the availability of food,climatic and ecological factors, habitat ,number of predator species and prey base. The phylogenetic scale suggest the complexity of system on evolutionary base.

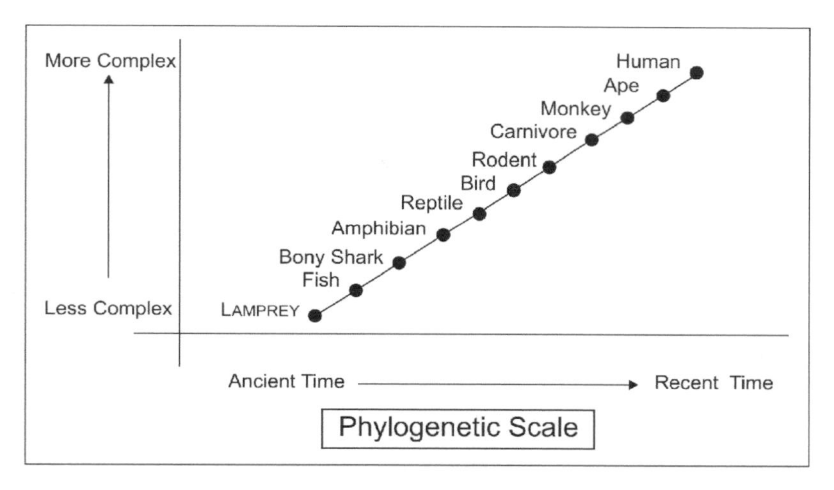

The energy,water,composion of food, species variation plays an important role in free living animals. The major food compositions are :

1. Water
2. Organic and inorganic substances
   (a) Carbohydrates
   (b) Fat
   (c) Protein
   (d) Vitamins and minerals
   (e) Enzymes
   (f) Hormones

ZOO NUTRITION

Very little scientific knowledge, whatsoever, is available on feeding and nutritional requirement of individual species of the wild animals, nevertheless fundamental principles of nutrition may be applied to the feral animals as well. Thus, food eaten by these animals in their natural habitats must form the basis for the formulation of dietary schedules. In the absence of natural food, wild animals in captivity may become emaciated or may even die. Careful study of the dietary habit is necessary for successful wild animal medicine, because nutritional deficien-cies in many instances may result into therapeutic failures.

BASIC FOOD INSTINCT

Australian koala feeds on tender eucalyptus leaves which can hardly be substituted; penguins thrive on fish and crustaceans found in Antarctica; birds like honey suckers regularly take worm diet while apparently sucking the nectar offlowers; a live fowl offered to a large feed suffering from anorexia may stimu-late appetite, elephants normally show tendency of geophagia and consume large quantities of saline soil either to replenish the deficit salts or for its laxative action, are a few examples which need consideration in zoo animal nutrition. Sometimes it may be necessary to recommend a radical change in the diet which may be accepted also by the animal.

The basic nutrients required by the feral animals are water, carbohydrates, fats, proteins, minerals and vitamins as in do-mestic and laboratory animals (NRC). Trace mineral salt blocks may be provided to the wild animals. A different variety of food may be provided to the animals so that the animal may choose and it may help to maintain a balance as well. Size and consistency of the food must also suit the species of the animal.

Small herbivorous animals may also be given small pieces of flesh or eggs in view of the fact that these animals in the free state may be consuming small worms which naturally maintains the nutritional balance. The food must be fresh and of high nutritional value, preferably fortified with vitamins and minerals including trace elements. Neonates of some species need colostrum immediately after birth, to acquire immunoglobulins, for resistance against diseases. Small herbivorous animals and carnivores may be given leafy vegetables like lettuce, spinach, cabbages and for large herbivorous animals, alfa-alfa, timothy, berseem and grains may be provided.

For zoo carnivores, good quality of muscle meat, liver, tripe, other viscera, bone-meal, green-bone paste, cod-liver oil, evapo-rated whole pasteurized or whole powdered milk (particularly in young growing animals) may be recommended. Horse flesh, beef (buffalo beef, chicken, rats, mice and rabbits etc. may be used according to the facilities. The meat may be sprinkled with vitamins (cod-liver oil) and minerals (Ca, P mainly). Bones may cause choke and hence may be avoided.

Diets for zoo carnivores consisting of meat fortified with other essential constituents, for animals such as wolves, lion. tiger, leopards, etc., have been formulated in many countries. However, all carnivores are very susceptible to the untoward effects of poly-unsaturated fats present chiefly in tuna fish and cod-liver oil. This diet in the absence of sufiicient quantities of tocopherol (Vitamin E) may precipitate yellow-fat disease (steatis) in carnivores, mainly in cats, mink and dogs etc., specially in captivity. Lactose in unsuitable for felids which may develop lactose intolerance characterized by diarrhoeal disorders and as such milk and milk products may be avoided in these animals.

Meat from animals given hormonal treatments (stilbestrol implants), glan-dular tissues containing thyroid and parathyroids and certain fish containing thiaminase enzyme, inhibiting thiamine (chas-tek paralysis) and other fish containing avidin, an iron binding factor may cause health hazards which may be overcome to some extent by cooking the food. Nursing sickness in minks is a dehydrating disease in lactating females under stress, proba-bly caused by the deficiency of salt. The food of the minks may be preserved by adding antibiotics

Carnivores animals also have a problem of anti metabolites in the feed such as chastek paralysis due to the presence of thiami-nase. Biotin deficiency occurring due to feeding of raw eggs may be prevented by cooking the eggs or adding biotin 1 mg per animal every three days. Hypovitaminosis A in foxes is charac-terized by neurological disorders such as circling which may be prevented by feeding of liver (conversion of carotine to vitamin A is poor in foxes as in minks and therefore, carotine precursors are not of much use). Riboflavin deficiency in foxes causes der-matitis specially if fed on high fish meal diets which may be prevented with vitamin B complex supplementation. Vitamin E acts as antioxidant and is recommended to protect foxes against yellow fat disease, caused by oxidation of unsaturated fats.

Among rodents, rabbits are fed on home grown feeds, com-mercial diets, feed supplements and pellets etc. Whole grains such as oats, wheat, barley, rye, corn,legume hay (dover, alf-alfa, cowpea) with salt bricks and plenty of drinking water are essential components of rabbit nutrition. Coprophagy is com-mon in the rabbits and is probably not a vice but a mode of self -supplementation. Wheat germ oil, trace elements, salt spools may be provided but cod or other fish oils are contra indicated.

Common diets of zoo animals are given as follows:

Marsulpials : *Opossum:* meat, raw eggs, milk, bread, banana,apple.

Kangaroo: greens, bread, apple, sweet potato, carrots, oat, hay, potatoes

Insectivores : *Hedge hogs:* meat, eggs, milk, dog food

Moles: meat, eggs, milk.

Carnivores : *Felines:* meat, bone meal, heart, liver

Hyena: meat, bone meal, liver, bones

Racoons: meat, eggs, bread, apple, oranges, banana, carrot bone meal

		Civet: meat, eggs, apple, banana, bone meal
		Panda: special mixture for mammals
		Fox: meat, apple, banana
		Bear: meat, bread, apple, carrot, fish
		Sea lion: fish
		Mink: fish, frogs, birds, insects, cereals, apple, meat
Rodents	:	*Squirrels:* greens, bread, apple, oranges, banana, sunflower
		Beaver and porcupine: Seeds, corn, dog food
Edentates	:	*Anteater and Armadillo:* meat, eggs, boiled ,sherdded milk
Primates	:	*Lemurs monkeys, kakoons, gorilla orangutan, etc*; greens, fruits, leaves, fl owers, milk, eggs, bread, monkey foods, etc.
Artiodactyla, Perisodactyla	:	greens, bread, apples, sweet potatoes, hay, oats, gram, bran, carrots, vitamins and minerals etc. The diet of elephants must be bulky consisting ofleafy fodder, bamboo, reeds and concentrates such as wheat flour 10 kg with salt andjaggary etc. These ani-mals must be given long time for feed-ing, including the nights.
Reptiles	:	*Iguannas:* Banna,pear,cabbage,rose petals,clover,calcium carbonate
		Chameleons: Live insects,meal worm,grass hoppers, locusts
		Tortoise: Vit.A rich diet,fish supplemented with 30 gm. of thiamine per kg.
		Crocodials: meat based diet with supplemention of Vit.E and with 30 gm. of thiamine per kg .
Bats	:	*Insectivorous:* mealworms, crickets, fruit flies, blowny larvae. The worms may be dusted with calcium and vitamines or artificail diets may be used.
		Frugivorous: banana, papaya, apple, pear, melon, grapes, carrot, sweet potato etc. with powdered milk, corn oil, eggs,dogs, feline diet,vit. and mineral mix
Birds	:	The captive birds requires more care during moulting and breeding. Amino acids, minerals, vitamines and carotenoid pigments forfeather colour needs special attention. The pesticides and other toxic level of ingredients must be tested periodically to avoid mortalities or magnification of their effect.

BALANCED FOOD DIET CHART OF WILD ANIMALS

Nutrition and Health

Nutritional status is intimately linked to health and reproductive output in numerous species. All nutrient categories - energy and protein, fatty acid, fat-soluble vitamin, and mineral imbalances - have been shown too directly, as well as indirectly, affect conservation goals of captive breeding.

Energy

Over-nutrition, with accompanying obesity, is a health issue for many captive psitticines, and may negatively impact reproduction in these species. Over condition can also be problematic for health of zoo primates, carnivores, and hoof stock, leading to problems with diabetes, respiratory and cardiac distress, hypertension, hypercholesteremia, foot/ hoof problems, and reduced reproduction. Obesity can still be considered a major problem in many zoo collections. We need to develop a series of species-specific body condition indices, similar to those developed for domestic livestock and pet species and/ or customized indices (for captive species) based on *in situ* populations as the guideline. These standardized scores can then be used to establish guidelines for amounts of diet to feed, based on energetic relationships, and will provide a tool for better understanding management variables, including the impact of increased activity through behavioral enrichment, on a global scale.

Protein

Excess protein can also prove detrimental to animal health. Within weeks of switching from a high (>40% crude protein on a dry matter (DM) basis) to a low (10%) protein nectar that better duplicated native diet composition hummingbirds at the Bronx Zoo successfully reproduced. A wide variability exists in commercial nectar product composition, with protein ranging from two to >20% of DM; these ranges appear to encompass levels that can prove harmful to captive populations of nectarivores. Similarly, the vulturine parrot *(Psittrichas fulgidus),* a fig specialist, experienced health problems associated with excessive dietary protein (kidney disease, poor reproduction and growth) on diets containing approximately 20% crude protein (DM basis). Nitrogen balance trials confirmed that adults of this species could maintain protein balance on diets containing only 2% protein. National Research Council recommendations for non-human primates, recommended diets in US zoological facilities have been recently altered to this lower (15% protein) level; true requirements and/ or nitrogen balance, however, has not been determined in this species. It is also possible that a genetic impact in limited captive populations may be involved in this health issue, but requires investigation.

Minerals and Vitamins

By comparison, vitamin E nutrition has been evaluated in much more detail in zoo species. Deficiencies in captive populations have been reported frequently, possibly due to improved detection techniques as well as a higher proportion of trained staff in many zoos compared to years past. Clinical signs of deficiency vary by species, but can include skeletal/ cardiac myopathies (in any species, but particularly in hoof stock), equine degenerative myoencephalopathy (in equids), microangiopathy (noted in swine and elephant species), erythrocyte hemolysis (primates including humans), exudative diathesis (avian species), and steatisis in carnivores.

Mineral imbalances have long been recognized in zoo species - the first documented nutrition problems in large carnivores fed meat - based diets, and primates and birds raised primarily on fruits - were reported from the London Zoo in the 1800s. Meat- and insect-based diets, lacking calcium, have been problematic in zoos for numerous species, resulting in supplementation recommendations as well as formulation of nutritionally complete commercially available products for feeding these species. Iron excess, leading to hemosiderosis, More recently, however, mineral imbalances in zoo hoof stock diets are being reported, resulting in the formation of gastroliths, enteroliths, uroliths, hypomagnesemia/ hypocalcaemia/ rumenitis/ and acidosis syndromes. In each instance, problems have been noted with hoof stock (both grazers and browsers) fed lucerne *(Medicago sativa)* as the primary forage. The mineral content of Lucerne compared with grasses (high in both Ca and Mg relative to P), may predispose ungulates to mineral imbalances and needs to be investigated in more detail.

Separate from these health aspects, mineral nutrition also has direct and indirect effects on animal reproduction. Lesser adjutant stork *(Leptoptilos javanicus)* chicks fed whole prey containing 2% Ca (DM basis) developed beak and leg deformities indicative of calcium deficiency / imbalance - even though the appeared adequate in both Ca and P. Further comparisons of semen characteristics in cats fed a nutritionally-complete commercial feline diet compared to chicken necks or red meat revealed improved motility (40 to 56%), % normal sperm (9 to 20%), and an increase in the amount of sperm per ejaculate (up to 10 times more

Mammals can be broadly classified into three groups according to their food habit i.e. herbivores, omnivores and carnivores. However, there is very thin lining between the species as far as their feeding habit is concerned. Some Artidactylids may be omnivorous, whereas, some carnivores may be actually omnivorous. Even some strict herbivores may occasionally indulge in some food of animal origin. For a general understanding, we will discuss orders artiodactyls, perissodactyla, proboscidae and lagomorphs as herbivores: orders primate and rodentia as omnivores and orders carnivora, pholidota and insectivora as carnivores. We will discuss order Chiroptera, as special order as there is wide variation in food habit among bats.

I. Order Artiodactyla

1. **Family *Suidae* :** A typical diet for wild boar should contain Concentrate mixture 1kg, seasonal vegetables 1kg, tapioca 250g and potato 3kg.
2. **Family *Tayasuidae* :** Peccaries can be maintained with swine ration supplemented with fruits and vegetables.
3. **Family *Hippopotamidae* :** Atypical diet for a Nile hippopotamus would comprise of 40-50 kg of hay, 5-6 kg concentrates (14% CP), 2 or 3 loaves of bread, 5-10 kg of vegetables including potato, cabbage, carrot, apple and onions. If available, green grass is refreshed greatly, but should be introduced carefully to avoid digestive disorder.

   An adult pygmy hippopotamus should be fed 7-8 kg of hay, 2-3 kg of concentrates, bread, potatoes, onions, carrots, cabbage and fruits.
4. **Family *Camelidae* :** A good quality grass hay (10-12% CP) or alfalfa hay is sufficient to *meet* all the requirement of an adult inactive camel, with activity this may be increased up to 3% of BW. Pregnant or lactating female or a growing young animal may consume up to 4% of BW.
5. **Family *Traguiidae* :** Mouse deer can be maintained on 100-150g soaked gram, 30-50g carrots, 20-25g tomatoes and 500g of leguminous fodder.
6. **Family *Ceruidae* :** Feeding schedule for some of the deer is presented in Table 1.

Table 1. Feeding schedule of deer

Species	Mash*(kg)	Green fodder (kg)	Tree fodder (kg)
Sambar	2.0	5	-
Swamp deer	2.0	3	-
Sangai	1.0**	2	5.0
Chi tal	0.75	2	-
Barking/hog deer	0.50	3	2.0

Mineral mixture added @1 %,** additional soaked gram O.5kg

7. **Family *Moschidae* :** In captivity musk deer must be fed grass 1.5 kg, wheat bran 0.5 kg, carrots 1.5 kg, tree leaves 3.0 kg green fodder 5.0 kg and salt 0.1 kg
8. **Family *Giraffidae* :** One typical giraffe diet should contain good quality hay 5 kg, Lucerne 1.5 kg, wheat bran 2 kg, horse gram 1.5 kg, Bengal gram 1.5 kg, decorticated oat 1kg, carrot 1 kg, cabbage 2 kg, onion 1 kg, apple 0.75 kg
9. **Family *Antilocarpidae* :** In captivity, musk oxen adapt to alfalfa and grass hays and will consume grains
10. **Family *Bovidae, Capridae* and *Ovidae* :** Feeding schedule of antelopes, and wild goat are presented in Table 2

Table 2: Feeding schedule of antelopes and wild goats

Species	Mash (kg)	Lucerne (kg)	Peepal leaves (kg)	Green grass (kg)
Gaur	3.0-4.0	5.0-6.0	20.0	Ad-lib.
Nilgai	2.0	3.0-4.0	-	Ad lib.
Chinkara	0.5-1.0	1.0-2.0	1.0-2.0	2.0-3.0
Black buck	0.5-1.0	0.5-1.0	2.0-3.0	2.0-3.0
Chowshiga	0.5-1.0	1.0-2.0	1.0-2.0	2.0-3.0
Goral	1.5	-	-	5.0

II. Order Perissodactyla :

1. **Family *Equidae* :** A ration containing 5.0 kg of dry fodder,10kg of green fodder and 5kg of concentrate mixture seem adequate for wild ass of Rann of Kutch. Similarly, a zebra can be fed with 3kg of horse ration and 20kg of green fodder.
2. **Family *Tapiridae* :** An average adult tapir require 3.5 kg of good quality alfalfa, 1-1.5kg of commercial monkey chow or herbivorous mash, 5-12kg of mixed fruits and vegetables.
3. **Family *Rhinocerotidae* :** A typical diet for Indian one-horned rhino should contain 2kg gram, lkg wheat bran, 1 kg mung bean, 10 bananas, 40g salt and 200kg of green grass. Black salt may be supplemented once in a month @500g.

III. Order Probiscidea

1. **Family *Elephantidae* :** Working elephants are fed with roti 6.0kg, oil 50g, salt 250g, and jaggery 500g, in addition to 400kg of green fodder. Non-working elephant are given only green fodder.

IV. Order Lagomorpha

1. **Family *Leporidae* :** Indian wild hare can be maintained with banana 4 numbers, vegetables 400g and dub grass 250g.

V. Order Primate

1. **Family *Cerocopithicidae* :**

   Rhesus monkey : fruits and vegetables 500 g, boiled rice 125 g, milk 250 ml, bread slices 2, groundnut 100 g and roasted gram 50 g.

   Bonnet monkey : Milk 100 ml, bread 200 g, apple 100 g, grapes/ chiko 100 g, banana 300 g, leafy vegetables 50 g carrot 50 g, cabbage 50 g, tomato 50 g, bean 50 g and ground nut 100 g.

   Assmaese macaque : fruits and vegetables 500 g, boiled rice 125 g, milk 250 ml, bread slices 2, groundnut 100 g and roasted gram 50 g.

   Stump-tailed macaque : Bread 200 g, apple 100 g, grapes/ chiko 100 g, banana 300 g, guava 50 g, carrot 50 g, onion 25 g cabbage 50 g, tomato 50 g, bean 50 g, green vegetables 50 g and ground nut 100 g.

Lion-tailed macaque : Fruits 250 g, vegetable 200 g, milk 50 ml, boiled egg 1, minced meat 150 g, insect/ honey 10 ml, nuts without shell 50 g, rice or bread 50 g and green vegetables 200 g.

2. **Family *Colobinae* : Common langur/capped langur:** Bread 100 g, milk 100 ml, orange/ mango 1/2 no. grapes/ chiko 100 g, vegetables 100 g, sweet potato 100 g carrot 50 g, onion 25 g, tomato 25 g, beet root 50 g, Bengal gram 50 g and ground nut 50 g.

   Spectacled monkey : Mixed fruit 150g, soaked gram 150g, banana 10, and raw vegetables 200g.

   Nilgiri langurs : Bread 100g, milk 100ml,egg 1 no. grapes/ chiko 100g, apple/ pear 150g,banana 3 nos. seasonal vegetables 50g, cabbage 50g, carrot 50g, onion 25g, tomato 50g, beans 50g, Bengal gram 50g and ground nut 100g

   Golden langur : Bread slices 4 nos. Bengal gram 250g, banana 6 nos., fruits 250g, vegetables 300g, green leaves ad lib.

   Capuchin/Red/PotashSpider : Fruits and vegetables 500g, Bread slices 2nos

   Baboon : Fruits 500g, bread slices 4 nos., vegetables 250g, milk 250ml, roasted gram 100g, boiled egg 1 no.

3. **Family *Lorisidae* :**

   Slow Loris : Boiled egg 1 no., banana 3 nos., milk 100 ml, cooked rice 100 g, minced mutton 100 g, oranges 2 nos.

4. **Family *Pongidae* :**

   Hoolock gibbon : Milk 100 ml, bread 100 g, boiled egg 1, gram 50 g, groundnut 50 g, banana 2 nos., apple 100 g, pears 100 g, sweet lime 100 g, beet root 100 g, tomato 25 g, cucumber 100 g, corn 100 g, potato 100 g, melon 100 g, onion 25 g, green leafy vegetables 100 g.

   Chimpanzee : Bread 200 g, milk 100 ml, boiled egg 1 no., sugar 50 g, banana 10 nos., carrot 250 g, sweet lime 400 g, apple 750 g, mixed fruits 650 g, mixed vegetables 850 g,coffee/ tea 100 ml.

   Orangutan : Milk 1 L, bread 150 g, banana 1 kg, mixed fruits 2.5 kg, sugar 50 g, egg 2 nos., tomato 150 g, maize cob 2 nos. bael 1 no.

   Gorilla : Bread 400 g, milk 150 ml, boiled egg 2 no., sugar 50 g, apple 1 kg, banana 750 g. gropes 1 kg, seasonal fruits 500 g , leafy vegetables 100 g carrot 250 g, cabbage 250 g, beans 50 g, coffee/ tea 100 ml.

VI. Order Rodentia

1. **Family *Hystricidae* :** Indian porcupine: Bread 100g, gram soaked 200g, maize cob 1 no., groundnut 100g, potato 200g, carrot 200g, leafy vegetables 200g.

2. **Family *Scuiridae* :** Giant squirrel: Milk 100ml, bread 100g, gram soaked 50g,banana 6 nos., groundnut 100g, mixed fruits 500g.

VII Order Chiroptera : Feeding habit wise bats are diverse group of animals ranging from frugivorous, nectarivorous, insectivorous to sanguinivorous. In zoos mostly frugivorous bats are kept.

Food items should be dated on arrival and should be used in a timely fashion.

VIII. Order Carnivora

1. **Family *Canidae* :**

   Wild dog/Dhole : Beef 2.5 kg, liver 100 g, chicken (bi-weekly) 500 g, milk 250ml.

   Jackal : 2 kg of buffalo meat

   Wolf : 4 kg of buffalo meat

   Fox : 0.5 kg of buffalo meat

2. **Family *Hyenidae* :**

   Hyena : buffalo meat 2.5 kg, 100 g liver, 500 g chicken

3. **Family *Ursidae* :**

   Sloth bear: Milk 500 ml, bread 2 kg, maize 1 kg, banana 3 nos., chiko/ grapes 500 g, beet root 250 g, sweet potato 250 g, Ber/ bael 100 g, papaya/ sweet melon 500 g, vegetables 250 g jaggery 100 g or alternately Rice 1 kg, milk 1 kg, wheat flour 1 kg, fruits and vegetables 3 kg, jaggery 200 g.

   Himalayan black bear : Bread 1 kg, milk 500 ml, banana 3 nos., fruits 500 g, carrots 250 g, roots/ potato 200 g, sweet potato 250 g, bel 100 g.

   Brown bear : Bread lkg, milk 500 ml, rice lkg maize 0.5 kg, tubers 1 kg, fruits 700 g

   Sun bear : Milk 500 ml, bread 1 kg, banana 3 nos. carrot 250 g, tomato 250 g, potato 250 g, cucumber 250 g papaya/ sweet melon 500 g, oral mau 300 g

2. **Family *Procyonidae* :**

   Red panda : Milk 11, sugar 20 g, carrot 200 g, papaya 200 g, protemax 10 g, apple 400 g, banana 3 nos., egg 2 nos., bread 150 g, bamboo leaves 2 kg.

3. **Family *Viviridae* :**

   Binturong : milk 100 ml, bread 100 g, egg 1 no., banana 10 nos., papaya 100 g, apple 100 g.

   Palm civet : buffalo meat 500 g, chicken 500 g, egg 1 no.

   Large Indian civet : milk 100 ml, bread slices 2 nos., meat 100 g, fish 100 g, and banana 1 no.

4. **Family *Mustelidae* :**

   Otter : Fish 1.5 kg, carrots 100-200 g.

   Hog badger : morning some fruits and bread. Evening 0.5-1 kg of boiled meat.
   Retell : bread 250 g, meat 500 g, milk 200 ml, banana 200 g.

5. **Family *Felidae* :**

Table 3. feeding schedule of some felids

Species	Meat(kg)	Milk(kg)	Egg (no)	Chicken (kg)	Remark
Tiger	8-10	0.5	1	weekly	Monday weekly off
Lion	8-10	-	-	-	-do-
Leopard	4	-	-	0.75	-do-,250g liver
Cheetah	3	-	-	1	-do-
Clouded leopard	2	-	-	-	-do-
Jungle cat	0.5	-	-	-	-do-
Leopard cat	0.5	-	-	-	-do-
Fishing cat	0.5	0.2	1	Fish 0.2	-do-
Golden cat	1	-	-	-	375g meaton off day

IX Order Insectivoira

1. Family Erinacidae :

Desert Hedge Hog : Egg 1 no., bread 50 g, milk 50 ml.

Pangolin : chopped meat 200 g, egg 1 no., carrots 100 g, termite mound 1,

SOME USEFUL TIPS FOR THE CAPTIVE FEEDING AND NUTRITION

1. Animal must be fed fresh, palatable, uncontaminated and nutritionally adequate food according to the species specific requirement of zoo animals.
2. The food must have optimum taste, nutritional value, palatability and preferred by the species.
3. If the food is required to be stored, it should be procured and stored as per the guidelines.
4. The perishable food and vegetables can be kept at 40-45 degree F.
5. Meat with its high level of protein and water presents an ideal medium for decomposing organism and must be cooled quickly after the animal is slaughtered.
6. After studying the digestive coefficient of various foods species requirement for the type, quantity and quality of food for species should be determined and provision, be made accordingly.
7. Feeding timings should be strictly adhered to until and unless there is some emergency Otherwise, the animals will pick up undesirable habits and vices.
8. Nutritional requirements and palatability are essential as animals requirements change from season to season and with age too. Changes in diets or dietary manipulations if made in overcoming the problems should be recorded with reason thereof. All the captive wild mammals require attention to stress because the benefits of nutritional support must be weighed against the duress suffered by wild animals under captive environment.

9. Individual diets should be modified to match the changing physiological state of the animal i.e. new born, young, growing, pregnant, lactating sick. recovering etc. For the new arrivals and also in case of convalescent animals nutritional support is always started gradually, no matter what the final caloric goal may be All are started at below maintenance levels for the first few days of refeeding and the total amount is increased gradually over time. If the nutritional support is started at the full amount, animals may have intestinal pain and diarrhea. If the first few meals are started slowly and diluted. There will be fewer problems associated with refeeding.
10. Starvation decreases metabolic rate producing a condition termed hypo metabolism. Starved, moribund and other sick animals require fewer calories than usual.
11. Any animal, rescued under extra ordinary circumstances and placed in the zoo, has to be provided good nutritional support for better thriving and also for safe health.
12. Ectotherms require attention to ambient temperature as food is digested incompletely. Reptiles such as python, king cobra, etc. are fed once a week. Most amphibians remain active and feed well at 20^0C to 25^0C. Snakes, turtles, crocodiles, lizards, etc. remain active and do well between 25^0C to 35°C. However, diurnal lizards and crocodiles are given voluntary access to the higher temperatures between 32° to 31"C (89 ^{0}F to 99^0 F) As the temperature goes down in winter, the metabolism of aquatic animals, reptiles, etc.is depressed and feeding requirements of many species are curtailed. In such cases forced feeding should be avoided.
13. Attention should be diverted to study abnormalities related to nutrition, metabolism, consumption of toxic substances, etc. Many disorders increase metabolic rate producing condition termed hyper metabolism. A vivid example is hyperthyroidism in cats, Occurrence of similar condition. In large felids can not be ruled out. Other examples include fractures, infections, burns, etc Animal with a disease that increases metabolic rate requires more calories than usual.
14. In case of deer, antelopes. wild asses and primates, to avoid conflicts and fighting among cage mates, the feed should be provided either on a long stretched platform or at several points so that each individual animal gets easy access to take its quota of ration, Trampling of weak neonates and intimidation of sub-adults by stronger ones must be. Aggressive individuals should be cared isolated immediately.
15. Big cats, canids and ursids, after weaning from their mothers, should be fed individually. Bears, apes and macaques are fed twice a day and small cats and puppies should be fed thrice a day. Never feed frozen cold meat until it is thawed to room temperature. The thawed meat, if not used, should not be refrozen to store.
16. Water quality for turbidity, salinity, oxygen, minerals and its availability in the enclosures should be checked regularly. The water troughs/pots should be cleaned regularly to avoid formation of algae and collecting therein of any foreign matter. Female's suckling offspring need more water. Water requirement also varies when there is change in the environment.
17. For hand rearing, special milk formulae have to be prepared which are different for each and every baby Some animals need milk with low fat and high protein content while others.

CHAPTER 24

ORPHAN MANAGEMENT

Hand rearing techniques are no more than emergency, when all else has failed. It is essential for the survival of some prime endangered species. In a model zoo, planned breeding, behavioural and medical record helps in managing pregnant animals. Past problems with exhibits, group social structures, dystocias or maternal neglects however should not be neglected. Quick decision under certain chaotic situation should be made for the young one and for the mother with possible enclosure modification, provision of supplemental heat and privacy, removal of aggressive cagemates and monitoring by birth watchers or closed - circuit televisions.

FACTORS ENFORCING HAND REARING

Maternal problem : Uterine inertia, dystocias, vaginal injuries, mastitis, agalactia and retained placenta can interfere with maternal rearing. When the dam dies and the neonate cannot be fostered, in such cases hand rearing becomes essential.

Neonatal problem : Congenital or hereditary problem, injuries or illness of neonates may need to rear them by hand. Aggressive males or overattentive mothers, pneumonia, enteritis or septicaemia are common neonatal illness.

Social problem : Social hierarchy and female's position in a group plays an important role in determining the need of hand rearing. An aggressive male or dominant female may interfere with the bonding, which should take place between mother and young one. In many carnivores the female must be isolated or both she and young one will be at risk. Infanticide problems of sub ordinate females in many packs are common.

Conditioning : For research , preventive model and educational purpose (touch and teach learning approach) some animals that are used are hand reared to condition them to captivity , close proximity of people and for captivity and confinement.

Why they should Hand Rear ?

Ideally all neonates and juvenile felids should be mother reared because: 1]. the natural mother does the job best;2].mother reared cubs usually are better adjusted behaviorally as adults; 3]. a mother and cubs make a terrific exhibit; and 4].it saves management costs. In the event that mother rearing is impossible on account of maternal neglect or for other reasons, cubs should be hand - raised.

Basic Needs to be Taken Care

Hand reared animals should be exposed to animals of similar species or same species.Some animals can be fostered to similar domestic animal females such as dogs, cats, goats, horses and cattle. Contact of sibling and tolerant females of the same species some times helps in hand rearing. A neonate without passive immunity of colostrum should be balanced with the infant's need. Reintroduction of weak, ill and injured neonates in short period with or without chemical restraint of females should be planned (Scott pers.comm.).

NEONATAL EXAMINATION AND NURSERY MONITORING

Routine examination of neonates to minimise the chance of maternal rejection (Meier and Willis, 1984) and to established normal serum values and developmental milestones (Read and Meier, 1998). Cannibalism of young one is a problem in carnivores especially if stressed; however, neonatal examination can be performed with selected individuals. The limited personnel (regular keeper and the examiner) should be allowed. Sterile surgical gloves should be worn. The keeper should enter the facility alone and shift the dam away from the young one, then remove the offspring from the den area. The examination should be done out of the sight and immediate hearing of the dam. It should be completed quickly. If the notch, ear- tag or chips are to be used it should be clubbed with this. The young one should be returned to the den and the dam allowed access to her offspring. Examination can be postponed for first few days of life if female and young one appear normal, although weekly examinations are possible with many species.

Housing

Housing must meet the physical, mechanical, psychological and social needs of the animals and individual species' characteristics must be taken into account. Environmental enrichment and the animal's use of space need to be considered. Use of cage furniture, logs, timbers and secluded hiding areas should be planned within the exhibit. Off exhibit holdings should have a space for medical treatment. A properly designed restraint cage which provides less stress should be planned. The walking off exhibit area should have sufficient traction, should not be hard and harmful the footpads. Regular record of animals should be planned with updates of all history and details of vital signs and the needs of neonates.

Feeding

Nursery personnel should use a consistent feeding schedule and pattern, allowing neonates to stabilise and become conditioned to a routine. Most animals learn to anticipate their feeding time and may wake up and become more active or restless, or cry or call just prior to feeding time. The growing felids appears to require higher levels of most essential amino acids, as well as more nitrogen from non essential amino acids to compensate for high obligatory nitrogen losses. The felids requires taurine, preformed vitamin A , niacin and long chained fatty acids as a species specific need. The muscle meat should be supplemented with vitamins and minerals, particularly calcium. Twelve gram of calcium carbonate (40 per cent calcium) can be added to 1 kg of muscle meat to provide approximately 1.5 per cent calcium (dry matter base).

Neonatal Examination

Evaluation of vital signs, (temperature, rate and character of respiration and pulse), regional examination of body and major organ systems, weight, sex, state of hydration (mucous membrane, skin fold test), vitality and activeness (response to stimuli and reflexes), palpation and auscultation of chest and abdomen, and observation of congenital abnormalities should be the part of the neonatal examination. Complete blood counts, serum chemistries may be run on as little as 3 ml. of whole blood. Serum gluteraldehydes is a good indicator of nursing. Hypo or a gamma globunemic or hypoglycemic young one should be considered for hand rearing or for the treatment and reintroduction to dam there after.

Nursery Monitoring

For supporting and development of young one, better husbandry approach through vital reference, past records and experiences should be shared among personnel and institutions. Resuscitate and stabilisations of neonates, managing shock if present and solving the emergencies with medical care are the three basic priorities of nursery monitoring. At the time of neonatal examination the navel should be disinfected with a betadine solution and ligated, if necessary. Each cub should be examined for congenital abnormalities including cleft palate. Another optional practice includes administering a prophylactic antibiotic, usually long acting penicilline (0.25-0.5 ml.). Use of colostrum either from domestic goat, mare, and cow should be used in neonates of hoofstock. If a deficiency in passive immunity is suspected, each cub should be given subcutaneous and oral species – specific sterile serum at the rate of 5 to 8 ml /kg subcutaneously for 2days and orally 2 to 5 ml/ feeding for 3 to 5 consecutive days on observing the response. Initially the cubs should receive 10 per cent dextrose for the first three feedings followed by milk replacer. The several milk replacers are available in the market can be selected which can be added with enzyme which break down the lactose thus reduces the digestive upsets (diarrhoea, blood in stool, gastric milk curds). For the first two days, cubs should be fed with a diluted formula in volumes of 8 to 10ml every three to four hours to enhance the appetite and minimise the digestive upsets. During subsequent days, milk volume is increased and frequency decrease, especially in cub continues to grow with weight gain and gastric impaction or distention does not occur. Concentration of milk replacer may be increased stepwise over several days to full strength. The cub should be stimulated to urinate and deficate at each feeding by massaging the ano- genital area with cotton moistened with warm water. If diarrhoea is evident, the formula should be diluted with oral electrolyte solution and the total volume decreased by 20 to 40 per cent for 8 to 12 hours. A stool culture before initiating antibiotics is recommended to check the pathogenic bacteria. If diarrhoea persist, all oral intake should be stopped for 12 to 18 hours and the cub should be supported with sub cutaneous fluids (~40 ml/kg/24 hours period).

Most hand-reared felids begin losing hair at 6 to 8 weeks of age probably because of unknown dietary deficiencies. Adding liver homogenate to the diet has been helpful in preventing and correcting this alopecia. Weaning cubs to solid food also usually enhances hair coat, growth and general appearance. Solid food (meat diet) should be introduced at 4 week of age. Weaning from milk should occur at 10 to 11 weeks of age. Hand raised cubs should be weighed regularly to monitor weight gain and calculate necessary food intake. Growth curves can be compared to available information to indicate the "normality" of hand rearing.

Emergency Care and Management

The initial assessment of cubs follows the ABCs taught in basic CPR (Cardiopulmonary resuscitation): airway, breathing, and circulation. An emergency kit containing basic resuscitation drugs and equipment may be life saving . Traumatic injuries, hypoglycemia and hypothermia should be managed with proper medicines and with fluid therapy. In hypothermia, use of incubators or box or cage with warm hot bottle or heat lamp, diaper bag cans serve the purpose. Extreme caution should be used when warming an animal with a blow dryer, electric blanket, and hot water bath or heat lamp. Moribund neonates can be monitored with electrolytes, fluid, caloric support, antibiotics, shock therapy and symptomatic therapy for any other problem. The respiratory upsets are common which can be handled with above ways and by use of decongestant, nebulisations. The digestive disturbances like parasites, vomitus, failure to pass muconium, bloat, colic and diarrhoea are very common. Use of electrolytes, fluid, caloric support, antibiotics, anthelmintics, gastric decompression for bloat, use of coating substances like kaolin and pectin, use of *lactobacillus* organisms, and symptomatic therapy should be done.

Immunisation

Prophylactic immunisation is good preventive medicine. However, claims of efficacy and safety for commercial vaccines can be made only for those species that have undergone for extensive vaccine testing. As a general rule, inactivated vaccines are safer for use in exotic animals than are modified live virus or bacterial vaccine. Fel-o-vax (Fort Dodge Lab.Inc.Fort Dodge, IA 50501) provides adequate antibody titres to feline panleukopenia, feline herpes virus and calicy virus diseases. This can be use at a 1-ml dose in adult felids. Cubs are vaccinated every 2 weeks from 8 to 16 weeks of age, with a booster vaccination given at 40 weeks to ensure that protective titres are maintained during the first year of life. A single annual vaccination appears to be adequate to maintain protective antibody titres in adults. The use of rabies vaccines depends upon local conditions. The serological periodical monitoring and checking of diseases of all felids are essential to provide sound data for a longitudinal disease prevention programme. The endo and ecto parasites should also be monitored and regular use of safe and effective anthelmintics and pest control drugs should be used. The sanitation and hygiene of premises and enclosures is necessary and should be planned out with basic sanitation rules. Phenolic compounds should be avoided while dealing with felid enclosure disinfection.

HAND-REARING APPROCHES AND MAJOR DISEASES

1. **Young Primates :** Natural rearing by the mother is preferable even though as mortality rates indicate-natural rearing in apes is, more critical than hand-rearing. Depending on the mother's behaviour towards the young within the first few days, her milk supply, attitude to handling and nursing her young, one must decide whether or not to remove the neonate. Risks encountered in natural rearing include infestation by gastrointestinal parasites, infectious disease, falls and injury. Direct feeding is not practical for apes or lower primates and besides the natural milk the only food available to the young is whatever they can reach with their own hands, but tasty treats are most often eaten by the mother. If deficiency diseases in young animals are to be avoided, it is necessary to provide abundant food for the mothers to assure sufficient leftovers for the young. Vitamins and solid food should possibly be given to the young directly.

   Numerous and comprehensive papers on the problems in hand-rearing and mental and physical progress in young apes have been published. Generally speaking, experience and methods in paediatrics are extensively used in the nutrition and medical care of apes and lower primates. Commercial baby foods can be used during first few months of hand rearing. Whipped banana or grated apple can readily be spoon-fed from the fourth week on. From the tenth week mild is gradually thickened with pablum or cheese and is fed with various types of commercial baby foods. Vitamins D3 drops are daily supplement to prevent rickets. Vitamin C drops and vitamin B12 are usually given from the sixth week up. Multivitamins and conditioning products are provided later on a regular basis. UV radiation ranging from 3-10 minutes is recommended once weekly.

   Iron deficiency anaemia clinically indicated in young apes by weakness and pale mucus membranes, is treated with iron that is supplied by way of the milk. Muscle spasms and tremors occasionally observed during hand rearing of young apes can be counteracted by high parenteral doses of Neurobion.

   Indigestion with diarrhoea during the first months of life in conjunction with teething problems occurs frequently. It is serious and requires appropriate attention. One of the most important therapeutic measures is a dietary program. Fennel or light black tea is five to the sick baby ape within the first 8-12 hours with artificial sweetener or in very weak animals. This diet can eventually be changed to baby diet food. In severe cases of diarrhoea 8% rice pablum or in somewhat older babies carrot soup and whipped banana are of value. Body warm faecal samples should immediately be submitted for bacteriological and parasitological examination. Infections are frequently caused by serous pathogens and antibiotic treatment should not be delayed. Oral antibiotic with adequate fluid therapy should be employed.

Infections of the upper respiratory system can be serious. Increased production of phlegm with coughing should be taken care with suitable therapy.

Great Apes: Septic arthritis occurs with some frequency in juvenile orang-utan as a result of septicaemia caused by umbilical and enteric infection. Bacteria pneumonia occurs commonly in juvenile animals as a sequela to human viral pathogens, such as rhinovirus, respiratory syuncytial virus, and influenza, bacterial pneumonias are caused by enteric organisms (e.g. Klebsella, Pseudomonas, Acinetobacter). Haemophilus influenza may cause primary pneumonia with subsequent death, as in children. Bacterial meningitis & miningoencephalitis have been seen in neonates of all species of great apes. These neonatal infections may result from the extension of ootitis and ethmoiditis.

Congenital or developmental anomalies: Mental retardation is perhaps more noticeable in affected great apes that in other species because of their higher mental functions. Mental retardation has been seen in orang-utans, possibly because of parental hypoxia or genetic defects. Perinatal hypoxia may occur, following deficient maternal behaviour, leading to hypothermia and hypoglycaemia. Computed tomography may be the diagnostic tool for documenting morphological changes in the brain.

Parasitic diseases: Sytrongylodiasis, caused by *storngyloides stercoralis* occur. Found in orang-utan, infection through milk or food. In young orang it causes vomition, diarrhoea leading to severe dehydration. In chronic case diarrhoea, may not be a feature of the disease but progressive weight loss and weakness may be the only indication. Treatment at this stage is often unsuccessful. Death is usually a result of pneumonia and peritonitis. Some of primate (rhesus) species have shown evidence of Hepatitis A virus. Bacterial Endocarditis in non-human primates in large primate colonies, approximately 1/3 of deaths are related to gastrointestinal diseases. Enterocolitis may account for 25% of infant rhesus monkey mortality, although it is much less frequent in squirrel monkey and other New world monkey infants. Bacteria associated with diarrhoeal disease in nonhuman primates include Shigella Sp., Salmonella Sp., Campylobacter Sp., Yersinia pseudotuberculoses, *Y.enterocolotica*, Aeromonas sp., *Mycobacterium avium* and enteropathogenic serotypes of *Escherichia coli*, Campylobacter and Shigella are the most frequent isolated pathogens.

Protozoal parasites of great apes: *Balantidium coli* may cause syndromes ranging form loose stools to watery diarrhoea without blood or mucous to dysentery. It is important to rule out other potential primary bacterial pathogens and some helminthic parasites such as Strongyloids stercorlis and Trichuris trichuris occasionally, *B.coli* may appear to be a primary cause of diarrhoea. Disease caused by B.coli may be self-limited and resolve without treatment or after appropriate treatment of primary bacterial or helminthic pathogens. Gorilla seems to be more susceptible to serious infections than other great apes.

2. **Carnivores :** Severe accumulation of ticks may cause anaemia in young animals. ticks may cause anaemia in young animals. Ticks may also transmit infections agents that cause such disease as piroplasmosis, borreliosis, talaremia ricketsiosis, and tick borne encephalitis.

   Demodex species the infection is transmitted form lactating female to young one, usually without causing clinical signs. Clinical disease is usually preceded by our immunodeficiency and is then characterised by erythema and alopecia, primarily in the skin of the head. Sarcoptic mange has been reported in canids, felids, mustelids and ursids.

   Isospora species (coccidia) and intestinal protozoan that cause clinical disease in young captive felids and canids. Clinical signs are watery diarrhoea, bloody faeces anaemia and emaciation.

   Toxoplasmosis caused by *T.gondi* was reported from Rostock zoo, where 16 to 20 deaths in 4 to 7 months old Kodiac bears were attributed to acute to sub acute septicaemia. Generalised toxoplasmosis with necrosis of the liver and pancreas was fond in a 6 day of polar bear cub. Infection of toxoscaris leonine, T. cati, Teania sp. and Anchylostoma have been observed in Asian lions (personal observation). In the wild, ascarid infections are most important in young animals. Young ones of canids felids and ursids are susceptible for Ascaris infestation.

   The infective eggs of *T. canis* and *T. cati* hatch in the carnivore stomach. The larvae invade the intestinal wall and within days, arrive in the pulmonary capillaries, tracheal migration of T.canis has been observed only in young carnivores. Trichuris vulpus in canids, felids and viverids may cause clinical disease in young animals.

   The resistance of neonatal canines depends largely on the presence of immunoglubulin in the young. Since canines have no diaplacental transmission of immunoglobulins it is vitally important for the neonate to receive colostrums. If neonates have to be removed from the mother before receiving colostrums, a domestic foster bitch, which has just given birth should be used as a substitute. Solely hand-reared pups must be bottled fed every 2-3 hours for the first 10 days. Suitable diets for this purpose are commercially available. For home make milk replacers the exact composition of the particular milk must be taken into consideration. Canine milk is rich in protein and fat. The carbohydrate part ranges form 3-4 %. For the prevention of rickets it is advisable to supplement vitamins and minerals and to provide UV radiation of 3 minutes daily from the sixth day on. The great susceptibility of canines to parasites makes an early faecal examination mandatory. Deworming can be initiated from the eighth day if necessary.

3. **Young Felines :** Good breeding results and healthy offspring is the aspiration of every person involved in zoo management. There has been continuous improvement throughout better knowledge of social behaviour, nutrition and health maintenance. Trying by foster dogs is an alternative if natural rearing is not possible due aberrant behaviour of the mother, lack of interest in the young, cannibalism, agalactia, insufficient

milk or other reasons. Best results in hand rearing were obtained by using Esbilac. This product contains crude protein 33 %, crude fat 40%, Mineral salts 6 %, various vitamins etc. The manufacturer recommends that the product be given in increasing concentration.

Additional feeding including minced meat and live by teaspoon between the second and eighth week is of value. Additional feeding at an earlier or later time has not been advantageous. The milk formula should be given at lest until the third or fourth month.

Hand rearing of large felines has proven successful by following these recommendations:

Food is first offered 12 hours after birth since the meconium is shed at this time.

Esbilac dissolved in boiled water is provided every 2-3 hours from 8:00 hours until midnight. The amount given is approximately 10 % of the cub's weight. A cub weighing about 5oo grams would thus receive 50 ml of Esbilac daily.

The milk must be at body temperature and must be kept at this temperature while being served.

Initially a nipple is used with a very small opening, which is gradually enlarged. Fast drinking caused bloating by swallowing excessive air, which in turn prevents the cub from 'playing with its front paws on the mammary system'. It is therefore important to hold the neonate in a physiological position with regard to the mammary gland. The neonate should have the opportunity to either cling to or push itself of the gland.

The daily quantity of Esbilac can be fully prepared and refrigerated. Each feeding requires a fresh, sterilized nipple and fresh, sterilized bottle.

Overfeeding should be avoided. The cub will show contentedness.

Daily weight control indicated milk quantity to be provided. Birth weight has usually doubled by the end of the second week.

Massaging the belly after each meal enhances digestion. The anogenital region is to be lightly massaged three times daily with a finger, which has been moistened in warm water. This usually triggers urination while defecation generally occurs only once daily.

Soft faeces or diarrhoea is first counteracted by feeding at shorter intervals (every 1-2 hours) with the daily quantity being unchanged.

Should this method fail, an antibiotic is added to the mild for 2-3 days.

The optimal ambient temperature for hand rearing of large felines is 29 ^{0}C during the first week, then is gradually lowered to room temperature by the fourth week. This can easily be accomplished by varying the distance form an infrared light.

The success of all these measures listed will depend entirely on the availability of skilled, patient and conscientious keepers.

Hand-reared cubs that may have missed colostrums should receive gamma globulin in the first 24 hours. Similarly cubs that have to be removed form the mother within the first few days should also receive gamma globulins. These prophylactic measures are maintained for all possibly arising disorders in cubs up to 4 weeks of age.

The occasional umbilical infection in cubs of large felines is characterized by redness, swelling and sometimes moisture in this region. The abdominal muscles appear to be tense accompanied by laboured breathing and diarrhoea. The animals become reluctant to nurse and often separate from the others. Treatment includes topical circumscribed injection of the umbilical region with penicillin streptomycin suspension, occasional additional antibiotics and spasmolytics.

We have not observed tuberculosis in the last 10 years, and prophylactic measures therefore have not been undertaken. If exposure occurs, BCG vaccination could be carried out at about 4 weeks of age (2 intradermal depots with 0.1 ml each).

A report entitled "Struma bei Lowen" (Goitre in lions) descries a disease in 10 young lions characterized by slow growth, short legs, relatively large heads and potbellies. The thyroid gland of a 5 month old animal was the size of a man's fist. Treatment was successful with the daily oral administration of 10-15 drops of Lugol's solution per animal over a period of 2-4 weeks.

4. **Young Bears :** Animals rejected by their mother are particularly vulnerable. Several hours may pass from the time of parturition untill rejection of the infant is recognized by the keeper and the unprotected, helpless young is subject to exposure. This may lead to subsequent respiratory noises and dysphagia. Advanced cases display serous and turbid nasal discharge. Prognosis of bronchopneumonia and pneumonia is unfavourable. Alleviation of respiratory distress and provision of oxygen should be the primary treatment. Administration of antibiotics over a period of 4-5 days is of value. Hand rearing is not particularly difficult once the young have accepted the bottle

5. **Young Elephants :** Colostrums produced within the first 10 days have a high protein content (up to 5%) and low fat content (around 2.5%). As secretion of mild continues the fat rises form 5.5-8.5% and peaks towards the end of the lactation period, after approximately 18 months, reaching values of 12-15% and occasionally as high as 20 %. The protein level is around 5% and lactose ranges from 3.5-6%. The composition of minerals resembles that of cow's mild. The high level of vitamin C is noteworthy at 77 mg%.

   For rearing of orphan elephants a number of mild formulas, which are close to elephant's milk, have been studied. However, rearing was even successful with regular, pasteurised cow's mild and with milk for children by Nestle with the addition of mashed banana and rice pablum right form the first week on. Cleanliness of all implements used is essential for a successful operation. Failure in rearing of many young orphan elephants in Africa is likely to be blamed on lack of cleanliness during feeding. One baby elephant fed on milk containing 15% fat experienced diarrhoea and skin rashes, which healed up immediately after reducing the fat to 5%. Another young elephant had severe stomatitis pronounced pruritus and yellow, pox like skin erosions which broke open after some weeks but the condition improved after correction of the diet.

6. **Young Equnes :** Occasionally hand rearing becomes necessary after loss of the mare. The proper composition of the milk as well as quantity, frequency and duration of suckling are important for successful rearing. A major problem is the lack of adequate colostrums. Frozen colostrums is usually unavailable and blood collection for serum transfusion is often not feasible. The alternative in hand rearing is to modify cow's milk prior to feeding to meet the requirements.

   Mare's milk contains 65% casein compared to 85% in the bovine. Untreated cow's milk forms clumps rather than a thin, flaky consistency in the foal's stomach and results in digestive disorders. The formula for modifying cow's milk is as follows: milk which has been allowed to stand for a while is slightly akimmed. To 1 litre of milk ass 3.5 ml of lactic acid or 0.4 grams of citric acid which is first dissolved in a spoon and slowly added to the mild; 0.2 litre of water to reduce the casein content and 20 grams of lactose. Saccharose from sugarcane or beets is unsuitable during the first 4 week since saccharose is only gradually formed in the neonatal animal. Vitamin A is low in cow's milk and must be supplemented. A foal suckles approx. 60-70 times during a 24-hour period, which means that at least 12-15 meals should be made available during this period. When the foal of a mountain zebra was hand reared, it was bottled fed three times during the first 6 nights, then twice and after 24 days only once. The maximum amount during the first few days was 180ml/meal, which was increased to 200 ml after 10 days, and was continuously increased until the 100th day when 400 ml were fed. The total amount consumed on the first day was approx. 600 ml, from day 2 to 24/3000 ml, and from day 25 to 100/4000-4500 ml daily. Animals reared without colostrums should be vaccinated against infection with Shigella equirulis (A.equuli). During the first few days attention must be paid to the presence of meconium.

7. **Young Rhinoceroses :** Breeding of rhinoceroses in captivity is a relatively recent achievement and consequently little information is available on hand rearing. Some cases are known, however, where young rhinoceroses wee caught in the wild and occasionally even reared. It is our opinion that the composition of the milk as we know it from the black and white rhinoceroses may be less critical than hygienic measures during feeding. Soiled milk immediately becomes "bacterial soup" and causes serious digestive problems. A young Indian rhinoceros was raised without any complications on pasteurised, homogenised cow's milk as marketed for human consumption. However, it receive colostrums form its mother for a total of 24 hours. The sterilized milk was fed to the young animal directly from the commercial bottle utilizing a sterile nipple. The consumption increased from the daily total of 2-9 litres to 20 litres by the time the young was 1 month old. At 6 weeks the animal began to accept greens and concentrate. Milk was completely eliminated from the diet at 9 months.

Fatal Salmonella septicaemia has been described in a juvenile southern white rhinoceros. It showed signs of disseminated intravascular coagulation, presumably secondary to endotoxemia.

Enteritis has also been associated with infection with *Pseudomonas pyocyanea* and coliform bacteria in young hand reared rhino. *Yersinia pseudotuberculosis* has been identified as the cause of enteritis and enlarged mesenteric lymph nodes in young rhinoceroses. Ompheloplebitis, in one incident associated with a valular endocarditis caused by E.coli was observed in captive white rhinoceros neonates.

Sarcoptes tapiri is a potentially pathogenic parasite of tapirs. In young animals i9nfectooon are severe and may become secondarily infected. Lesions generally begin on the belly and inner limbs and progress to the sides of the tarso and back, becoming large, erythemetous hairless patches, with skin thickening and curst formation. Purities may be severe and can lead to secondary trauma. The death of a young captive greater one horned rhinoceros have been attributed to F. gigantica.

8. **Young Hippopotamuses :** Hand rearing of hippopotamuses has repeatedly been attempted, but has rarely been successful rearing was achieved at many zoos. The young was isolated from the mother immediately after parturition and was then raised on goat's milk. Leipzing zoo succeeded in hand rearing a young hippopotamus to the age of 7 weeks. The initial diet consisted of two third of goat's mild with the addition of vitamin A and D. Bottle-feeding, however, created difficulties and the method was changed to feeding from a crock. The attendant stuck two fingers into the animals mount to facilitate drinking form the crock that worked perfectly after a while. The daily quantity supplied increased form ½ kg to 8-½ kg and decreased to 3-2 kg during episode of diarrhoea. Diarrhoea began after 4 days and lasted for about 10 days. The diet was changed form cream and milk to cream and tea and suitable treatment was employed to cure diarrhoea. After slight improvement diarrhoea recurred 3 weeks later with rapid deterioration of the animal and finally death. The cause of death could not be clearly established at necropsy, but was believed to have been due to circulatory failure. Not to be ruled out is the possibility that the diet of cream, milk and tea may have been inadequate resulting in debility.

   Several reports describe losses of newborn. Necropsies indicate they died of drowning, injury to the head and neck, and starvation. A pigmy hippopotamus died after 4 weeks. Based on observations made in the zoos at Basle, Washington and Zurich where a total of seven pigmy hippopotamus had lost their first young, it was concluded that the first young born does not seem to survive.

   Two hippopotamuses died at the age of 6 and 8 months. Subcutaneous abscesses and broken ribs were found at necropsy and were thought to be due to crushing and butting by the mother.

9. **Young Cervids :** Hand rearing of cervids is generally not difficult since mother's milk can readily be replaced by goat's milk. Young animals often die form starvation. Neonates tend to lie away form the mother which makes observation of suckling

difficult. They may only suckle when not being watched. Mortality in the young can be reduced by providing vitamins, antibiotics or even serum. Supplementation of mother's milk 2-3 times daily during first few days, especially in reindeer is of value.

OVER ALL MANAGEMENT AND HAND REARING OF ORPHANED ANIMALS

Hand raising of orphaned wild young one may be necessary when the natural raising, process fails. Because of many practical difficulties in hand rearing and development of in appropriate behavioural trait in its relationship with animals of its own kind due to close association with its human foster parents, the decision to put the animal on hand rearing should be made after critical assessment of the situation.

Whenever the situation has aroused for hand rearing, following primary steps should be taken. Bath the animal if it is dirty in warm water, dry it with clean towel and put it in a confined, dry place. Provide thick dry bedding and an appropriate heat source. Provide sufficient space to enable the animal to move away or closure to the heat source. The temperature should be around 30-32 °c during 1 st week of age; which should be adjusted downwards to around 24-26 °c at the age of 4th week.

If there is dehydration, Ringer's lactate solution or glucose/saline can be used depending upon the state of animal. Fluids can be administered subcutaneously or intraperitoneally if the animal does not take fluid orally. Total of up to 40 ml/kg may be given at a time.

Thorough examination of the animal for any physical injury, physiological status, disease etc. should be made and suitable measures should be makes to correct it.

When the condition of an animal is stable and the decision has been made to undertake the task of hand raising the animal, following may be used as guidelines for further care and management of the orphaned animal.

Selection of feeding equipments depends on the species, the size of the individuals and the expected volume of mild per feed. In a smaller species dolls bottles are suitable, 20-30 ml capacity bottles (per mature human baby bottles) may be suitable for the species up to 500 g and standard baby's bottles can be used for other species. Selectio of teats is also important and the opening should be just sufficient which gives adequate mild without causing strain to the animal or aspiration in lungs. in some cases feeding tubes may be required. All equipments must be sterilised properly and leaking of teats should be prevented.

Colostrum feeding during first 12-24 hours, after birth is essential for transfer of maternal immunity. if colostrums is not received, collect 30-50 ml. of blood from an adult animal of the same species allow blood to clot, harvest serum after centrifuge, inject 5-10 ml subcutaneously and remaining serum is given in first feed. Repeat if possible after 12 hours.

First feed after birth should be of electrolytes and glucose. Use a glucose/saline infusion or oral electrolytes solution. If young one is not in a position to suckle, then do not force feed, use dropper until strong enough to suckle.

Nasogastric tube usually well tolerated. Use local anaesthetic 1-2 drops only in nostril or apply anaesthetic cream on tip of tube. Insert in ventral part of nostril, lift the head and pass the tube gently into pharynx, wait for the animal to swallow and pass the tube gently into the stomach. A small amount of saline injected in to the tube will indicate if it is in the trachea, inject milk mixture slowly to avoid regurgitation and do not exceed 20-30 ml/kg per feed. The maximum volume of the stomach in new born animal is approximately 50 ml/kg body weight.

Table 1: Showing normal milk value of some animals:

Species	Total solids g/100 ml	Fat g/100 ml	Protein g/100 ml	Lactose g/100 ml	Ash g/100 ml
Cow	12.7	3.7	3.4	4.8	0.7
Jackal	22.00	10.5	10.0	3.0	1.2
Leopard	19.4	6.5	11.1	4.2	0.8
Lion	30.2	17.5	9.3	3.4	-
Antelope	-	9.2	8.1	Carbohydrate2.4	-

Looking to above mentioned milk values it is obvious that cow's mild alone may not be suitable for hand rearing of many other species. Bellow given is a table showing suitable milk substitute for some species.

Table 2: Mild substitute suitable for raising wild carnivores:

Formula I		Formula II	
Cow's milk	200 ml	Cows' milk	210 ml
Cream	100 ml	Egg yolk	15 ml
Calcium caseinate	20 g	Multivitamin drops	2 drops
Multivitamin drops	2 drops		

Suitable milk substitute for raising herbivores : If cow's milk is used as a base, it should be enriched to increase the protein and fat content. Cow's milk can be enriched adding 65 ml of fresh dairy cream and 10 ml of egg yolk to 1000 ml of milk.

No supplementation of vitamins and calcium is require for herbivores unless the deficiency is manifested clinically.

Volume of Feed : Herbivores: The total amount of mild substitute should be equivalent to 10-15 % of body mass, divided for 406 feeds in 24 hours.

Carnivores : Total volume of mild fed varies from 15-25% of the body mass per day.

Feeding Posture : Hold the animal under its chest, with its head lifted slightly, and place the teat in to its mount. Assess the animals ability to suck and adjust the teat opening if necessary.

Anogenital stimulation : Young animals will not defecate or urinate voluntarily until approximately four weeks of age. Excretion must be stimulated after each feed by gentle massage using warm, damp cloth or cotton wool for several minutes until defecation and /or urination occurs. Gentle rubbing of the lower back or stomach works for some other species, suitable stimulus should be found and used.

Weaning : Milk mixture should be fed to an orphan animal until estimated time of natural weaning slowly the animal should be taught to accept solids along with milk mixture and solids can be gradually increased. Vitamin A and calcium supplementation is extremely important factors during hand rearing of carnivores. Meat must be balanced with extra calcium; add 2.5 g calcium carbonate for every 500 g of meat in feed. Herbivores usually does not require vitamin and mineral supplements.

Clinical problem of hand rearing :

1. Diarrhoea : Over feeding, bacterial infection incorrect formula may feed to diarrhoea. If diarrhoea persists, replace the formula with a limewater and saline mixture for 24 hours. After 24 hours start including the formula in a mixture with saline and limewater. Take the next 36 hours to get the animal back on to the full formula. In severe cases antibiotic can be given. If severe dehydration is noticed subcutaneous fluid therapy may be indicated.
2. Constipation : Adjust the formula, decrease the amount of supplementary protein in the formula and add glucose to the formula. Administer a few drops of liquid paraffin if the above changes do not result in an improvement. Saline enema may be given in severe cases.
3. Pneumonia : Aspiration pneumonia occurs if the animal is weak and does not suckle well, or if the opening is too big. Distension of abdomen, constipation, increase respiration rate, excessive rib cage movement, and high temperature are common manifestation. Such cases should be treated aggressively with antibiotics, fluid therapy may be given if dehydration noticed. Other clinical problems are prolapsed rectum, tympany, rumen dysfunction, colic, naval ill etc, which should be corrected if present.

Importance of Record keeping in hand rearing : Details record about date, age in days, body mass, feed intke per feed, feeding time, urination, fefaecation, any other observation make, change in milk mixture. Such information is very useful for successful rearing of animal.

CHAPTER 25

DISASTER AND VETERINARY MANAGEMENT PLAN

The natural or manmade calamities either due to earthquake, flood, cyclone, drought /famine, cold famine, land slides, fire, industries and technologies, pollutions, gas release, epidemics are some of the examples. The impact of disaster is obviously high and the vulnerability to the animals has been not much highlighted since long. Looking to the present scenario of Tsunami in Japan (March, 2011) and the heavy rains in Gujarat and Mumbai (July-August, 2005, 2010) and hurricane, Katrina, Rita in different States of America sensitized the human being to think about the calamities ,its impact and planning to avoid, reduce the after effect. Many Developing countries are largely disaster prone. The peninsular Indian states (22 out of 32 states and Union Territories) are vulnerable to some disaster or other. The cyclone in the state of Andhra Pradesh, of November 14 to 20, 1977, caused an estimated loss of 2, 30,146 cattle and 3,44,056 other livestock, as against 8,515 human deaths. Similarly in Orrisa cyclone (in India) during the 4th of June, 1982 there had been 11,468 cattle lost against 243 human deaths. (Loss of other animal is not being projected). Earthquake has comparatively less impact on animals; but in Uttarkashi earthquake (of India), 3100 cattle heads were lost as against 770 human lives. Between 1953-1990 we lost 1,02,905 cattle against 1532 human lives because of earthquakes. It is apparent that even though animals are the main source of livelihood to the poorest of the poor and to the landless, concrete steps towards disaster management of livestock and other animals are yet to be taken.

I. PREPADENESS :

A. Event Preparedness : It should contain a checklist of basic material and expertise required. For instance, a veterinary surgeon could provide inputs on trauma injuries that might occur in an earthquake, and a list of basic first aid equipment could be provided for reference.

B. Institutional preparedness : The following aspects are important under institutional preparedness:

Hazard identification

Listing of potential impacts, outlining methods of mitigation, and control, using professional expertise.

Drawing up of specific action plans

Rehearsals of plans

II. DISASTER MANAGEMENT PLAN FOR ANIMALS : The most important features of such plans would be:

1. A clear, valid and viable outline of specific emergency action necessary in specific situations (related to specific hazards), along with the designation of a coordinator, or lead agency.
2. Documentation of material and human resources required and accessible during emergency situations.

III. OUTLINE : A basic outline of disaster management plan for zoo animals should essentially include:

1. Retrospective epidemiological survey study : Retrospective epidemiological study of the disasters in the area and this shall include,

(a) Data Collected interpreted & analysed (i.e. information), on the basis of which some prediction can be made.

(b) Disaster Vignetting: is a means by which mapping is done on the basis of incidence frequency magnitude, epicenter, vulnerable areas.

(c) Zoo profile: the total captive animal population (number), vulnerable animal population as per their species, age, sex and physiological status etc.

(d) Community Profile, the total zoo staff population, their age, sex, socio-economic status, education, cultural distribution etc.

(e) Animals at risk; the nature of hazard, the intensity of impact and mortality rate (immediate or delayed).

(f) Risk factor analyses: is the analysis of type of risk (identification & analyses)

2. Management and Action Plan : Since following disaster, animals are to be rescued and collected in relief camps (a suitable space or alternative area), the immediate priority would be controlling and combating disease. The animal health component of disaster mitigation will include

(a) Promotional and priority health care such as nutrition, pregnant animal care, care of new-born and young animal etc.

(b) Prevention of risk is through removal of waste disposal, hygiene, vaccination, pest/ vector, control, sanitation etc.

(c) Specific therapy by way of early diagnosis and treatment.

(d) Rehabilitation: help animals to recover from any trauma/stress or fear.

(e) Disposal of dead animals: Carcass utilisation is one method. Many animals in which treatment is unlikely to be beneficial may have to be plan looking to the condition of animal and schedule to put to sleep i.e. Euthanasia ("mercy killing")

3. **Resources Planning :**
   (a) Assessment of available man power i.e. Veterinary doctors, Para veterinary staff, ancillary staff, voluntaries.
   (b) Store and equipment include the medicine, surgical and essential medical appliances, diagnostics, life saving equipment etc.
   (c) Logistical needs: that is the need for zoo animal food, water, fuels, lighting equipment, tents, sheds, grass bedding, trolleys, material for sanitation, storage of feed and fodder and water for subsequent days.
   (d) Ambulance and out reach facility:- transporting animals is very cumbersome when they are sick, injured and non-ambulatory.
   (e) Veterinary medical facilities as veterinary hospital, mobile veterinary units etc.
4. **Planning of Training :**
   (a) Training veterinary personnel, para veterinary staff, zoo keepers, attendants etc.
   (b) Training administrators like T/B.D.O., telephones & Fire Service personnel, Civil defense, personnel, Collectors, Municipal/ Forest officers, administrators
   (c) Animal Health awareness for trainees such as social workers, volunteers (NGO).
5. **Allied Supportive Planning (Organisation) :**
   (a) Augmenting political and administrative support
   (b) Involving N.G.O.'s, media, animal welfare organisations and other volunteer groups.
   (c) Eliciting commitment and allocation of funds
   (d) Formation of veterinary service groups at State and Central Government level.
   (e) Organisation of district/ area level bodies, assigning specific tasks and responsibilities to State A.H. department.
   (f) Establishing communication channels, alternate channels like ham radios.
   (g) Establishing alternate source of power, energy etc.
   (h) Plan for monitoring and supervision.
   (i) Publicity and public relation activities (Vet-PR)
   (j) Plan for mitigation and rehabilitation of animal owners along with animals
6. **Execution of short & long term plan :** Short term and long term plan implementation includes estimation of animals involved, damage assessment by local and central level committee. One may make use of media to help locating remote areas where often organizers' attention may not easily reach. Monitoring is part of implementation.

7. **Impact assessment :** Experience gained from similar hazards of nearby areas is positively helpful. This is especially useful in animal disaster management which is still in its infancy.
8. **Post execution evaluation of plan :** Follow-up of disaster management especially that of animal management are valuable experience that need be shared among disaster prone areas. Unfortunately this is not forthcoming.
9. **Linking with allied agencies for uniformity in all regions :** It is a part of disaster management programme as unbiased and uniform measures help transparency and hence would instill confidence of the community in the system.

IV. BASIC COMPONENTS OF A VETERINARY DISASTER MANAGEMENT FACILITY :

1. **Campsite Vety. Hospitals :** The equipment and other infrastructure are provided along with stock pile of emergency, equipment and drugs. All veterinary personnel may not be used at a time. It would be important to plan shifts to avoid mental and physical fatigue. Early phases would need more work force and in all departments of activity. Supervision and monitoring would be important and demands all qualities of leadership (like confidence, competence, experience, patience, communication skill etc.). As search and rescue would be over, relief and rehabilitation would acquire importance.
2. **Recovery and Rehabilitation :** The impact of disaster being calamitous one may come across a number of cases which are beyond mere physical injuries or contagious disease. The mental trauma due to the loss of a sibling or a partner can be encountered and will have to be handled at appropriate stage. Many of the cases of extensive trauma that cannot be handled may have to be put to sleep (mercy killing). Doing so under full public view where the public is already surcharged with losses of life and property may set in depression or even public outcry. Overenthusiastic and ill trained animal activists often can blow things out of proportion, specially because in many parts of India religious sentiments are attached with some animals.
3. **Control Rooms :** For the information exchange and co-ordination of veterinary support, control rooms are used. These handle co-ordination information among directorate of Animal Husbandry, district Vet. hospitals, rural hospitals of the affected area etc. Control rooms keep link with and co-ordinate supplies from agencies. The control room would be working on feed back from affected area on the extent and nature of emergency, the need for special equipment, emergency equipment (pain killer, sedatives, antibiotics, fracture equipments etc.) The roles of a veterinary public relation man (Vet PR) assume significance.
4. **Control of epidemic/diseases :** Immediate task after the first emergency, treatments are completed would be taking care of diseases that may spread. Armed with retrospective epidemiology, authorities would identify emergency diseases and sanitation measures. This would include digging manure pits, drainage facilities (more important in floods, cyclone etc.), drinking water and water troughs.

This will have to be simultaneously established in hospitals, temporary shelter camps as well as in the affected areas from time to time and as per priority.

For health care operations there is a need for discipline; public co-operation is also essential. Animal welfare agencies working in the areas can help arranging stockpiles. Media can help by providing feed backs from field.

5. **SWOT Approach :** **S**trength, **W**eakness, **O**bjectives/outcome and **T**hreats of planning such programme should be outlined. Some factors that may hamper health & Veterinary support are,
   (i) Lack of communication among officials.
   (ii) Shortage of space for operating various activities like laboratory, Medical store etc.
   (iii) Uncontrolled demand for additional materials (this would be more if it is felt that any material is in short supply)
   (iv) Improper or inefficient materials like equipments that do not operate on battery.
   (v) Lack of staff trained (Veterinary extension and Public relation) to handle anxiety.
   (vi) Lack of after-care facilities.
   (vii) Lapses in co-ordination, duplication.
   (viii) Absence of overall training for disaster management.
6. **Post-Disaster phase and Reconstruction :** Apart from reconstruction of the damaged or lost veterinary and health facilities one may have learned from the experience of disaster handling, the pitfalls in the zoo husbandry and veterinary services. This experience would help during rebuilding some of the damaged or lost facilities. Some equipments or medicaments would also get enlisted as per priority.
7. **Role of old structures :** With the sentiments of Indian community to animals, there are already many zoos established by Nawabs or Maharajas with past experience and local need base thought. Vadodara zoo having the central elevated area with tree plantation is the best and life saving example of open moated system for herbivorous which saved many lives of deer's and antelopes in the recent flood of July 2005 in Gujarat.
8. **Carcass disposal :** The methods of animal waste disposal/ recycling vary with animals. Improper disposal can enhance pest or vector problems. Preparation of compost or digging the manure pit be considered out side the zoo area. Manure pits should be layered with dry waste and lime regularly. During prolonged stagnation of flood water, duck rearing and fish farming can be considered as the means of pest control as naturalistic way but not suitable for zoos. The disposal of dead animals poses acute problems during floods and cyclone, as the number of animals dying would be enormous.

V. ACTION-ORIENTED RECOMMENDATIONS:

1. **Instant availabilities and awareness of SOS Dockets :**

   (a) In most of the zoos the ready reckners for disaster management plans, dos and don't does list of emergency items, alternatives, guidelines, materials, records, resource persons, keys storage and important telephone numbers, addresses of important institutes, associations should be handy and in a written form.

   (b) They should be kept in a safe, fireproof, quickly accessible place with other important documents and taken along if it becomes necessary to evacuate the farm.

   (c) Each member of the farm family and herd personnel should know of and practice the plan so that action may be taken even in the absence of key management personnel.

2. **Creating a Reference Document/Manual :** It is possible to tap the expertise of veterinary professionals to create a document that would contain information on:

   (a) Hazard identification in different situations, e.g. research institution, zoo, etc.

   (b) Potential impact on Animals in a specific situation e.g. epidemic, fire, floods

   (c) Veterinary aspects of these impacts from the point of view of a surgeon, epidemiologist, animal nutrition expert, ethologist, and zoologist expert.

3. **Application oriented and Instant/on line(Internet) information :** Information on disaster preparedness should be more application-oriented, with inputs from experts. If a sustained flow of information is available, it may be possible to consider publishing a regular column. The online resources, list of important web sites and addresses should be known by the zoo officers. It could also provide the much-needed forum for victims/eye-witnesses and workers to discuss real-life situations, and provide information on the practical problems encountered by citing examples and problem faced by the zoo persons.

CHAPTER 26

SURGICAL PROBLEMS AND INTERVENTION

Like domestic animals wild animals (in captivity and free range) also often affected with surgical diseases and they require surgical interventions. Some of these diseases are categorized as general surgical affections like traumatic injuries (laceration, gunshot wound and horn-gore wound, fracture, sinus and fistula), punctured wound, abscess, thermal injuries (burn and scald). Sometimes surgeries are considered necessary as elective options (e.g. onychectomy, debeaking, vasectomy, castration, laparoscopic sex determination in birds etc.) from management point of views. Species wise common surgical problems are listed below :

ELEPHANTS

Abscesses

The thickness and toughness of elephant skin contribute to the very common problem of subcutaneous abscess formation. Abscesses can form under the skin as a result of contusions, chafing, wounds, parasitic invasion, and general debility. Owing to the thickness of the skin abscesses may go undetected for long periods of time, to the point that they either has become quite large of have spread and undermined a significant area of skin. Because abscesses tend to spread beneath the skin rather than come to a head and rupture externally, serious sequelae can occur. Work elephants that develop abscesses over the spine have had dorsal spinous processes of vertebrae eroded owing to the spreading nature of the abscesses. Diagnosis of abscess formation is made by observation of an initial hot, hard swelling or later of a fluctuant swelling underneath the skin. Treatment consists of surgical exploration and complete drainage of the abscess followed bay flushing with an antibiotic and antiseptic solution. Large abscesses may be packed with gauze that is soaked with an antiseptic or antibiotic, which is gradually withdrawn as the abscess heals. Hot compresse may help bring an abscess to a head for drainage.

Trauma

Shallow skin wounds and abrasions usually heal rapidly in elephants with little or no treatment. Deep wounds may form abscesses secondarily and should be vigorously flushed with antibiotic or antiseptic solution; if the wounds do not penetrate the entire dermis, they may be treated with topical antiseptic or antibiotic ointment.

Ulceration

Skin ulceration can occur as a sequelae to abrasions or superficial wounds that have been three types of ulcers in the skin of elephants have been described: granulating, spreading, or stationary. Treatment of ulcers consists of debridement of the outer surfaces and application of an antibiotic ointment. It has been reported that situations that create great anxiety or stress in elephants can cause the rapid (e.g., overnight) formation of pockets of ventral edema. This may possibly be a type of angioneurotic.

Routine Foot Care

Foot problems are common in captive elephants. The major causes of these problems are chronically wet and dirty conditions and inadequate exercise and wear.

Cracked Sole

Cracks usually occur in the soles of the feet of elephants that are exposed to wet conditions and poor sanitation. The cracks penetate into the hoof and expose deep tissue to dirt and infection. Signs associated with cracked sole can include lameness, pain, exudate, erosion and ulceration of the edges of the crack, and exuberant granulation formation and development of a hole in the bottom of the foot.

Cracked Heel

Cracked heel also is caused by poor sanitation and wet conditions. The crack occurs between the junction of the skin of the leg and the sole at the posterior of the foot. Treatment is identical with that used for cracked sole.

Ingrown Nails

Even elephants whose nails are regularly trimmed can occasionally develop ingrown nails. Signs associated with an ingrown nail include lameness, pain, inflammation, and exuberant granulation. Treatment consists of trimming back the ingrown nail and surgically removing the exuberant granulation when present, followed by application of topical antibiotic or antiseptic ointment until healing is complete.

Overgrowth of Cuticle

Overgrowth of cuticle is usually seen concurrently with overgrown of sole and overgrowth of nail. Overgrown cuticle appears as a roughened, split area of skin just proximal to the nail. Roughened cuticle is quite sensitive, and most elephants will not readily tolerate its removal with a blade. An old circus method for maintaining healthy cuticle works very well, however. This method requires daily application of vegetable oil or mineral oil to each cuticle. The oil treatment gradually softens the cuticle to the point that the elephants will rub off the excess themselves. After the cuticle has been restored to normal, a periodic application of oil will help maintain it in good condition.

Trunk Injuries

The trunk of the elephant is subject to many different types of injuries, such as crushing, laceration, penetration by foreign bodies and damage to motor nerves. The many specialized functions of the trunk dictate that every effort is made to ensure that healing takes place with minimal loss of function. The blood supply to the trunk is massive and a deep incised wound or laceration high up on the trunk can produce shock and death. Trunk paralysis can occur as result of parasite migration in the area of the motor nerves, bacterial infection from wounds near the motor nerves, a crushing injury that causes massive damage to the trunk or tumors that place pressure on the motor nerves; it can also be caused by tetanus or rabies. Treatment of trunk paralysis in which motor nerve damage from parasitic or bacterial infection is suspected includes use of an appropriate anthelminthic, high doses of steroids, and therapeutic levels of broad-spectrum antibiotics.

Tusk Problems

Animals that repeatedly rub their tusks on bars or that always sleep on the same side will wear down the ivory on a particular area of one or both tusks. One can prevent excessive wear by placing a metal ring with a lag screw on the portion of tusk being worn. It the tip is being worn, a steel ball with a lag screw may be applied. Tusks that break distal to the root canal are no immediate problem, except that the ivory gets weaker as the pulp cavity is neared and that tusks broken near the pulp cavity is neared and that tusks that occurs over the root canal causes hemorrhage and distress and may result in the loss of the entire. Common sequelae to nerve exposure are infection abscessation, nerve death, sinusitis, and loss of the tusk. Treatment flushing of the pulp canal with an antiseptic or antibiotic solution to prevent infection while reactionary dentin is laid down to fill the defect. Splitting tendency in a tusk can be halted by application of a tight metal ring, which is pounded down along the tusk and may be held in place with a lag screw. Tusks that spilt along their entire length, exposing the nerve, should be treated similarly to fractured tusks. Infection of the tusk sulcus in usually due to the presence of a foreign body. If sulcus infection remains untreated, it can lead to tusk damage or loss. Treatment of sulcus infection consists of removal of the foreign body, if still present, and regular flushing with antibiotic or antiseptic solution until healing is complete parenteral antibiotics, chosen on the basis of culture and sensitivity, are also given.

HOOF STOCK

Hoof Problems

Hoof abnormalities of wild animals are similar to the hoof problems of domestic horses, cattle and sheep. Wild animals are plagued with overgrown hoofs, excessive wearing of the hoofs, hoof cracks, cuticle overgrowth, with contusions, penetrating wounds, laminitis, thrush, gravel and infectious pododer-matitis. Essentially, these are managed in wild animals as they are in domestic animals. Hoof trimming is so often required that basic principles adopted for domestic animal.

Most hoof trimming can be done with an equine hoof nipper, a pair of pruning or food-rot shears, a hoof knife, and an equine hoof rasp. The work can be made much easier with the use of a hoof trimmer for larger hoofed stock; a drawing knife for elephants, camels, and their relatives; a miter-box saw for wild equines; and a power sander for final smoothing and shaping. In hoof trimming, the most important consideration is the operator's perception of the normal appearance of the hoof, which varies from species to species. For most species, one can assume that in the normal animal the angle of the outside of the wall should be the same as the angle of P-3. Some exceptions to this include elephants, camels, and related species. In cloven-hoofed animals it is common for one claw to be markedly longer asymmetrical growth is likely to result from conformational faults of the animal. Such faults may be genetically controlled or caused by a change in gait as a result of lameness or injury to a limb or foot. Trauma to a claw would automatically result in the animal's favoring that part of the foot and continued lack of wear would result in overgrowth of the claw. The most severely overgrown or deformed claw should always be trimmed first. If the corium has been deformed by the overgrowth and hemorrhage appears before the normal hoof shape is attained, foot balance can be restored by allowing the accompanying claw to remain a bit longer in length. Foot balance is important; otherwise the basic problem will be perpetuated, not solved. The wall and sole should be trimmed parallel to the angles of P-3. In horses, deer, swine, cattle, sheep, goats, and related species, the wall should bear the brunt of the pressure when the foot is placed on the ground. If the wall is trimmed so that the sole extends beyond the wall, disproportionate weight is borne by the sole, predisposing the sole to contusion.

Choke

This has occurred in captive exotic ruminants and free ranging ruminants that try to swallow large pieces of carrot, apple, hedge, potato, beet, turnip or ears of corn. Clinical symptoms include bloat, salivation and extended neck, palpable object within the esophagus, cyanosis, and dyspnea. Treatment should include trocarization of the rumen, when indicated. Object in the cervical esophagus may be massaged anteriorly and removed with an oral speculum. A stiff stomach tube of relatively large caliber may be used for thoracic obstructions. Obstructions located near the rumen may be removed via a rumenotomy. Esophagotomy should be employed only as a last resort, as poor healing usually results in an esophageal fistula.

Traumatic Reticulitis

This occurs when a sharp foreign object, such as a piece of wire or nail penetrates the reticulum. The clinical symptoms are chronic debilitation, "humped-up", stiff gait, reluctance to move, elevated temperature, and increased WBC with a shift to the left. Chronic cases may show intermittent recovery and relapse. Treatment consists of large doses of parenteral board spectrum antibiotics or oral sulfonamides. Rumenotomy, with removal of the aggravating object, is usually indicated.

Metritis and Pyometra

These occur in exotic ruminants, usually in post parturient animals. The clinical signs include anorexia, depression, and a mucoid purulent discharge from the vulva. Diagnosis is based on the observation of clinical signs and a history of recent brith. Treatment should consist of stilbestrol therapy (10-50 mg I.M.), broad spectrum antibiotics and infusion of the uterus with sulfamethazine or tetracyclines, either as solutions or boluses.

Prolapse

Prolapse of the vagina or uterus, or both, occurs with some frequency in exotic ruminants. Vaginal and uterine prolapse is a common sequela to dystocia. Ruptured uterine ligaments have also occurred. Therapy is similar to that used in cattle and sheep. The standard method of reduction and restrain of the prolapse with a purse-string suture requires that the animal be restrained a second time to remove the sutures from the vulva.

Orchitis

This has been reported in the African cape buffalo and caribou as result of infection with brucellosis. The affected testicles wre enlarged and contained abscesses measuring 1.5x0.7 cm.

Scrotal Hernia

Scrotal hernia has been reported in a white-tailed deer. The hernia was discovered at autopsy. Should the occasion arise, the hernia may be repaired in the same manner as in the domestic ruminant.

Urethral Calculi

Urethral calculi occur most frequently in exotic ruminants the clinical sings are abdominal pain, anorexia, and uremia. Calculi may be in the from of "sand" or a large single stone.

Calculi in the urethral process of sheep and goats may be treated by amputation the urethral process after the animal the exotic equine, zebra, and pachyderm. These traumatic syndromes are usually accompanied by soft, painless swellings. Close confinement and rest of the joint and limb will often relieve the inflammation. In chronic cases of bursitis the accumulated fluid should be aspirated and cultural for pathogenic organisms. If infection has become established, the appropriate antibiotic should be injected into the bursa if the lesion is sterile, a corticosteroid injection may be value. The oral administration of phenylbutazone at the rate of 0.5 to 1 gm per kg of body weight for three to five days will relieve the animal's discomfort and allow the joint and associated soft tissues to recover. Special corrective orthoses have been applied to elephant legs.

Osteoarthritis

Occurs in the older equine, zebra, and pachyderm. The articular cartilage is usually destroyed leaving a raw, painful bony surface. Diagnosis in usually made by clinical signs and a history of recurring lameness becomes reduced as the animal "warms up". The treatment of choice is rest and oral administration of phenylbutazone as required at the rate of 0.5 to 1 gm per 100 kg body weight luxations have occurred in elephants. Degenerative arthritis has been described in the elephant, zebra, and black rhinoceros. Pedal rotation following laminitis is a common occurrence in the exotic equine and zebra.

Orchiectomy (Castration)

A pronounced decline in the rate of fighting and uring spraying occurs in many of the cats shortly after surgery. Predatory behavior and fear induced aggression, unlike sexual aggression and inter-male fighting are not affected by castration. Ketamine and xylazine combination will provide effective anesthesia for orchiectomy. The hair is plucked or shaved from the scrotum and the surgical site is prepared by scrubbing with surgical soap and applicaltion of a surgical disinfectant. One testicle is grasped with slight tension to immobilize it in the scrotum. A lengthy incision is made through the skin and tunica vaginalis, parallel to the median raphe on the central ventral aspect. The testicle is then expressed through the incision and the scrotal ligament separated from the tunica vaginalis, allowing the testicle and spermatic cord to be retracted.

CARNIVORES

Nail Problems

Nails are not likely to be overgrown. They are less massive than hoofs and usually break of if growth extends much past the end of the digit, however, nails are subject to trauma (foreign bodies, lacerations and contusions), onychia, paronychia, cracks, onychocryptosis (ingrown nail) and excessive cuticular growth (hangnail). These are managed in wild carnivores in the same way as they are in dogs and cats.

Claw Problems

In captive animals, claws are subject to overgrowth, abrasions, digital paralysis, contusion, penetrating wounds, avulsion and cracks. Overgrowth interferes with retraction and therefore with normal function of the claw. The overgrown claws of a domestic cat are usually simply cut off. This is not sufficient in wild felines. Aside form the trimming and management procedures already described or referred to, the most commonly requested procedure is onychectomy, the removal of claws.

Periodontitis and Dentoalveolar Abscessation

These are fairly common in aged exotic ruminants. The cause appears to be mineral depletion which allows movement of the teeth in alveoli, causing the accumulation of the feed alongside and under the root of the teeth. The clinical symptoms are unmasticated roughage in the feces, collection of "cud" in the buccal cavity, anorexia, fetid breath, chronic weight loss, and salivation. Diagnosis requires visual inspection of the teeth. Treatment includes the administration of board spectrum antibiotics, removal of loose teeth, curettage of the alveoli, and correction of the diet.

Vasectomy

This operation is desirable when the zoo veterinarian or the owner of a male exotic cat wishes to sterilize the animal yet maintain the behavioral patterns and secondary sex characteristics of an intact male, Vasectomy will meet these requirements, and is currently a very popular surgical procedure. The spermatic cords may be palpated when caudal traction is applied to the testicle. Ketamine (10 to 15 mg/kg) combination provides a good surgical plane for vasectomy. The patient is placed in dorsal recumbency and the public area is scrubbed with surgical soap for sterile surgery. A 5 to 10 cm incision is made over the spermatic cords. The tunica vaginalis is incised longitudinally, exposing the spermatic vessels and the ductus deferens. The ductus deferens is dissected free, and about 2 to 4 cm of the isolated ductus is double ligated with non-absorbable suture material the section of ductus deferens between vaginalis is closed. The procedure is then repeated on the second spermatic cord, through the original skin incision.

Overiohysterectomy (Spaying)

The removal of the ovaries and uterus in not requested as often in the exotic as in the domestic cat, as reproduction is usually desirable in the zoo situation and the animals rarely become accidentally pregnant. Spaying is most often performed on the exotic felid as a treatment for chornic life threatening pyometra.

Hairball (Gastric Trichobezoar)

A partial obstruction may result in a chronic wasting syndrome of increasing cachexia. Abdominal palpation and radiography are not useful diagnostic procedures. Definitive diagnosis requires and exploratory gastrotomy. Often a true trichobezoar is not found. There may be small multiple hairballs or a less discrete aggregate of hair forming a small plug that occludes the narrow pyloric lumen. Laxative preparations for cats are occasionally successful in the treatment of rabbit hairballs; otherwise, surgical removal with careful pre-and postoperative care is the only successful therapy.

AVIFAUNA

Very few surgical problems are unique to Galliform Birds and they are described below:

Debeaking : Removal of the tip of the upper beak may be indicated to control cannibalism (such as toe and nose picking among young birds) and fighting among male birds. This procedure must be done so as not to interfere with normal eating or drinking. A guil-lotine-type dog nail clipper or a human toe nail clipper is useful for this purpose.

Amputation of the Comb or Wattles

Surgical removal of these appendages may be indicated after extensive injury, edema, or infection the combs of birds are highly vascular; thus, adequate hemostasis is necessary to prevent fatal hemorrhage. Either under general anesthesia or after the use of a local anesthetic. The wattles may be removed with surgical scapel or scissors. Hemorrhage should be controlled with hemostatic forceps and suitable astringent.

Crop Impactions

The ingestion of too much bulky, dry, fibrous food may result in over distension and cessation of normal activity of the crop. In some instances, it is possible to pass a tube into the crop by way of the esophagus, inject water or oil, and manually massage the organ until the impaction is broken down. If this technique proves unsuccessful or if the crop is too pendulous, surgical intervention may be necessary. Feathers are removed from the skin over the crop. The skin is cleansed with a surgical antiseptic and swabbed with tincture of iodine. The skin is incised and reflected from the crop, and then the crop is incised. This incision should be no larger than is necessary to evacuate the crop. Closure of the surgical wound should be by two independent lies of closure-crop, then skin. A topical antibiotic powder may be used. Food and water should be withheld for up to 12 hours, after which the bird should be fed sparingly for several days.

Foreign Bodies in the Crop: Frequently foreign bodies lodge in the crops of birds. Surgical removal is usually indicated. The surgical approach is as described for crop impactions.

Sex Identification

For methods of sexing gallinaceous birds, using endoscopes or otosecope general or local an aesthesia.

Egg Retention

These cases must be handled individually with respect to the bird's general health status and the value of the bird. When the egg is retained within the oviduct near the posterior abdomen, one may be able to deliver it by applying gentle pressure on the abdomen while at the same time dilating the vent with a finger or suitable speculum. The vent and posterior part of the oviduct should be well lubricated to facilitate passage of the egg. Before any surgical manipulation is at tempted, the bird should be placed in a humid, warm environment at 85°P. If the egg will not pass through the oviduct, it may be possible to manipulate it posteriorly so the shell can be seen through the vent. Using a large-gauge needle, penetrate the shell and aspirate the contents. After aspiration of the contents, the egg can be collapsed and the shell fragments removed with a suitable forceps. After the egg has been removed, the cloaca may be irrigated with a mild antiseptic solution.

ELECTIVE SURGICAL INTERVENTIONS

These include surgeries which are conducted electively.

1. **Dentistry :** Dentistry in zoo animals present unique problems. The roots of canine teeth in monkeys and carnivores are more extensive than the exposed crown and dislodging with a dental elevator is essential. There have been numerous reports of injuries in ursids, felids, primates and elephants. Traumatic dental disorders resulting in exposure of the dental pulp should not be ignored. Dental therapy may range from the routine removal of a calculus to performance of a root canal. The rapid development of restoratives with adhesive properties (glass ionomer cement, composite resin cement, composite amalgam) has provided, the veterinarian with promising new applications in veterinary dentistry.
2. **Bone Plating :** Bone plating is one of the internal fixation methods for immobilization of fractures in wild animals. it is mostly done in case of long oblique fractures ,multiple fractures, comminuted fractures etc. extra strength and support is given by using orthopaedic wiring also.

   Clinical repair of a metatarsal fracture by internal fixation with bone plating in a Sambar
3. **Endoscopy :** The non-invasive nature of endoscopy and the clarity of the structure seen make it an important tool in the zoo and wildlife clinicians armamentarium. Flexible and ultrasound endoscopy systems are prevalent. Endoscopy is one of the most diagnostically useful tools available. It affords the clinician a token invasive method to investigate cause of recurrent chronic gastrointestinal disorders, small bowel tumors and inflammatory disease of stomach and intestine, which could not be

determined by current diagnostic motalities. Minimal invasive surgery is collection of surgical techniques designed to minimized the extend of anatomical approach while still maintaining precision and efficiency. Endosurgery/ endoscopic surgery means involves performing a minimal invasive surgical procedure with visualization provided by an endoscope.

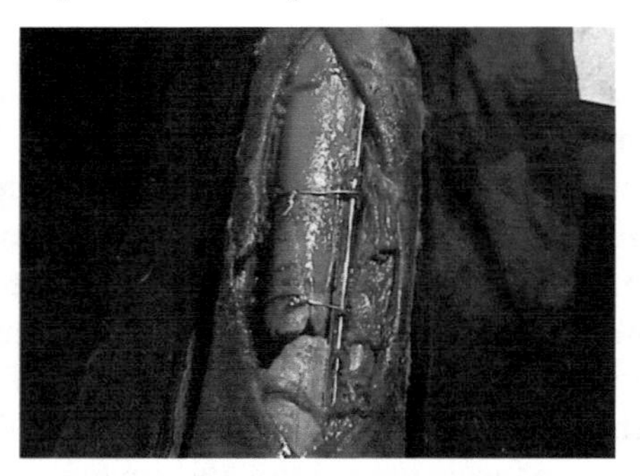

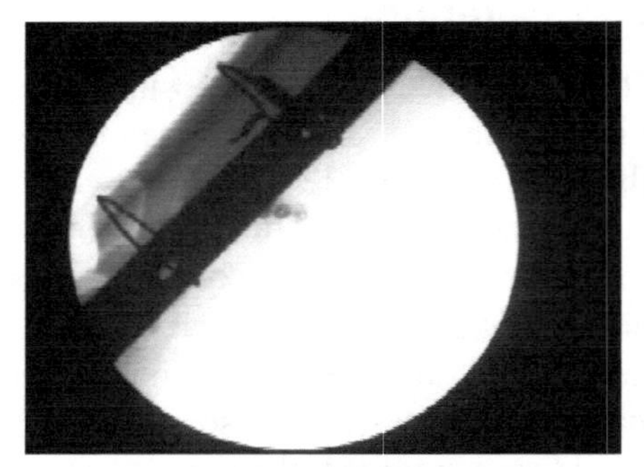

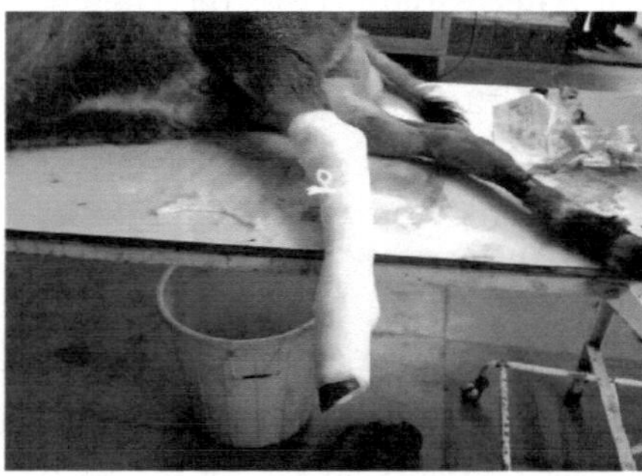

A realistic and common sense appraisal of the animals future should direct the medical efforts, and the animal should, not be subjected to surgery and treatments simply to massage the ego or pad the release statistics. The veterinary profession should play the leading role in establishing guidelines both for the medical and surgical treatment of injured wildlife and for the process of responsible and humane decision taking.

4. **Laparoscopy :** Laparoscopic surgery, keyhole surgery, or pinhole surgery is a modern surgical technique in which operations in the abdomen are performed through small incisions (usually 0.5 - 1.5 cm) as compared to larger incisions needed in traditional surgical procedures. Laparoscopic surgery includes operations within the abdominal or pelvic cavities, whereas keyhole surgery performed on the thoracic or chest cavity is called thoracoscopic surgery. Laparoscopic and thoracoscopic surgery belong to the broader field of endoscopy.

   Rigid endoscopes are routinely used in veterinary practice for laparoscopic examination.

   Indications – visual examination of the abdominal organ and also biopsy of the organs showing lesions of neoplasia or even normal. Acute or chronic colic are diagnosed and as well as treatment in horses. Affections and surgeries of gonads like ovariectomy, cryptorchidectomy, and castration are performed in small animal birth control program and also in horses. Surgeries for repair of inguinal hernia are also performed by this technique. Embryo transfer with aid of laparoscopy is performed in equines, sheep, goats and cattle. The new addition in this field is of correction of abomasal displacement. The procedure of which is still under standardization, it can be of turn over point in large animal abdominal surgeries. Contraindications are diaphragmatic hernia, peritonitis abdominal wall malignant tumors

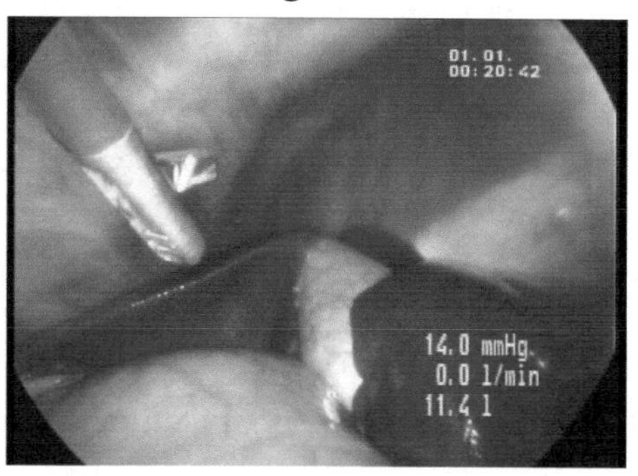

Conceptually, the laparoscopic approach is intended to minimize post-operative pain and speed up recovery times, while maintaining an enhanced visual field for surgeons. Due to improved patient outcomes, in the last two decades, laparoscopic surgery has been adopted by various surgical sub-specialties including gastrointestinal surgery (including bariatric procedures for morbid obesity), gynecologic surgery and urology. Based on numerous prospective randomized controlled trials, the approach has proven to be beneficial in reducing post-operative morbidities such as wound infections and incisional hernias and is now deemed safe when applied to surgery for cancers such as cancer of colon.

5. **Application of Pop Cast :** Cast application is the most practical method of external fixation of lower long bone fractures. For an adequate stabilization, joints proximal and distal to the fracture site should be included in the cast. Plaster casts are also used to immobilize frac-tures proximal to the knee and hock joints. Many distal radial and tibial fractures can be managed successfully by this method. Cast is routinely used in combination with various internal fixation and trans-fixation devices. A combination of external support and internal fixation is advantageous because the latter provides rigid stability at the fracture site and the former provides strength to bear the weight of the animal. A cast should incorporate the foot of the animal to avoid pressure sores around the coronary band or on palmar/plantar surface of the fetlock area. The rim of the cast around the wall and toe of the foot should be of adequate strength to prevent its breakage' during use.. A cast should not end at the mid shaft of radius or tibia. These bones are surrounded by little protective soft tissues and the proximal rim of the cast invariably causes pressure sores if the am ends over the diaphysis of the long bone.

For application of a POP cast, the animal is sedated and restrained in lateral recumbency with the affected limb uppermost. The affected limb including the foot is cleaned and dried. Any wound, if present, should be cleaned, debrided, covered with antiseptic dressing and bandaged. The interdigital space should be padded with cotton. A piece of wire is fixed through the predrilled holes in the wall of the medial and lateral claws. The wire allows a slow and steady traction to facilitate reduction and proper alignment of the fractured fragments. Apart from facilitating reduc-tion, traction wire also helps in maintaining the limb in normal position during application of the cast.

6. **Onchyectomy :** The surgical amputation of the Distal phalanx (last knuckle) along with the accompanying claw, by traditional cutting.
7. **Vasectomy :** Mass vasectomy is carried out for population control in case of deers like Chital, Sambhar and antelopes like Nilgai.
8. **Laparotomy :** It is done in case of intestinal obstruction, intestinal anastomosis, urinary obstruction, urinary bladder rupture, intussusceptions etc.
9. **Cataract :** It is the focal or diffused opacity of the crystalline lens or its capsule. Severe trauma to the globe may result in cataract. Trauma may result from a crushing nonpcnetraling blow from impact with an automobile. Traumatic wound may penetrate the lens capsule and disrupt the cortical material. With tearing of lens capsule a variable amount of lens protein material which is foreign to the immune system leaks in to the anterior chamber. The resultant inflammatory response may be quite reverse leads to lens induced uveitis.

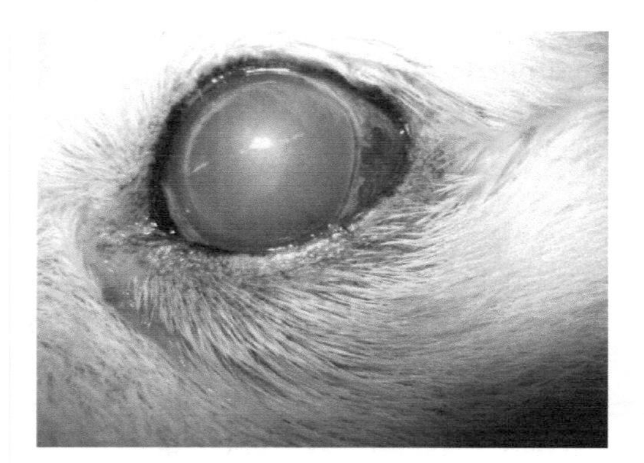

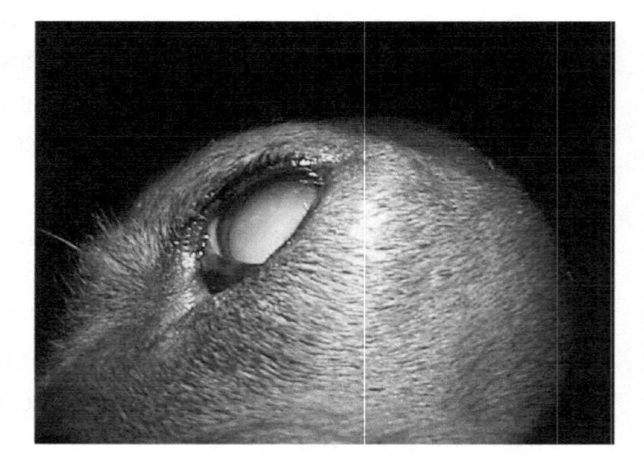

10. **Tumour Resection :** Veterinary oncology is becoming a speciality of ever increasing relevance to the small animal practitioner. No topic in veterinary oncology generates more interest than the treatment of mammary tumours. Mammary neoplasms comprise between 25 to 42 per cent of total neoplasms seen in the animals.

    The ultimate evaluation of malignancy of disease is metastasis, either by local infiltration or spread of disease to distant sites. Metastasis of mammary tumors of the dog has been reported most frequently in the lungs and regional lymph nodes, but they can metastasize to any organ including bone, brain, liver, spleen, kidney and skin. These sites of metastasis suggest that both vascular and lymphatic routes are involved in the spread of this disease

11. **Hernioplasty :** This is mainly done in case of herniation.umbilical hernia is more common among all other hernias.The hernioplasty with centilene mesh 2 X 4 is performed and ring is sutured with prolene taking overlapping sutures and additional reinforcement sutures.

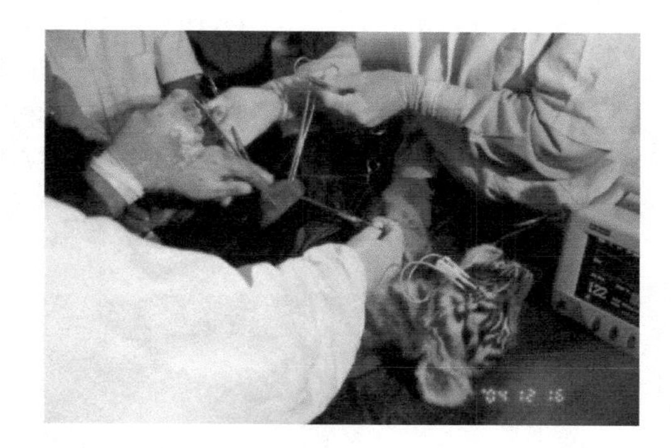

12. **Cryosurgery :** Cryosurgery or cryotherapy may be defined as controlled application and destructive effect of profound cold (freezing) m performed with the aid of special instruments (cryoprobe) for local freezing of diseased or excisable tissue without any harm to normal tissue. It is said to be easy, safe, rapid and least disturbing to patient's physiological integrity. The end result of freeze is to produce local necrosis of tissue and to allow the devitalized area to slough off or absorbed from body. In recent past, development and availability of cryosurgical equipment, increased understanding of tissue response to freezing and successful clinical reports have brought about a renewed interest in cryosurgery. The cryosurgery offers two unique features of cryodestruction (cryonecrosis) and cryoadhesions. The first effect i.e. cryonecrosis is used for treatment of various neoplastic and non-neoplastic conditions. The principle of cryoadhesion between the occular tissues and proves enable us to treat the various intraoccualar disorders like retinal detachments, glaucoma, cataract and intraoccular neoplasms.

13. **Interlocking Nailing :** The goal of fracture treatment is to achieve complete functional recovery of the injured limb as quickly as possible by obtaining and maintaining reduction of fractured fragments in as near normal an anatomical position as possible. Good reduction and stability at fractured site are very much important for the bone to heal. There are various fracture fixation techniques being used in small animal practice which include both internal and external fixative devices. Intramedullary interlocking nail, which is a modification of kuntscher nail, is a more recent addition to the fracture fixation devices developed in 1968. Interlocking nail differs from the Steinmann pins commonly used in veterinary practice because the nail is secured in position by proximal and distal transfixation screws that pass through the IM nail and engaged both the near and far cortices. In human medicine, orthopedic surgeons

routinely use fluoroscopes to control placement of the nail and to accurately place the locking screws through the bone and nail. But most veterinary practices do not have intraoperative fluoroscopy available, so an implant system for veterinary use needs external guides to line up the screws with the nail screw holes.

14. **Image Intensifier Television (C-arms) :** An image - ultraviolet, visible light, or near infrared - is projected onto he transparent window of the vacuum tube. The vacuum side of this window carries a sensitive layer called the photocathode. Light radiation causes the emission of electrons from the photocathode into the vacuum, which are then accelerated by an applied DC voltage towards a luminescent screen (phosphor screen) situated opposite the photocathode. The screen's phosphor in turn converts high energy electrons back to light (photons), which corresponds to the distribution of the input image radiation but with a flux amplified many times.

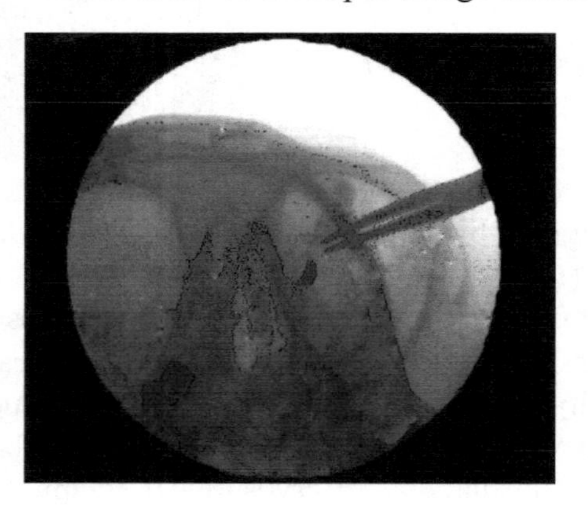

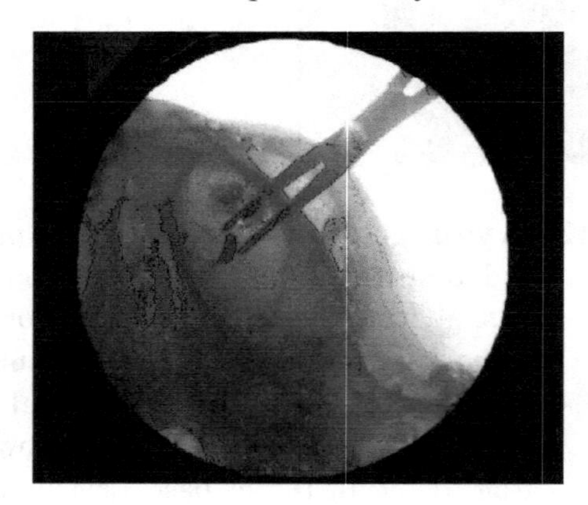

A device used to convert X-rays that have passed through the object or patient into a light image which can be recorded with a television camera or photographic camera. The image intensifier (below named II) was introduced in the 1950s and has since then been subject to massive improvements, making it one of the most important components of modern radiology. Schematically, the II consists of an evacuated glass or metal bottle, having an input screen onto which the X-rays enter and an output screen where the light image is created. The input screen consists of a supporting base, commonly made of 0.5 mm aluminium, a scintillating layer and a photocathode. For all image intensifiers, one has to consider the conflicting demands on efficient absorption of the X-ray photons and high spatial resolution. As in the case with the intensifying screen, the higher the resolution, the thinner the scintillating layer, which in its turn leads to a high noise level. In the early days of the II, the scintillating layer was made of ZnCdS: Ag. This material was not optimal, since the compromise of thickness and resolution was very unsatisfactory. Since the beginning of the 1970s, this material has been replaced by CsI: Na, having a much higher X-ray absorption efficiency. Furthermore, the scintillating layer of CsI:Na crystals can be manufactured as an array of crystallites, small rods of crystals, 150-400 mm high and about 5 mm in diameter. This arrangement prevents light from being scattered inside the input screen and improves the spatial resolution dramatically. The efficiency of converting X-ray photons to light photons is quite high. A 60 keV photon that is absorbed in the CsI: Na will produce around 3 000 light photons at about 420 nm wavelength.

The light produced in the input screen is converted into electrons, which are produced in the photocathode. The spectral sensitivity of the photocathode should match the emission spectrum from the input screen. The photocathode is commonly made of $SbCs_3$, having a good spectral matching to the scintillator and a conversion efficiency (number of electrons generated per light photon entering it) of 10-20%. Therefore, following absorption of a 60 keV photon, around 400 electrons are emitted from the photocathode. Attempts made to retrieve bullet located near the left orbit under image intensifier television.

15. **Intramedullary Pinning :** Use of intramedullary pins for fracture fixation is the most commonly practiced technique, especially in small animals. The most commonly used IM pin is stainless steel Steinmann pins, they are available as not threaded, partially threaded, or totally threaded pins with one of several types of points (trocar, chisel) on one or both ends of the pins. The trocar pointed pin is three sided and has the ability to penetrate cortical bone whereas the chisel point is broad, flat and two sided with poor cortical penetrating ability. The currently available threaded pins offer no additional advantage. Insertion is more difficult and it does not improve fracture stability. Pin failure at the thread non thread junction has been reported.

    Intramedullary pins are inserted by hand using a Jacob's chuck and it provides a better 'feel' as to pin location in the bone. Pin can also be inserted with the help of power equipment. It has been claimed that, use of power equipment eliminates the pin "wobble" that may occur when pins are placed by hand. Pins may be placed in the bone by normograde or retrograde insertion, and it can be inserted either by closed or open method. Closed (blind) pinning is indicated when the fracture is quite stable and recent one with no over riding (transverse, short oblique or incomplete fractures) - especially the fracture of femur, tibia, humerus and ulna, as these bones have prominent land marks or minimal soft tissue covering. However, fracture reduction and immobilization are less optimal, there may be interposition of soft tissues and additionally, it will not neutralise all forces acting at the fracture site, as no other ancillary form of fixation (circlage wires) can be used. On the other hand, open reduction and pinning has the advantage of direct visualization and anatomical reduction can be achieved by direct observation. Retrograde pin placement is also possible. Single intramedullary pins are used in transverse or short oblique diaphysial fractures of long bones. When the pin occupies the entire medullary canal and fixed at both ends, both bending and shearing forces would be eliminated. However, this does not happen in clinical cases, as the pin cannot occupy the entire medullary canal. Similarly, complete fitting of medullary canal is not desirable, as it adversely affect the blood supply. So, it is recommended to select a pin that occupies approximately 60-70% of the medullary canal. In this case, fragment apposition and resistance to shear forces cannot be guaranteed. Rotational forces can, to some extent be neutralised by using external fixators or orthopaedic wire fixation.

16. **Electroacupuncture :** Acupuncture was found to be beneficial in cases in which analgesics and anti-inflammatory medications have been ineffective or have demonstrated side effects, and in cases in which surgery was not recommended. It appears that acupuncture not only provides long-term analgesia but also increases circulation to the affected areas and decreases inflammation. Techniques and selection of appropriate acupuncture points depend on the condition treated. This therapy gave excellent result and recovery in the posterior paresis was observed after three weeks onwards. The acupuncture therapy was routinely practice in other animal for posterior paresis and it gives excellent result in other animal.

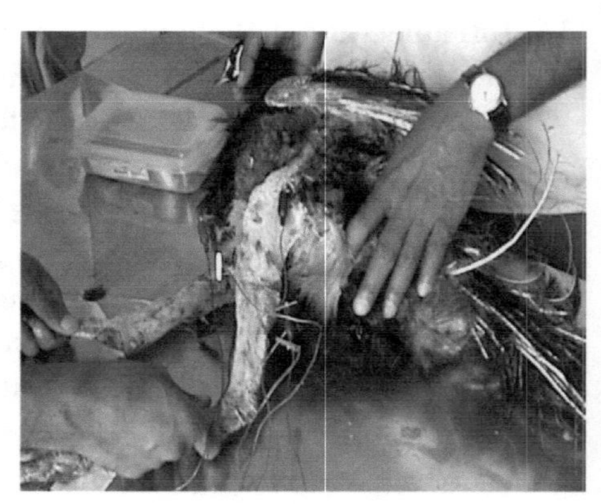

SURGICAL EMERGENCIES

Medical emergencies encountered in the immobilization of wild animals in general do not differ from those in human medicine. Respiratory depression, hypo- and hyperthermia, cardiac arrest, and shock are treated like in humans. Special emergency situation in exotic animals include the following:

1. **Trauma**

    Capture wounds (bandaged animals cannot be released)

    Contusions

    Abrasions

    Lacerations

    Fractures: necks and legs

    Wound Management : Wild animals unlike domesticated ones are more prone to self-inflicted injury and may also inflict injury to fellow animals in the enclosure. The initial step of providing first aid and emergency treatment to wild animals is to capture and/or restrain the animal and if possible to transport it to separate ad hoc enclosures/

cages where veterinary assistance can be provided. As such, there must be provision for such enclosures and/or cages where the animals requiring emergency veterinary care can be kept until complete recovery. Every possible precaution should be taken to minimize struggling of animals during restraint, capture and transport. The cages and/or enclosures should be spacious and comfortable.

Wound is defined as any break in continuity of skin or tissue cause by injury or surgical intervention. Trauma also literally means physical injury. However trauma is more deeply distressing or disturbing condition resulting from any stressful event or physical injury. It may be associated with physical shock, sometimes to long-term neurosis. Wound can broadly be grouped as open or closed wound.

Open Wound, usually contaminated mixed with many objects.

Clean open wound or aseptic wound, which seems practically non-existing.

Contaminated wounds, which are inflicted by clean objects such as puncturel wounds of compound fracture.

Infected wounds are most common form of wounds under field conditions. Presence of debris and exudates and poor blood perfusion due to infection, infected wounds are difficult to heal.

An abrasion is a wound in which the skin is not completely punctured and only outer border is lost. These wounds heal up rapidly after cleaning, as the epithelial layers are still fresh.

Protocol for Wound Management in Wild Animals : Following general guidelines are recommended for management of wound:

Always wear the protective clothing including head and face masks and gloves while handling an injured or wounded wild animal.

The basic objective of all wound/trauma management is to clean the wound, so that it ran be closed and further bacterial invasion is prevented.

Contaminated wound should never be closed. It may lead to abscess and latter septicemia.

Carefully remove wound inflicting agents or contaminants like piece of wire, glass, pebble pieces, etc. by using clean/ sterilized forceps. This will help in better healing.

Collect bacteriological swab for culture and antibiotic sensitivity test. This helps in deciding appropriate course of antibiotics.

Clip the hair of the area surrounding the wound. In case of birds, pluck the feathers. Clean the surrounding areas with antiseptic like povidone iodine, cetrimide (1:1dilutions), chlorhexidine (1:1000 dilution), etc. Avoid contaminating wound during the process.

After cleaning the surroundings, clean the wound. Mechanical cleaning of wounds with swabs or cotton wool will damage healthy tissues and may force debris further deep into the wound. Loose strands from swabs or cotton may also get deposited as additional debris.

The safest way to clean the wound is to flush it with sterile saline. No antiseptic or disinfectants is required, as some of these may be cytotoxic and destroy healthy tissues except chlorhexidine 0.05% solution. Plain water can be used for flushing out bulk of the foreign matter in very dirty wound. But being isotonic, best option is saline.

Chemical debriding agents such as mixture of benzoic and salicylic acids will facilitate removal of necrotic tissue from the wound. This is very useful in wild animals , as some times it is not easy to dress the wounds.

Then apply antiseptic dressing. The aim of wound dressing is to keep the wound in moist condition, so that the necrotic tissue can be removed by animal defense mechanism. Use of silver ions and silver mixed with sulpha drugs is very effective in treatment of chronically infected wounds or heavily contaminated wounds.

The wound should be packed with some antiseptic swab, to prevent the ingress of more debris during cleaning. This practice also continues to keep the wound moist, which is essential for good, rapid healing.

If wound is assumed to be cleaned of contamination, apply antiseptic dressing at daily interval.

Removing dressing may dislodge any clots that might have formed. Any resulting bleeding can be controlled with sterile pads of preferably haemostatic dressing.

Management of Maggoted Wound : Any wound during warm and moist period of the year is susceptible to invasion by flies and Maggot infestation. Animals with diarrhoea or faecal contamination around the hindquarters are at high risk. Flies lays egg on these wound and they hatch quickly into tiny maggots that starts eating necrotic and healthy tissues. Removal of the maggots either mechanically (picking with forceps) or chemically (by applying turpentine oil or diluted cypermethrin solution) is the first step towards management of maggoted wound. Injection of ivermectin can also be used to control maggots. Inject non-steroidal anti-inflammatory agents such as flunixin or meloxicam to counteract inflammatory effects of toxin produced by the maggots.

2. **Respiratory Depression or Arrest :** Tissue hypoxia resulting in necrosis or damage caused by inadequate oxygenation of blood hemoglobin

Causes :

Drug-induced depression of respiratory center

Airway obstruction (nose/trunk/tracheal occlusion, edema/vomiting blocking airway)

Pressure of the diaphragm (bloat, intestinal contents)

Clinical Signs :

Few or no respirations

Cyanosis

Noisy breathing

Oxygen saturation <70% for more than 2 min SO2

Treatment :

Cease further administration of immobilizing drugs

Establish patent airway

Begin artificial ventilation (mouth to mouth, mouth to nose; endotracheal tube/air sac tube; tracheotomy if laryngeal block)

Administer 1-2 mg/kg doxapram (Dopram) IV

Administer appropriate antagonist IV

3. Hyperthermia :

Causes :

Metabolic heat from physical exertion

Heat absorption from environment (direct exposure to sun, confinement in poorly ventilated space)

Drug-induced alteration of thermoregulatory centers

Bacterial infection

Treatment :

Cease all further administration of immobilizing drugs

Cool the animal (move out of direct sunlight)

Whole body immersion: most effective

Spray entire animal (groin & belly)

Pack ice or cold water bags on groin, head

Douse with isopropyl alcohol (rapid evaporation cools quicker)

Cold water edema

Cold lactated Ringer's solution IV or IP

Administer appropriate antagonist IV

4. Hypothermia/Frostbite :

Causes :

Drug-induced (decreased metabolism/endogenous heat production; alteration of thermoregulatory center

Cold ambient temperature

Loss of insulation (wet, soaked coat)

Oiled fur or feathers

Malnutrition (decreased fat)

Long recumbency (compression)

Inadequate circulation (shock, foothold trap)

Treatment :

Warm the animal

Containers of warm water

Blankets

Foam pads

Hand warmers

Body heat

Electric heat pads, lights

Comments

Antagonism of immobilizing drug is not recommended. You may have to give more drugs until animal is warmed

Chilling below 24 C invariably results in death

5. **Shock :** Clinical syndrome characterized by ineffective blood perfusion of tissues resulting in cellular hypoxia. It is often seen in animals that underwent a stressful or strenuous capture.

Causes :

Prolonged physical exertion

Prolonged physiological/psychological stress

Severe blood loss

Clinical Signs :

Rapid heart rate :

Low blood pressure (slow capillary refill)

Muscle weakness

Depressed sensorium (masked by drugs)

Hyperventilation

Treatment :

Cease all further administration of immobilizing drugs

Administer 30 ml/kg Lactated Ringer's solution IV - correct hemorrhage

Administer 5 mg/kg dexamethasone IV Prednisone at 10 mg/kg IV or methylprednisone at 20 mg/kg can also be used

Assist ventilation if necessary

6. **Bloat :** It is a common occurrence in ungulates when excess gas resulting from normal fermentation accumulates in the rumen. The rumen enlarges, compressing the diaphragm and lungs, and impairs respiration. Treatment and prevention include correct sternal body position and insertion of a stomach tube to release intestinal gas. If the bloat cannot be relieved, the bloated rumen has to be trocharized in the left flank with a large-bore needle.

7. Vomiting/Aspiration :

Drug-induced (Xylazine)

Stress, excitement

Head positioned lower than stomach/rumen

Clinical Signs :

Gurgling sounds during respiration

Choking, gasping

Cyanosis

Presence of foreign material in larynx, trachea

Respiratory arrest

Treatment :

Cease all further administration of immobilizing drugs

Clear airway

Begin artificial ventilation if necessary

Administer 1-2 mg/kg doxapram (Dopram) IV

Administer long-term antibiotics to prevent pneumonia

8. Dehydration :

Causes :

Decreased water intake

Hyperthermia- loss by transpiration

Fever

Chronic vomiting

Wound drainage

Polyuria

Clinical Signs :

Skin lacks pliability

Mouth, gums dry

Weak pulse

Depressed sensorium

Signs of shock

Treatment :

Cease all further administration of immobilizing drugs

Administer fluid therapy

Determine the volume deficit (4-6-8-10 rule)

Loss of 4% in body weight is determine by fluid loss or history

Loss of 6% obvious fluid deficits- mouth mucosa red and dry

Loss of 8% severe fluid loss - weak pulse, depression

Loss of 10% life threatening- shock

9. Seizures/Convulsions : Transient disturbance of cerebral function characterized by a violent, involuntary contraction or series of contractions of the voluntary muscles

Causes :

Drug-induced (ketamine or ketamine combinations)

Trauma

Hypoglycemia

Clinical Signs :

Uncontrolled muscle spasms- whole body spasms

Rigid extension of limbs

Mouth gaping

Treatment :

Administer 10 mg diazepam (Valium) IV slowly

Monitor temperature - endogenous heat, keep below 41 C

10. Cardiac Arrest : Loss of effective cardiac function resulting in cessation of circulation

Causes :

Drug-induced

Hypoxia -respiratory failure

Acid-base imbalance

Electrolyte imbalance - hyper/hypokalemia, hypocalcemia

Autonomic nervous system imbalance - increased sympathetic/ parasympathetic tone

Hypothermia

Clinical signs :

Weak or absent heart sounds or pulse

Poor capillary refill (> 2 sec)

Cyanosis

Increased respiratory rate, abnormal pattern or apnea

Dilated pupils

Cold skin

Loss of consciousness

Treatment :

Cease all further administration of immobilizing drugs

Make sure that the animal can breathe -artificial respiration/doxapram

Begin external cardiac massage - press at one release at one with 60-100 cycles/min - detect femoral pulse

Inject 0.2 mg/kg of 1:10,000 epinephrine IV or IC and continue massage. Insert needle between 4th and 6th ribs, above and behind elbow. NOTE: many epinephrine concentrations come as 1:1,000; dilute each ml with 9 ml of physiological saline or lactated Ringer's

If no response to the above, inject 0.1 ml/kg calcium chloride solution (10% or 100 mg/ml) IV or IC

If still no response, repeat epinephrine and calcium chloride doses plus inject 10-20 mEq sodium bicarbonate IV or IC

11. **Capture Myopathy :** CM is a complex condition affecting animals undergone a stressful or strenuous capture or handling.

Causes :

Prolonged physical exertion

Prolonged physiological and/or psychological stress

Ataxia, weakness

Paresis or paralysis

Myoglobinuria

Treatment:Administer 5 meq/kg sodium bicarbonate IV. Administer slowly (4-5 ml/min) to avoid cardiac arrhythmias

SURGICAL INTERVENTIONS IN BIRDS

Pre-surgical Evaluation

PCV < 25-30 consider transfusion : Loss of five drops of blood in small birds (canaries) is about 1 5%of total blood volume producing hypotension and cardiac arrest. PCV > 55-6O Rehydrate with fluids because i.v. catheters pose formidable problems, subcutaneous administration is indicated for dehydrated birds. Best sites arc the inguinal and axillary areas or the back between the wings. A rule of thumb is to provide 0.05 ml of fluids/g body wt. /day (50 ml/kg/day) in divided doses at multiple sites.

Preoperative Preparation and Precautions

Fasting

1. Small birds needs energy - reserve, should not be fasted.
2. Medium size -High metabolic rate fast for 2 hours,

3. Large birds: may regulate crop contents resulting in aspiration so fast for 6-12 hrs.

   Intramuscular injections are administered in pectoralis muscle. Avoid leg muscles because of the possibility of nerve damage and first-pass effect due to the renal portal system.

MINOR SURGICAL PROCEDURES

Ketamine (i.m.)

<250 g - 10mg/kg

>250 g - 30 mg/kg

Telazol

Mallard 5-10 mg/kg i.m.

Parakeet 15-20 mg/kg i.m.

Bumble Foot Treatment in Raptors

Bumble foot is a pododermatitis, characterized by excoriation, ulceration, and cellulitis or abscessation of the plantar epithelium. Left untreated culminates into septic Osteomyelitis. Secondary problem may arise including infection of other joints, valvular endocarditis and tetanus. Conceptually bumble foot is equivalent to a bedsore in a bedridden human. Loss of integrity of the epithelium of the plantar surface of the foot from various contributing factors, followed by bacterial invasion is the usual mode of development. Diagnosis is by physical examination, microbiological examination of exudates and radiography. Goals of therapy are to reduce swelling and inflammation, debride necrotic tissue and establish drainage in cases of abscesses filled with purulent exudates, eliminate pathogens and promote wound, healing and granulation and remove the underlying cause before returning a recovered bird to its former environment. Treatment consists of DI130 cocktail and enrofloxacin 10-15 ml/kg b.i.d. Skin respiratory factor preparations (haemorrhoidal ointment) is used promote wound healing.

SURGICAL INTERVENTIONS IN REPTILES

Anaesthesia in Reptiles

Intravenous

1. Turtle : Ventral abdominal vein after drilling through plastron
2. Crocodiles : Caudal vein or into hemocanal 'in ventral spinous processes of coccygeal vertebrae
3. Snakes : Large central abdominal vein or buccal vein

Intramuscular

Snakes : In longitudinal musculature along the dorsum, inject small volumes at multiple sites.

Crocodiles : In the base of the tail

Ketamine

Turtle : 44 mg/kg for minor surgery i.m.
66-88 mg/kg for major surgery i.m.

Reptiles : Ketamine and Telazol

Telazol

Turtle : 5-10 mg/kg i.m.

Snakes : 50-100 mg/kg i.m.

Iguana : 10-25 mg/kg i.m.

Fatalities

1 Prolonged recovery time up to 6 days

2 Permanent aggressivity of snakes after recovery from ketamine anaesthesia.

CHAPTER 27

MEDICAL MANAGEMENT OF REPTILES

Reptiles have been on this planet before mammals were born. Crocodiles, lizards, turtles, tortoises, and snakes now comprise some of these animals on the earth. Snakes belong to the class Reptilia, and are classified under the order Ophidia.

The Snake : It is important to obtain a through clinical history from the owner. This should include feeding, weight, ecdysis (process of shedding of skin), faecal and urate production.

Urate : The non-fecal part of the excreta. Depending on the species, this may be a clear thin or viscous fluid with or without thicker white parts. In carnivorous and omnivorous reptiles, the white part of the urates may be semi-soft pellets which harden after deposition into a chalk-like substance. In herbivores, the white part of the urate may be laced through the urates and around the fecal pellet or hidden within the feces.

The following need to be assessed :

1. Obtain accurate length (Rostrm-Cloaca) and weight.
2. Inspect rostrum (the nose and area surrounding the nostrils and front of the top lip.), nostril and infra-orbital pits (Discharge, occlusion, trauma)
3. Check eyes clear, inspect spectacles under magnification for abnormalities.
4. Ensure tongue flicking normally and snake moving normally, able to support head etc.
5. Examine oral cavity (Mucous membranes shlould be normally clear, note petechiation, excess salivation, oedema, fluid from respiratory tract.
6. Body should be rounded (triangular if emaciated).
7. Skin elasticity? Check hydration and palpate swellings.
8. Check scales for haemorrhage, blisters, loss (Check ventral scales too).
9. Check cloaca for oedema, erythema, discharge, swellings.
10. Make faecal smear, if faeces produced.
11. Smaller snakes and hatchlings may be transilluminated.

Average organ position in boas and pythons

Organ	Position
	(expressed as % of total length rostrum-cloaca)
heart	22 - 33
lungs	33 - 45
air sac	45 - 65
liver	38 - 56
stomach	46 - 67
intestines	68 - 81
right kidney	69 - 77
left kidney	74 - 82
colon and cloaca	81 - 100

Cranial 1/3 Oesophagus, Trachea, Heart

Middle 1/3 Lungs, Liver, Stomach, Cranial air sac

Caudal 1/3 Pylorus, Duodenum, Intestines, Kidneys, Gonads, Fat body,Cloaca

The Lizard : As above where relevant plus :

1. Check medial temporal joint (orally) for white deposits (= uric acid deposits)
2. Assess limb strength and locomotion, and bone/soft tissue swelling (see Metabolic Bone Disease)
3. Check digits and tips of tail (dry gangrene with dysecdysis)
4. Smaller lizards e.g. geckos can be transilluminated
5. Wrap in damp towel to auscultate lungs/heart
6. Faecal smear if deposited

The Chelonian* (* A collective term referring to turtles and tortoises)

Turtles" belongs- to the order Chelonia. This order also includes (1) Marine turtles (2) the fresh water tortoise (3) Terrapines (4) the fresh water turtle and (5) the land tortoise. Main difference between the turtle, tortoise and terrapines are:

Turtles : They are water dwelling species and its limbs have evolve into flattened flippers.

Tortoise : They are land dwelling species. Their roes are short and have no welrbing unlike tire aquatice species of turtles.

Terrapines : They are found in ponds and river and have webbed feet.

As above where relevant plus:

1. Assess body condition using Jackson's ratio
2. Check locomotion (does eventually come out of shell when left alone?) and strength.

3. Check shape of shell for evidence of metabolic bone disease, hemorrhage,trauma.
4. Tympanic membrane should be flat/concave
5. Peak should be evenly apposed.
6. Inspect mouth, note clear eustachian tubes

CLINICAL ASPECTS OF REPTILE MEDICINE

Injection Sites

The renal portal venous circulation in reptiles means that injection into the hind limb musculature may be eliminated via the kidneys before reaching the rest of the body (however, recent work disputes this; JWD Jul 95).One should avoid injecting into the tail of those animals that can shed their tails (a process known as autonomy). These animals include most geckos, the green iguana. Although the tail will regrow, it will be a different shape and often colour than the original. It is also embarrassing to explain to the owner!

i/m : snake intercostal muscles of body
lizard fore- and hind-leg muscles, tail muscles
chelonian as lizard, plus pectoral muscle mass at angle of forelimb and neck

s/c : in loose skin (over ribs in snake/lizard)

i/v : see blood sampling veins

Force Feeding

Give oral fluids e.g. lactated Ringers, Hartmanns, daily equal to 4%- 10% body weight. Mix up feed at 5g protein and 500kJ/Kg bodyweight daily.

e.g. BuildUp (Carnation foods), Protinaid (VetDrug).

1. **Snake :** Manually restrain animal, open mouth and insert gag e.g. folded piece of radiograph with hole cut into centre. Hold anterior of snake vertically. Insert well lubricated end of French catheter into oesophagus to level of stomach. Syringe in fluid.
2. **Lizard :** The stomach is positioned just behind caudal edge of ribs. Proceed as for snake.
3. **Chelonian :** The stomach is positioned midway down plastron. Measure stomach tube from caudal end of abdominal shield to just beyond gular notch. Hold the chelonian upright, sitting on caudal shields. Extend neck and hold head behind mandible. The neck must be fully extended to ensure the oesophagus is straightened.

   Prise open mouth and insert gag. Insert lubricated, prefilled tube to correct depth and slowly infuse liquid. Withdraw tube slowly.Take care not to overfill the stomach. If that occurs, you will see food wellling up into the mouth.

Fluid Therapy

Most sick reptiles will present dehydrated, requiring fluid therapy or force feeding. Consider these if; the reptile has continued weightloss, dehydration (PCV<.25l/l) with an associated hypoglycaemia (blood glucose < 5.2 mmol/l). Fluids given s/c, i/p or i/v.

Care : As reptiles lack a diaphragm, the administration of large volumes of fluid i/p may impair respiration.

Faecal Examination

Reptiles often deposit urates/faeces when being examined. If not, a colonic wash may be performed as follows:

1. Insert lubricated French catheter attached to syringe filled with sterile saline into the cloaca and colon (it should slide in easily with the right size)
2. Flush in saline and aspirate several times.

The following examination techniques may then be used:

1. Direct wet mount : dilute small sample with saline and examine under high power (x400) to see spinning flagellates and static nematode ova
2. Mix faecal sample with eosin for background staining to show encysted entamoeba
3. Gram Stain. Care : Gram negative bacteria are frequently recovered from clinically healthy captive reptiles. However, most infections are caused by gram negative pathogens e.g. Pseudomonas, Aeromonas, Proteus, Providentia, Arizona, Salmonella
4. Examine sediment (following centrifugation) for protozoa and trematode ova
5. Flotation technique for nematode ova

Care : Reptiles may pass prey parasites e.g. mouse nematodes which are non-pathogenic to the reptile.

Blood Sampling

It is imperative that the maximum blood volume that may be safely withdrawn is accurately calculated as it is easy to overestimate in small animals. In Reptiles the total blood volume varies with species but is approximately 5-8% bwt (70 ml/Kg). Of this, 10%may be the maximum withdrawn safely. Thus a 100g reptile can only have 0.7ml safely taken.

It is obviously important that the reptile patient is weighed accurately and the calculations made before blood is withdrawn!

Collect blood into lithium heparin tubes (EDTA tends to lyse cells)

Blood Sampling Sites

1. **Snake :** The palatine vein, ventral tail vein, or cardiocentesis

(a) Ventral vein, the only method not requiring sedation, - identify cloaca, insert needle distal to this, into tail at midline at 45 0 angle,advance to vertebrae, aspirate as slowly withdraw.

(b) Cardiocentesis requires sedation. Palpate/ visualise beating heart, stabilise with finger and thumb. Use 23/25G needle on 3-6ml syringe. Slide needle under ventral scale and aspirate syringe. If only clear fluid is withdrawn, this is pericardial fluid.

2. **Lizard :**

   Large animals : Ventral tail vein

   Smaller lizards : Clip toe nail and collect blood with capillary tube.

3. **Chelonian :** The following sites may be used: cardiocentesis,jugular vein,brachial vein, ventral coccygeal vein, orbital sinus and toe nail clipping.Atempts to venepuncture limb veins often results in collection of lymph only, as these vessels are large and the veins cannot bevisualised. Collection from a jugular vein; hold animal between your knees and extend the neck towards you. the jugular vein will be seen as a bulge between the tympanic membrane and the base of the neck. Swab the site with 70% ethanol. Insert 23/25G butterfly catheter.In Mediterranean tortoises, the dorsal tail vein is the most convenient site for venepuncture.

Haematology

Measure number RBC, WBC, differential WBC count, PCV, and the haemoglobin concentration.

Biochemistry

Use plasma (gain greater volume from blood sample and serum tends to clot). Take blood sample and centrifuge immediately,remove plasma.Measure Na, Cl, Ca, P, Gl, urea, uric acid, creatinine, cholesterol, AST, ALT, ALP, total protein.

Blood Smear

The following may be carried out: Differential WCC, morphology of cells, level of toxic changes, inclusion bodies, blood parasites, bacteria

Radiography

Positioning is important when radiographing reptiles. Animals can be taped down, radiographed through a box or bag if not sedated. Assess organ position, shape, size, density and homogeneity. Check state of reptile nutrition; skeletal density, gastrointestinalorgans and contents.

Snake

Radiograph regions of suspected lesion only. If radiographing whole snake, take sequential sections along length of snake using lead markers every 10 - 20 cm.

WHOLE BODY COILED RADIOGRAPHS ARE ALMOST USELESS

Barium Meal : Studies can be performed as follows; A 2Kg snake requires 10ml barium sulphate suspension by oesophageal tube followed by 90ml air for double contrast study. 15mins later you should see oesophageal folds, gastric rugae, pyloric sphincter and duodenal villi. 5mg metoclopramide reduces the GIT food transit time from days to 12 hours.

Lizard : Poor skeletal density most common finding, if suspected, reduce the kV. Normal lizards show similar bone/soft tissue contrast to mammals. Dorso-ventral and lateralviews as mammalian positioning.

Chelonian : DV View; care; healthy animal can move very quickly off the Table!

Take exposure between expiration and inspiration. Placing animal on a raised column with the feet off the table to aid restraint Lateral -tilting chelonians onto their side distorts the diaphragm and lungs, thus horizontal beam required; centre beam on 6-7th marginal shield at right angles to vertebral column cranio-caudal view is useful for contrasting two lung fields. Centre horizontal beam on nuchal shieldhead, neck, limbs General anaesthesia required for optimal positioning of extremities

Barium meal : 2ml barium sulphate by stomach tube followed by 18ml air for double contrast studies in animals 1Kg

Ultrasonography

Lizards and Snake : 7.5 and 10MHz transducers with stand-off for suitable resolution in small reptiles. 5 and 3.5 MHz transducers for larger reptiles,linear array transducers are used to view the internal organs via the ventral body wall using an aqueous gel.

Chelonian : The only sites available for the access of ultrasound are the soft tissue areas known as the femoral fossa (cranial to the hind limb) and the cervical fossa (at the base of the neck).

CHAPTER 28

EXOTIC BIRD HEALTH CARE AND MANAGEMENT

Since time immemorial the hobby of keeping companion birds for their singing, talking, beauty and colour by the human society is universal. Out of major 27 orders of birds, having 8700 species of which Passeriformes (finches, sparrows, mynah and perching birds) and Psittaciformes (Parakeets, parrots, cockatoos, macaws and lorikeets) are commonly preferred as companion birds by most of the pet lovers.

The geographical distribution of the order Psittaciformes includes the Central and South America, Afro-Asian and pacific areas.

Psittacines : The major birds under this group include budgerigars, conures (Aratinga spp.) Amazones (Amazones spp.), macaws (Ara spp.), cocktails (Nymphicus spp.), Indian ring necked Parakeet (Psittacula spp.) blue winged parrot (Neopnemachrysostona), Australian king parrot (Alistern spp.), Princess parrot (Polytelis spp.), Baraband parrot (Polytelis spp.), Pileated parrot (Pinopsitta spp.) as well as Kea (Nestor spp.) .

Biology : The size of these birds ranges from 10 cm. (pygmy parrot) to 100 cm. (hyacinth macaw). They have large, hooked upper mandible with thick, heavy muscled tongue and zygodactyl foot, which help them in climbing and to grasp the food. The general biological information is presented in Table-1.

Table 1. General biological information of parakeets.

Companion bird	Body length (Inch)	Body weight (Gram)	Age of sexual Maturity	Maximum life span (Years)	Average life span (Years)
Budgerigars	7	30	4 months	18	6
Cockatiels	12.5	75-100	6-12 months	32	5
Cockatoos	12-28	300-1100	Small: 1-2 Yrs. Medium: 3-4 years Large: 5-6 years	80	15
Love birds	5-7	38-56	8-12 months	12	4
Macaw	12-40	200-1500	Mini: 4 - 6 yrs. Large: 5 - 7 yrs.	50	15
Amazon Parrot	10-18	350-600	4-6 years	80	15
Eclectus Parrot	12-14	380-450	3-5 years	20	8
African gray Parrot	13-22	300-33	-	50	20

Basic clinical biology : Body temperature of psitacian birds ranges from 40-42^{0} C, heart rate ranges from 200-290 per minute and respiration rate ranges from 55-130 per minute. For the blood collection, larger anterior toenail is preferred. In birds having 30-200g.body weight jugular puncture is preferred with 26-gauge needle and with tuberculin syringe. 0.2 ml. blood is enough per 50 g. of bird.

Housing : They may be kept in cages, indoor bird rooms or in out door aviaries. Cages of parakeets should be square rather than round. They should be enriched with plenty of perches of large diameters, food and fresh water receptacles.

Depending upon the species of pet psitacian birds, they can be kept in such cages that are available commercially. It is advisable to have a space to fly free in the cage. The cages can be enriched with dropping pans, perches, toys, ladders, bells and mirror to avoid the boredom of birds. Its provides a good **"occupational therapy"** to the caged bird but care should be taken to avoid over crowding of the cages. For the larger birds wood blocks and other indestructible objects can be used. The bills and claws of parakeets occasionally become too long in captivity, which can be trimmed with nail clipper.

Approximately 12 hours of darkness should be provided for the caged birds. For nesting of such pet birds, commercial nest boxes, large leaves, strips of papers are used .For budgerigars and love birds minimum of three pairs is advisable to keep in cage for breeding. The lovebirds colonised in large 'gang'. Cockatiels require a ladder or hard ware with cloths inside the nest box is good. For macaws, hardwood nesting boxes and perches can be used.

Feeding : They are reported to eat a wide variety of food stuffs which includes fruits, berries, tree branches, legumes insects, larvae and seeds. It is advisable to provide them a variety of foods in the form of "boredom relievers"(tree branches, whole nuts, berries, spray millets, corncob, bones, sprouts and whole fruit) in captivity. Monkey biscuits or dog biscuits, sprouts, fresh vegetables, dark leafy greens, carrots, sweet potatoes, red chili peppers, yellow sweet corn, etc. are commonly preferred food items of parakeets. Large psittacian enjoy chewing on chicken bones occasionally. Small quantities of citrus fruits, vitamin and mineral supplements are needed.

Watering : Fresh water for drinking should be available at all times and fresh water for bathing should be provided at periodical intervals. Parakeets enjoy sitting in rains and bathing by rolling about on wet floor. It is advisable to provide freshwater in earthenware or copper bowls at periodical intervals. Use of nipple bottle for watering has been tried in some zoos.

Major infectious diseases : Psittacosis, New Castle disease, pox, avian influenza, pacheco's disease, salmonellosis, tuberculosis, mycoplamosis, E.coli, aspergillosis, candidiasis and erysipelas have been documented.

Non infectious diseases : Beak abnormalities (over grown, mal occlusions, tumors, deformities, deficiency problems), claw deformities (over grown), mites (Knemidocoptes spp), trauma, dermatoses, feather follicle cyst, self mutilation, diabetes, gout etc. have been observed in parakeets.

Budgerigars : This is also known as 'budgies' (*Melopsittacus undulatus*) or "Parakeet". The wild form is green with black bars on the wings and back, narrow bars on the crown, nape and cheeks with a yellow fore head and bib. The male and female are identical except that the male cere and feet are blue and the female's cere and feet are brown or pink. By crossing several colour varieties of white, yellow, blue and different shades of green coloured have been produced by hobbyists.

Social structure : They are gregarious and originate from very arid portion of Australia. They can taught to "talk" or mimic from the very young age. They are tame and can learn variety of circus type tricks.

Most common diseases : Bacterial or viral infection, tumors, polioma virus carrier, over growth of beak and nails, reproductive disorders like egg bound condition, foot problem, feather disease, feather picking, self-mutilation, leg paralyisis, scaly face mites, chalmidiosis and internal parasites.

Cockatiels : The cockatiels (*Nymphicus hollandicus*) are medium sized, relatively quite, non-destructive and entertaining birds that are easy to care for. Because they are considered so gentle, they are excellent as companion birds for children.. Immature gray cocktiels have yellow stripes under the primary wing feathers. A male loses these stripes around nine month of age Head and facial marking are usually brighter on males. Vocalization is the easiest method of sexing these birds- male has melodious call, female have more of monotonous chirps.

Social structure : Cockatiels are playful, easily amused by simple toys. They love to chew the toys. They originates from the arid region of Central Australia. They are easily housed in pairs or in colonies.

Most common diseases : Obstetrical problems (egg bound condition),eye problems, chlamydiosis, yellow coloured feathers in lutions, gasping, lead poisoning, intestinal parasites, injuries, incoordinations , feather picking, self-mutilation, obesity, lipomas and upper respiratory diseases.

Cockatoos : Cockatoos are appreciated as companion birds as they enjoy "cudding". A great care is required to manage these birds to avoid noisy and destructive habits if improperly socialized. They are the most reluctant to change their eating habits to a healthy diet. Most free ranging species are considered as endangered or threatened. The white species (e.g. Umbrella, sulphur crested, and citron) are in high demand. They can be trained for mimic and to perform circus type tricks. They are affectionate and highly intelligent birds.

Social structure : The gender of the cockatoos can be determined by the colour of iris; those of adult males are black and adult females are reddish brown. However, this does not hold completely true in all species or an individual, there fore endoscopic method is more reliable. The cockatoos are originates from the Australia, New Zealand and South Pacific islands.

Most common disorders : Psittacine beak and feather disease virus, feather picking, self-mutilation, pododermatitis, obesity, lipomas, metabolic bone disease, upper respiratory disease, anti social behaviour, oral abscesses, idiopathic liver cirrhosis, parasites, prolapse of cloaca.

Lovebirds : They are small colourful parrots with short tails and relatively broad bodies. They are moderately common as a companion birds. There are several species like peach faced, black masked and Fisher's love birds. The Fisher's lovebirds are very popular in United States of America. The Nyasaland birds are highly prized.

Social structure : Sex determination is difficult by simple observation. They rarely learn to mimic. Their clutch size is 3-8 white eggs. The free ranging birds are found in central and southern Africa. They are non-destructive, mischievous birds that love to hide such as under paper, in shirt pockets or in long hairs. They are territorial and may kill new addition or weak birds in their colonies however named as love birds. They are prolific breeders.

Most common diseases : Psittacine beak and feather disease (PBFD virus), feather picking, self-mutilation, bacterial infection, chlamydiosis, cannibalism, heat stress, pox virus, fungal infection, epilepsy and obstetrical problems like egg bound condition.

Amazon Parrot : Their primary colour is green with varying amount of yellow on the head. The red and blue spots on the wings. .they are stout, short tailed. their sub species includes Panama amazon, Yellow napped amazon, Yellow fronted Amazon, Mexican double yellow head and Levaillan's Amazon.

Social Structure : They originated from the rain forest of central and South America. They become good pets and can talk on training. They don't reproduce well in captivity.

African Gray Parrot : African gray parrot (*Psittacus erithacus*) are commonly kept parrot in high society and considered to be the best 'talker' of all the parrots. Both sexes are gray with a radish orange tail.

Social structure : They originate from the forested areas of Africa.

Most common diseases : Psittacine beak and feather disease (PBFD virus), feather picking, self-mutilation, bacterial infection, chlamydiosis, cannibalism, proventicular dilatation syndrome, pox virus, fungal infection, epilepsy and obstetrical problems like egg bound condition.

Macaws : Macaws are some of the most intelligent and beautiful coloured parrots. The subspecies of macaw includes blue, scarlet, and golden. The blue and golden macaw is blue on the neck and back, dorsal wing and tail. The crown is green. There is a black chin and yellow body. The large hooked bill and feet are black. The scarlet macaw is bright, scarlet over the head and body with yellow and blue on the wings. The tail feathers are red and blue. The upper mandible is white or gray in colour and the lower mandible is black. There is no colour dimorphism. The hyacinth macaw is the largest macaw. It is deep blue and has a massive, hooked with yellow ring around both eyes and at the base of bill.

Social structure : They originate from the forested areas of Mexico, Central and South America. It is difficult to differentiate male and female birds based on physical characteristics, therefore, endoscopy or laboratory methods must be used for sex determination in breeding facilites.

Macaws that are allowed unrestricted access in home can encounter numerous physical dangers or toxins, therefore, wing clipping is recommended which helps to prevent it from developing rapid and sustainable flight and to prevent escape. Trimming also helps the birds more owner dependent and less aggressive.

Most common diseases : Neuropathic gastric dilatation, reproductive disorders, feather picking and cyst, papilloma, mutilation syndromes, chlamydiosis, bacterial and viral infection, sunken eye sinusitis, toe deformities and mal coloured feathers.

CHAPTER 29

WILDLIFE DISEASE SURVEILLANCE

Conservation of a wildlife needs monitor and survey several factors in a given period of time for understanding the trend for positive and negative effect on a given population. Several epidemics and pandemics records justify the need of survey of individual entities. Recent pandemic of corona virus (COVID-19) in human being from more than 213 countries with 90,81,145 confirmed cases along with 4,71,308 total deaths (June 22, 2020) with highest cases of 23,57,667 in USA followed by countries like Brazil, Russia, India, United Kingdom, Spain, Peru, ,Chile, Italy, Iran, Germany and China (arcgis.com) alarms and suggest needs of preparedness for the wildlife pathogen surveillance program on top priority. New York Zoo tigers and lions were also reported positive for COVID-19 along with domestic cat in USA should not ignore the future threats to wildlife population.

For planning any programme for pathogen or disease surveillance, it is necessary to understand the difference of monitoring and surveillance.

Mon·i·tor·ing (mon'i-tŏr'ing)

1. Performance and analysis of routine measurements aimed at detecting a change in the environment or health status of a population.
2. Ongoing measurement of performance of a health service.

Sur·veil·lance (sŭr-vā'lănts)

1. The collection, collation, analysis, and dissemination of data; a type of observational study that involves continuous monitoring of disease occurrence within a population.
2. Ongoing scrutiny, generally using methods distinguished by practicability, uniformity, or rapidity, rather than complete accuracy.

"Wildlife disease surveillance" may also refer to pathogen surveillance in wildlife, given that infection with pathogen(s) may not always produce visible clinical signs associated with disease in a given species or at a given point of time. The objective of a surveillance programme should be clearly defined as to whether it is aimed at disease or pathogen detection.

Advantages of wildlife disease surveillance

1. It can be a useful and complementary component of human and domestic animal disease surveillance.
2. It helps in prevention and control measures of diseases.
3. It boost up the conservation efforts.
4. It may provide information of domestic and wild animal morbidity and mortality.
5. It helps to identify changes in patterns of disease occurrence over time
6. It will assist in early detection of disease outbreaks, including those linked to emerging diseases.

Limitations of wildlife disease surveillance

1. Wild animals do not have close observational vigilance and monitoring, which can limit detection and reporting of diseases in wildlife as well as access to data collected from other sources .
2. Some diagnostic tests may not be validated for wild species in terms of specificity and sensitivity.
3. There are also different stakeholders and participants; for example, wildlife biologists and ecologists should be engaged in the development, analysis, interpretation and communication of results for a wildlife disease surveillance programme which may not have enough knowledge of patterns of animal diseases.
4. Additionally, hunters, wildlife managers or rehabilitators, protected area managers and other stakeholders may be key collaborators in acquiring specimens. While there are not always clear solutions for management and control of diseases in wildlife detected by surveillance efforts, knowledge of the occurrence of specific diseases and pathogens in wildlife can be used to reduce health and economic risks to domestic animals and people.

Planning of wildlife disease monitoring

With a few modifications based on the limitations specified above, much of the diagnostic, information management, and communication capacity in existing animal health surveillance programmes can be used for wildlife disease surveillance programmes. Like domestic animal disease surveillance, wildlife disease surveillance programmes should be implemented as an ongoing, continuous activity providing actionable information.

Core Components wildlife disease monitoring

There are four essential core components of all disease surveillance programmes

1. **Detection of pathogens and diseases:** These efforts may require broad participation from many stakeholders to gain access to samples. Training of stakeholders can greatly improve detection.

2. **Identification of pathogens and diseases:** Many pathogens infecting wildlife are readily identified by diagnostic capacity of well-equipped veterinary diagnostic laboratories established for domestic animals. Some wild animal pathogens or diseases may be rare or new to science, and their identification may require follow-up analysis (e.g. genetic sequencing). Detection and identification of pathogens of importance in wildlife may justify investment in targeted surveillance efforts to acquire more detailed information.
3. **Analysis and communication:** Review of information obtained from surveillance and analysis in various ways requires input from epidemiologists, wildlife biologists and ecologists . The validity and accuracy of test results should be carefully considered, especially if the sensitivity and specificity of the diagnostic tests used have not been validated in wildlife .
4. **Information Management:** At least a minimum level of data should be collected; for example, data should be recorded on the disease incident or sampling event, date, latitude and longitude coordinates, observation of mortality or sickness, specimen identification numbers, animal species, laboratory identification numbers, and diagnosis with associated detection method .

Surveillance Programme Strategies

These critical components are independent activities carried out by different groups of people. Therefore, constant coordination across all four critical components is crucial; roles must be clearly designated, with frequent communication across the surveillance network.

There are two main categories of wildlife disease surveillance. Both are designed with the same four essential components, but have the following distinctions, which largely affect sample collection methods:

I. **General or Scanning wildlife disease surveillance (sometimes referred to as "passive" surveillance):** It is aimed at detecting disease and pathogens in wild animals, rather than obtaining statistical data on one or a few pathogens, such as pathogen prevalence estimates. A wide range of stakeholders (such as wildlife researcher, biologist, forest workers, rangers, conservation organisations, etc.) might be involved in an opportunistic disease detection network for general surveillance. Anatomical pathology is an especially important capacity for general wildlife disease surveillance to determine cause of death and disease which requires a trained veterinarian for providing details of post mortem findings.

II. **Targeted wildlife disease surveillance (sometimes referred to as "active" surveillance):** It is focused on one or more particular pathogens in one or more wild animal species, typically is used to obtain statistical data on prevalence, age and sex distribution of infection, or geographic distribution of the pathogen. Although there are often challenges in getting a representative sample base, this approach can more precisely estimate prevalence or incidence, unique field methods (such as radar tracking or mark-recapture efforts) may be necessary to estimate population size and structure. Specific decisions must be made regarding, sample size, sampling times and places, specific species, and number and type(s) of samples to collect in targeted surveillance programmes.

The determination of whether to use general or targeted surveillance depends on the goals of each programme, as well as the resources available. Programmes may also employ a mix of general and targeted wildlife disease surveillance.

Risk-based surveillance approaches may also be used, which may be informed by targeted surveillance data. This may be an especially important priority for initiating wildlife disease surveillance in settings where resources are limited.

The lack of validation of some diagnostic tests in wild animal species may present unique challenges in selecting tests for specific pathogens. In these cases, tests should be selected on a species-specific basis, in addition to considering the capabilities of laboratories available to conduct the testing, cost, and recommended sample type.

OIE References Laboratories can be consulted to advise on such considerations for pathogen-specific screening methods in wildlife.

The specific goals of a wildlife disease surveillance system should be clearly defined.

Planning of wildlife disease surveillance:

1. **Flexibility of time frame:** However, wildlife disease surveillance systems, like any surveillance system, also benefit from flexibility. This is especially important as more information is generated that can improve understanding of wildlife disease risks and help refine surveillance strategies. Flexibility is also important given that priorities may change; for example, an influenza outbreak in poultry originating from a wild bird strain may demand enhanced wild bird surveillance. Having wildlife disease surveillance capacity in place that can be scaled up rapidly as needed can help achieve early detection, and inform response, and control measures.
2. **Collaborative approach:** Where possible, surveillance of wildlife at sites where human or domestic animal surveillance is also occurring may help provide information on cross-species disease transmission risks.
3. **Conservation:** Non-lethal sampling of wildlife is encouraged to support biodiversity conservation goals (and killing of certain wildlife species may be prohibited by national or regional endangered species listings). However, this should not be to the exclusion of samples provided by hunters, or samples from wildlife mortality events, where available and appropriate. For a select number of diseases, animals exhibiting suspected disease may require culling to obtain samples for disease screening (for example, as seen with infection with rabies virus).

4. **Communication:** printing materials, teleconference lines, websites, hosting of or travel to meetings with stakeholders, including for coordinated planning, data review and interpretation.
5. **Training:** Capacity building resources (such as information workshops, hands-on training, textbooks).
6. **Budget and financial resources:** Additional costs may include fees for laboratory certification, continuing education, and consultations with reference laboratories. Budgets can be developed for a general national wildlife disease surveillance plan, or for disease-specific targeted surveillance (for example, a specific programme targeted for highly-pathogenic avian influenza in wild birds).

Budget Considerations

Many existing resources often can be used for wildlife disease surveillance, which may provide substantial cost-efficiencies. For example, specimens collected from wildlife disease surveillance may be tested at existing human or animal health laboratories, rather than developing a separate laboratory. Such an integrated approach is highly encouraged to reduce duplication of efforts and unnecessary investments, and to promote collaboration between wildlife and domestic animal and/or human health authorities. In many countries collaborative wildlife health regional centers are doing good and can be used for such surveillance programme.

Practical Considerations

1. **Optimization of survey with modern technologies :** Information management is critically important to pathogen surveillance. It requires dedicated full-time personnel and continuous modification as the standards and tools of computing and data management change over time. The information management system usually can be designed to serve the needs both of general surveillance and also of targeted surveillance. Through the Internet, it now is feasible and affordable to create a central information management system that can be used by all participants in the surveillance programme in all parts of a country.
2. **Analysis of Data and Communication of Results:** The important component of pathogen surveillance is analysis of the data produced by detection and diagnosis, and communication of those results to those who need this information. Who can do this work? Analysis and interpretation of wildlife pathogen surveillance data requires the combined expertise of wildlife biologists, wildlife pathogen and disease specialists, epidemiologists and communications specialists. Each of these areas of expertise is required to correctly interpret the results of wildlife pathogen surveillance and to transmit the information to others. Thus, the surveillance programme must include a small team of people expert in these fields and who understand the purpose of the surveillance programme.

Information on wild animal pathogens generally is required in four areas of public responsibility:i. public health ii. domestic animal health iii. wildlife conservation and management iv. environmental management. Analysis of surveillance data must serve all four of these areas, and the concerns and interests of each often are very different. For example, public health agencies will want to know about zoonotic diseases and food safety. Veterinary Services will be concerned about pathogens shared with domestic animals and potential implications for food production, agricultural economics and international trade. Wildlife conservation agencies will be concerned about potential effects on wild animal populations and potential conflicts between wild animal populations and human activities. Environmental managers will be concerned about the stability and resilience of ecosystems and detection of toxic chemicals or other environmental contaminants. In addition, the public will expect to be informed accurately and immediately, whenever pathogens in wild animals create a significant risk to themselves, their animals or their environment, including wildlife.

Interventions to manage pathogens and diseases in wild animals

In human medicine and in veterinary medicine applied to domestic animals, it is standard practice to use pharmaceutical agents, vaccination, sanitation, food inspection and other actions to prevent, treat and reduce the impact of infectious pathogens and diseases. However, this is not so for pathogens and diseases in wild animal populations. Standard medical techniques are difficult, often impossible, to apply to wild animals. When techniques such as vaccination or treatment with drugs are successfully applied to wild animals, each programme is preceded by years of costly research to develop and validate the techniques used and each requires years of very costly implementation to achieve the desired results. Most attempts to control pathogens and diseases in wild animal populations have failed; only a very few have succeeded.

There are four strategies that can be applied to management of pathogens and diseases in wild animals.

I. **Before health issues arise from wildlife pathogens:** Prevent new health problems from arising.

II. **After health issues from wild animal pathogens have emerged:**

   Take no action or response to the health issue;

   Intervene to control the health issue to some degree;

   Intervene to eradicate the pathogen of concern.

III. **Many approaches have been taken to control or to eradicate pathogens from wild animal populations.**

   These approaches have included:

   treatment with drugs delivered in oral baits or by remote injection

vaccination: oral baits, remote injection, or trap-vaccinate-release

reducing animal populations: reduced reproduction, translocation, killing

changing animal distribution: fences, deterrents, attractants

altering the environment: drainage, flooding, burning, insecticides.

IV. Decisions on whether or not to attempt to control or eradicate pathogens in wild animal populations should be informed by a complete review of the control methods available and of the rationale and objectives of a control programme.

Most often, there is little that can be done to control pathogens in wildlife populations, and the best choice will be to attempt to reduce the impact of such pathogens by actions that target the affected domestic animal or human populations.

Separate domestic animals from infected wild animals;

Vaccinate people and domestic animals;

Focus on human behaviour: - cook meat - purify drinking water - prevent insect bites - control rodent populations around people.

Such programmes can significantly reduce transmission of wildlife pathogens to people and domestic animals, but do not require that disease management be attempted in wild animal populations.

1. **Prevention:** Countries should strive to have active programmes to prevent health issues arising from wild animal pathogens. Actions to prevent emergence of new health issues associated with wild animal pathogens are feasible and cost-effective. These should focus on the greatest risk factors associated with wild animal pathogens. One such risk factor is the transportation, or translocation, of wild animals from one geographic area to another geographic area. The OIE Working Group on Wildlife Diseases has identified wild animal translocation as a particularly high-risk activity. Such wild animal translocations are carried out with high frequency world-wide. Health risk assessments carried out for such wildlife translocations are powerful preventive actions that can identify health risks and propose mitigation strategies to reduce or eliminate these risks.
2. **Health risk assessment in wild animal translocations:** Wild animals are moved from place to place for many different reasons. Most often, they are captured in the wild, transported, held in quarantine, and released again into the wild for conservation or wildlife management purposes. Sometimes this also is done for commercial purposes.

   There are potential health risks associated with all such movements of wild animals.

   The principal risks are:

   - That the animals will carry pathogens into the destination environment that will cause harm to the destination environment.
   - That the animals being moved will encounter pathogens in the destination environment and will be harmed by these new pathogens.

Health risk analysis can be carried out prior to the translocation of wild animals in order to determine:

a) whether or not such risks exist, and

b) the magnitude of the potential consequences, to the economy and ecology of the destination area and to the success of the translocation programme.

3. **Risk analysis process:** Health risk analysis is a rigorous application of common sense to determine whether or not there are important health-related risks associated with a proposed activity, such as animal translocation.

   Risk analysis can be qualitative, in which risk is estimated as being negligible, low, medium or high, or it can be quantitative, in which mathematical models are used to give numerical estimates of the probability of a negative outcome and the economic, ecological and social harm that would occur as a result.

   Wild animal health risk analysis usually will result in a qualitative assessment of risk. This is because, most often, there is not enough reliable numerical information about wild animals and their pathogens to support a reliable quantitative assessment of risk.

   Qualitative risk assessments are extremely valuable and can contribute as much or more to decision making and risk mitigation as can quantitative risk assessments.

   The product of a risk analysis is a written report that documents all steps followed, all of the information considered and the way that information was evaluated.

   Basic steps in health risk analysis in wild animal translocations includes:

   a) Translocation plan: A complete, detailed description of the translocation is made. This clearly defines the activity for which health risks are to be analysed.

   b) Health hazard identification and selection for assessment: complete list of potential hazards, the hazards that appear most important are selected for detailed consideration. Often, only a small number of hazards can be fully assessed.

4. **Information requirements:** Many different kinds of information are required for risk analysis: species and populations of animals, pathogens and their mechanisms of transmission and spread, transportation and quarantine facilities and procedures, and general information about the source and destination environments, including their human economies and cultures. If sufficient information is not available, it is not possible to carry out an analysis of health risks. Yet, very often, sufficient information is available to permit a health risk assessment that will contribute importantly to reducing risks and preventing wildliferelated disease problems.

5. **Decision making:** Decisions whether or not to proceed with wild animal translocations, or with other programmes that include wildlife health hazards and risks, may be determined by the results of health risk analysis, but they also may be influenced by a variety of other factors. Risk analysis informs decision makers regarding health risks and provides them with options to reduce risk if it is decided to proceed with the translocation or other programme.

6. **Using Digital Tools for Disease Surveillance**

Mapping Innovations: advance scientific approaches are now available and should be used for planning the wildlife health care management and surveillance.

Health maps, Geographical Information system (GIS), Remote sensing (RS) and several advanced software can be helpful for long term and quick wildlife surveillance programmes.

CHAPTER 30

ADVANCE SURVEILLANCE TECHNOLOGIES

Data availability and required actively generated information in relation to time, place and updated status of agent for any disease outbreak is a basic need which makes the baseline surveillance strongly. Strong surveillance is now become essential so that it can alert authorities for new outbreaks, provides real-time information here to improve situational awareness of existing out breaks and which can identify geographic areas where surveillance needs to be strengthened.

Broadly speaking, such strong surveillance needs digitalized information with use of advanced technological tools and soft wares.

The existing innovations can be broken down into three different categories:

1. **Passively generated information:** The data already exists; we just need to use it.
2. **Actively generated information:** Using new methods to get data and
3. **Information aggregation and visualization**

1. Passively Generated Information

This type of information is generated as a result of our day-to-day activities. Information about what people say and do is captured in online media like news, advertising, or social media, and their use of financial, communication, or information.

Since this information is already being produced, it is typically low cost, having high volume/frequency, noisy and digital driven by technology users like Germ Tracker or Global Public Health Intelligence Network (GPHIN) programmes etc.

a. **Germ Tracker:** Germ Tracker analyses Tweets using a combination of machine learning and human computation to find meaningful trends about the spread of illnesses.
b. **The Global Public Health Intelligence Network (GPHIN):** It identifies outbreaks from websites, news wires, local and national newspapers retrieved through news aggregators. It uses machine translation and manual analysis. Examples like social networking sites were used in the surveillance of an avian influenza A(H7N9) outbreak in 2013.

c. **Sick weather:** It scans social networks for indicators of illness so you can check for the chance of sickness in your geographic area.

d. **Google apps:** Several Google apps uses certain search terms as indicators of COVID-19,dengue and influenza activity. They use aggregated search data to estimate activity around the world in real-time.

e. **Doctor Me:** is a mobile app developed to provide health information to users in Thailand.

2. Actively Generated Information

Information is generated by collecting information from public, through media or from authentic sources which will be compiled. Examples of such information

a. **Citizen reporting**

Flu Near You and Outbreaks Near Me both use information from HealthMap to enable the community to report new outbreaks and receive current outbreak information. 'Flu Near You' is a website, 'Outbreaks Near Me' is an android application.

Wildlife Health Event Reporter is a USA-based online tool to gather information from citizen scientists about sick, injured and dead wildlife.

Dengue na Web is a tool used in Brazil to enable citizens to contribute information on dengue cases. This data is used for real-time analysis, and to simulate disease spread in the city and intervention scenarios. This was based on the influenza project Gripe net in Spain. Influenza net is a Europe-wide, web-based flu surveillance system that monitors the activity of influenza-like-illness with the help of citizen volunteers. Here, Daniela Paolotti discusses their experiences implementing this project, including some of the considerations that they made when implementing this system across different countries and cultures.

In India Government has also launched "Aarogya Setu" app for COVID-19.Similarly Google and Apple is also set up different apps .

b. **Mobile data collection through surveys:** For mobile collection through surveys, multiple reporting tools with different characteristics have been developed and the range can be quite overwhelming. The NOMAD Project is a really valuable source of information for navigating the available options. Their online selection tool is a quick survey that matches your needs with appropriate solutions.

3. Information Aggregation and Visualization

No single data source can capture the full picture of the complex environments that we work in. These new data sources should be considered as complementary to existing sources. Aggregation is important to pull the different sources together, visualization so that the information can be understood and used.

ADVANCED DIGITAL TOOLS FOR DISEASE SURVEILLANCE

i. HealthMap

It brings together disparate data sources, including online news aggregators, eyewitness reports, expert-curated discussions and validated official reports, to achieve a unified and comprehensive view of the current global state of infectious diseases and their effect on human and animal health. These disparate data sources include crowdsourcing from 'Flu Near You' and 'Outbreaks Near Me' (as discussed above).

HealthMap, a team of researchers, epidemiologists and software developers at Boston Children's Hospital founded in 2006, is an established global leader in utilizing online informal sources for disease outbreak monitoring and real-time surveillance of emerging public health threats.

The freely available Web site 'healthmap.org' and mobile app 'Outbreaks Near Me' deliver real-time intelligence on a broad range of emerging infectious diseases for a diverse audience including libraries, local health departments, governments, and international travelers. HealthMap brings together disparate data sources, including online news aggregators, eyewitness reports, expert-curated discussions and validated official reports, to achieve a unified and comprehensive view of the current global state of infectious diseases and their effect on human and animal health. Through an automated process, updating 24/7/365, the system monitors, organizes, integrates, filters, visualizes and disseminates online information about emerging diseases in nine languages, facilitating early detection of global public health threats. Download our brochure to learn more.

ii. OpenStreetMap

OpenStreetMap is a project to create a free and open map of the entire world, built entirely by volunteers surveying with GPS, digitizing aerial imagery, and collecting and liberating existing sources of geographic data. In response to the recent Ebola outbreak in West Africa, the humanitarian community worked online to map the outbreak area. As of April 15th (20 days in total), 363 contributors had mapped 1.65 million objects, 150,000 buildings, 5100 places, 9,900 landuse polygons and 22,200 highway sections.

iii. Epi SPIDER

It takes free text from outbreak/epidemic summaries and visualizes it on a map using natural language processing and visualization algorithms. Although the visualization is automated, the data source is Pro MED-mail it is curated manually by experts. Malaria Atlas Project works to generate new and innovative methods of mapping malaria risk.

iv. Global Public Health Intelligence Network (GPHIN) programmes

Accurate and timely information on global public health issues is key to being able to quickly assess and respond to emerging health risks around the world. Health Canada, in collaboration with the World Health Organization (WHO), has developed the Global Public Health Intelligence Network (GPHIN). Information from GPHIN is provided to the WHO, international governments and non-governmental organizations who can then quickly react to public health incidents.

GPHIN is a secure Internet-based "early warning" system that gathers preliminary reports of public health significance on a "real-time", 24 hours a day, 7 days a week basis. This unique multilingual system gathers and disseminates relevant information on disease outbreaks and other public health events by monitoring global media sources such as news wires and web sites. The information is filtered for relevancy by an automated process which is then complemented by human analysis. The output is categorized and made accessible to users. Notifications about public health events that may have serious public health consequences are immediately forwarded to users.

What types of surveillance does GPHIN conduct?

GPHIN has a broad scope. It tracks events such as disease outbreaks, infectious diseases, contaminated food and water, bio-terrorism and exposure to chemicals, natural disasters, and issues related to the safety of products, drugs and medical devices.

Who uses GPHIN?

Users include non-governmental agencies and organizations, as well as international government authorities who provide public health surveillance. GPHIN is used by the global public health community in its efforts to minimize health risks by developing appropriate risk management, control and prevention measures and responses.

Who manages GPHIN?

GPHIN is managed by Health Canada's Centre for Emergency Preparedness and Response (CEPR), which was created in July 2000 to serve as Canada's central coordinating point for public health security. It is considered a center of expertise in the area of civic emergencies including natural disasters and malicious acts with health repercussions. CEPR offers a number of practical supports to municipalities, provinces and territories, and other partners involved in first response and public health security. This is achieved through its network of public health, emergency health services, and emergency social services contacts.

v. PH surveillance

Public health surveillance is the ongoing systematic collection, analysis, and interpretation of data, closely integrated with the timely dissemination of these data to those responsible for preventing and controlling disease and injury. Public health surveillance is a tool to estimate the health status and behavior of the populations served by ministries of health, ministries of finance, and donors.

Because surveillance can directly measure what is going on in the population, it is useful both for measuring the need for interventions and for directly measuring the effects of interventions. The purpose of surveillance is to empower decision makers to lead and manage more effectively by providing timely, useful evidence.

vi. Informal Networks as Critical Elements of Surveillance Systems

WHO and other agencies frequently receive telephone calls or informal reports about urgent health events. WHO publishes an informal list of these "rumors," which allows public health workers to respond to health risks promptly rather than waiting for formal reports (http://www.who.int/csr/don/en/).

vii. Global Surveillance Networks

Globally, infectious disease surveillance is implemented through a loose network that links parts of national health care systems with the media, health organizations, laboratories, and institutions focusing on particular disease conditions. WHO has described a "network of networks" that links existing regional, national, and international networks of laboratories and medical centers into a surveillance net work.

viii. Global Infectious Disease Surveillance Frameworks

Government centers of excellence (for example, CDC, the French Pasteur Institutes, and FETPs) along with WHO country and regional offices also contribute to disease and health condition reporting. Military networks, such as the U.S. Department of Defense's Global Emerging Infectious Disease System, and Internet discussion sites, such as Pro Med (http://www.promedmail.org) and Epi-X (http://www.cdc.gov/epix), also supplement the reporting networks. In 1997, WHO started the Global Outbreak Alert and Response Network, and it was formally adopted by WHO member states in 2000. The network has more than 120 partners around the world and identifies and responds to more than 50 outbreaks in developing countries each year .

Future of Surveillance:

Public health agencies, ministries of finance, and international donors and organizations need to transform surveillance from dusty archives of laboriously collected after-the-fact statistics to meaningful measures that provide accountability for local health status or that deliver real-time early warnings for devastating outbreaks.

This future depends in part on developing consensus on critical surveillance content and developing commitment on the part of countries, funding partners, and multilateral organizations to invest in surveillance system infrastructure and to use surveillance data as the basis for decision making.

This vision of the future assumes a coherent, integrated approach to surveillance systems that is based on matching the surveillance objective with the right data source and modality and on paying attention to country-specific circumstances while maintaining global attention to data content needs.

Public health surveillance is an essential tool for ministries of finance, ministries of health, and donors to effectively and efficiently allocate resources and manage public health interventions.

To be useful, public health surveillance must be approached as a scientific enterprise, applying rigorous methods to address critical concerns in this public health practice.

Although the surveillance needs in the developing world appear to differ from those in the developed world, the basic problems are similar.

There is need to set up system for wildlife disease surveillance program in collaboration among practitioners, researchers, nations, and international organizations is necessary to address the global needs of public health surveillance.

CHAPTER 31

WILD LIFE PROTECTION ACT

WILDLIFE PROTECTION ACT, 1972

It covers the following aspects in details as definitions, Authorities to be appointed or constituted under the Act, Hunting of Wild Animals. Prohibition of Hunting Maintenance of records of wild animals killed or captured, Hunting of wild animals to be permitted in certain cases, Grant of permit for special purposes, Suspension or cancellation of licence, Appeals, Hunting of young and female of wild animals, Declaration of closed time, Restrictions on hunting, Protection of Specified Plants, Sanctuaries, National Parks and Closed Areas, Sanctuaries, Declaration of Sanctuary, National Parks, Declaration of Game Reserve, Game Reserve, Declaration of closed area, Power of Central Government to declare areas as Sanctuaries or National Parks, Central Zoo Authority and Recognition of Zoos, Trade or Commerce in Wild Animals, Animal Articles and Trophies, Restriction on transportation of Wildlife, Purchase of captive animal, etc. by a person other than a licensee, Prohibition of trade or commerce in trophies, animal articles, etc., Power of entry, search, arrest and detention, Penalties, Attempts and abetment, Punishment for wrongful seizure, Power to compound offences, Cognizance of offences, Operation of other laws not barred, Presumption to be made in certain cases, Offences by companies, Officers to be public servants, Protection of action taken in good faith 60A Reward to persons, Power to alter entries in Schedules, Declaration of certain wild animals to be vermin, Power of Central Government to make rules, Power of State Government to make rules, Rights of Scheduled Tribes to be protected.etc.

For the detail information of wildlife protection act, the reader is requested to go through the book on WLP act. Here are some of the amendment bills and draft bills up to 2010, information provided for current status on act.

The scheduled animals under different categories are specified as I, II, II, IV, V groups. Mostly under Schedule-I covers most of endangered species like lion, tiger, black buck etc. The list is attached separately.

THE WILD LIFE (PROTECTION) AMENDMENT BILL, 2010

Preamble

A Bill to further amend the Wild Life (Protection) Act, 1972

BE it enacted by Parliament in the Sixtieth Year of the Republic of India as follows:—

1. **Short title, extent and commencement**

   (1) This Act may be called the Wildlife (Protection) Amendment Act, 2010

   (2) It shall come into force on such date as the Central Government may, by notification in the Official Gazette appoint.

2. **Amendment of Section 2 of the Act :** In the Wildlife Protection Act, 1972 (hereinafter referred to as the principal Act), in Section 2 the following amendments shall be made:

   (1) In sub-section (15) the following shall be added after the words "wild animal": "or specified plant"

   (2) In clause (b) of sub-section (16), the word "electrocuting" shall be inserted after the word "trapping".

   (3) The following shall be inserted as sub-section (17A): ""Leg hold Trap" means a device designed to restrain or capture an animal by means of jaws which close tightly upon one or more of the animal's limbs, thereby preventing withdrawal of the limb or limbs from the trap."

   (4) In sub-section (24) the words "a body corporate, any department of the Central or the State Government or any authority or any other association of persons whether incorporated or not." shall be inserted after the words "a firm".

   (5) In sub-section (35) the word "chainsaw" shall be inserted before the word "firearms".

   (6) In sub-section (36) the words "and found wild in nature" shall be replaced by the words "or found wild in nature"

   (7) The following shall be inserted as sub-section (30A): "Scientific research" means any activity carried out only for the purpose of research on any wild animal or plant listed in any Schedule to this Act or discovered in the wild in India and the habitats of the same."

3. **Amendment of Section 8 of the Act :** In clause (c) of Section 8 of the Principal Act, the words "of the Act" shall be inserted after the words "and schedule".

4. **Insertion of new Section 9A**

   (1) After Section 9 of the principal Act, the following section shall be inserted: "9A. Prohibition on Leg hold Traps.- (1) No person shall manufacture, sell,purchase, transport or use any Leg hold Trap.

   (2) No person shall keep in his custody, control or possession any Leg hold Trap except with prior permission in writing of the Chief Wild Life Warden.

   (3) Every person having at the commencement of this Act, the possession of any Leghold Trap, shall within sixty days from the commencement of this Act, declare to the Chief Wild Life Warden, the number and description of Leg hold Traps in his possession and the place or places where such Leg hold Traps are stored.

(4) The Chief Wild Life Warden may, if he is satisfied that a person will use a Leg hold Trap in his possession only for educational or scientific purposes, issue such person with written permission to possess such trap subject to such conditions that he may see fit to impose.

(5) All Leg hold traps, which have been declared under sub-section (3) of this Section and in respect of which permit has not been granted in writing by the Chief Wild Life Warden under sub section (4) of this Section, shall be property of the State Government.

(6) In the prosecution for any offence under this section, it shall be presumed that a person in possession of a Leg hold Trap is in unlawful possession of such Trap, unless the contrary is proven by the accused.

5. **Insertion of new Section 12A :** After Section 12 of the principal Act, the following section shall be inserted: "12A. Grant of Permit for scientific research.-

   (1) Notwithstanding anything contained elsewhere in this Act, the Chief Wildlife Warden, shall:

      (a) On application, grant a permit, by an order in writing to any person, toconduct scientific research on any animal specified in Schedules I to IV or any specified plant found wild in India or habitat of the same.

      (b) The Chief Wildlife Warden shall ensure that all permits for scientific research are processed and granted in accordance with the rules and regulations as may be prescribed, from time to time, by the Central Government in this behalf.

   (2) The Central Government shall, by notification in the Official Gazette, make rules and regulations regarding the conduct of scientific research including but not limited to:

      (a) the minimum qualifications of the persons and organizations who will be eligible for the grant of permits under this Section;

      (b) the time frame in which proposals for scientific research must be disposed off; which shall in no case exceed 120 days.

      (c) the conditions subject to which permits for scientific research may be granted.

6. **Amendment of Section 22 of the Act :** In Section 22 of the principal Act, the words, ", and of the Gram Sabha" shall be inserted after the words "the records of the State Government"; and the words," acquainted with the same" shall be replaced by, "acquainted with such right".

7. **Insertion of new Section 26B :** The following shall be inserted after Section 26A as the new Section 26B:"26B. Compliance with Forest Rights Act.- In the settlement of rights for all scheduled tribes and forest dwellers in sanctuaries and National Parks for which the notification under sub-section (1) of Section 18 or sub-section (1) of Section 35 has been issued after the commencement of The Scheduled Tribes and Other

Traditional Forest Dwellers (Recognition of Forest Rights) Act 2006 (Act No. 2 of 2007), the Collector shall ensure that the provisions of that Act are complied with."

8. **Amendment of Section 28 of the Act :** In clause (b) of sub-section (1) of Section 28 of the principal Act the words "and documentary film-making without manipulating any habitat or causing any adverse impact to the habitat or wildlife" shall be inserted after the word "photography".

9. **Amendment of Section 29 of the Act :** The Explanation to Section 29 of the principal Act shall be deleted and the following shall be inserted in its place: "Explanation – For the purposes of this section, an act permitted under section 33, or hunting of wild animals under a permit granted under section 11 or under section 12, or the exercise of any rights permitted to continue under section 24 (2) (c), or the bona fide use of drinking and household water by local communities, shall not be deemed to be an act prohibited under this section."

10. **Amendment of Section 32 of the Act :** In Section 32 of the principal Act, the words "or equipment" shall be inserted after the words "other substances".

11. **Amendment of section 33 of the Act**

    (1) The words "in accordance with approved management plans" shall be inserted after the words "manage and maintain all sanctuaries".

    (2) In clause (a) of section 33 the words "or Government" shall be inserted after the word "tourist".

12. **Amendment of Section 35 of the Act**

    (1) In sub-section (2) of section 35 of the principal Act, the following shall be inserted after the words "declared as a National Park": "The notification shall, wherever feasible, include forest compartment numbers and relevant details of forests, revenue and other Government records pertaining to the area proposed to be declared a National Park."

    (2) In sub-section (8) of Section 35 of the Act, the word "18A," shall be inserted before the words "27 and 28".

13. **Amendment of Section 36D of the Act :** In sub-section (2) of Section 36 D of the principal Act, the word "five" before the word "representatives" shall be deleted; and the following Explanation shall be inserted after sub-section (2) of Section 36D: "Explanation – Where a Community Reserve is declared on private land under section 36C(1), the Community Reserve Management Committee shall consist of the owner of the land along with a representative of the State Forests or Wildlife Department under whose jurisdiction the Community Reserve is located."

14. **Amendment of Section 39 of the Act**

    (1) In clause (a) of sub-section (1) of Section 39 of the principal Act, the words "or specified plant picked, uprooted, kept, dealt with or sold" shall be inserted after the words "bred in captivity or hunted"

(2) The following sub-sections shall be inserted after sub-section

(3) of Section 39 of the principal Act:

(4) Where any such Government property is a live animal, the Central Government, or the State Government as the case may be, shall ensure that it is housed and cared for by a recognized zoo or rescue centre when it cannot be returned to its natural habitat."

15. Insertion of new Chapter VB : After Chapter VA of the principal Act, the following Chapter shall be inserted, namely: "CHAPTER VB: REGULATION OF TRADE IN ENDANGERED SPECIES OF WILD FAUNA AND FLORA"

(1) The following shall be inserted as the new Section 49D "49D.- In this Chapter, unless the context otherwise requires:

(a) "artificially propagated" shall have the same meaning as given to it in Conference Resolution 11.11(Rev. CoP15) of the Convention, as it may be amended from time to time.

(b) "bred in captivity" shall have the same meaning as given to it in Conference Resolution Conf. 10.16 (Rev.) of the Convention, as may be amended from time to time.

(c) "Convention" means the Convention on International Trade in Endangered Species of Wild Fauna and Flora signed at Washington, D.C., in the United States of America on the 3rd of March 1973, and amended at Bonn on the 22nd of June 1979, its appendices, decisions, resolutions and notifications made thereunder, to the extent binding on India;

(d) "import" with its grammatical variations and cognate expressions means bringing into India from a place outside India;

(e) "exotic species" means species of animals and plants not found in the wild in India and not listed in the Appendices to the Convention, that are notified by the Management Authority under sub-section (3) of Section 49F for the reasons mentioned in clause (b) of Section 49E;

(f) "export" with its grammatical variations and cognate expressions means taking outside India from a place in India;

(g) "introduction from the sea" means transportation into India of specimens of any species which were taken from the marine environment not under the jurisdiction of India;

(h) "Management Authority" means the Management Authority designated under section 49 F;

(i) "plant" means any member, alive or dead, of the plant kingdom, including seeds, roots, and other parts thereof;

(j) "readily recognisable part or derivative" includes any specimen which appears from an accompanying document, the packaging or a mark or label, or from any other circumstances, to be a part or derivative of an animal or plant of a species included in Schedule VII;

(k) "re-export" means export of any specimen that has previously been imported;

(l) "Scientific Authority" means a Scientific Authority designated under section 49 G;

(m) "scheduled specimen" means any specimen of a species listed in Appendices I, II and III of the Convention and updated from time to time.

(n) "specimen" means

(i) any animal or plant, whether alive or dead;

(ii) in the case of an animal: for species included in Appendices I and II, any readily recognisable part or derivative thereof; and for species included in Appendix III, any readily recognisable part or derivative thereof specified in Appendix III of Schedule VII in relation to the species; and

(iii) in the case of a plant: for species included in Appendix I, any readily recognisable part or derivative thereof; and for species included in Appendices II and III, any readily recognisable part or derivative thereof specified in Appendices II and III of Schedule VII in relation to the species;

(o) "trade" means export, re-export, import and introduction from the sea."

(2) The following shall be inserted as the new Section 49E: "49E.- The provisions of this Chapter shall apply to:

(a) specimens of animal and plant species listed in Appendices I, II and III of the Convention, and incorporated in Schedule VII; and

(b) exotic species of animals and plants, that is to say specimens not covered by the Convention, which require regulation:

(i) to protect the indigenous gene pool of the wildlife found in India; or

(ii) because such species maybe invasive in nature and may pose a threat to the wildlife or ecosystems of India; or

(iii) because such species are, in the opinion of the Scientific Authority,critically endangered in the habitats in which they occur naturally."

(3) The following shall be inserted as the new Section 49F: "49F.-

(1) The Central Government may designate, by notification in the Official Gazette, an officer not below the rank of Additional Director General of Forests as the Management Authority for the purposes of this Chapter.

(2) The Management Authority shall be responsible for issuance of permits and certificates regulating the import, export and re-export of any scheduled

specimen, submission of reports, registration of institutions and other documentation as required under this Chapter.

(3) The Management Authority shall, by notification in the Official Gazette, and on the advice of the Scientific Authority, notify the exotic species of animals and plants not covered by the Convention.

(4) The Management Authority shall prepare and submit annual and biennial reports to the Central Government for forwarding it to the Secretariat of the Convention.

(5) The Central Government may appoint such officers and employees as may benecessary to assist the Management Authority in carrying out his responsibilities under this Chapter, on such terms and conditions of service including salaries and allowances as may be prescribed.

(6) The Management Authority may with the prior approval of the Central Government, delegate its powers, except the power to notify exotic species under subsection (3), to such other officers not below the rank of Assistant Inspector General of Forests, as it may consider necessary for the purposes of this Chapter."

(4) The following shall be inserted as the new Section 49G: "Section 49G.- The Management Authority shall, while implementing the provisions of this Chapter, be guided by the following principles, namely:

(i) The export or re-export or import of a specimen under Schedule VII shall be in accordance with the provisions of the Convention;

(ii) Specimens for export or import shall not be obtained in contravention of the laws of the country concerned relating to protection of fauna and flora;

(iii) Any living specimen for export or re-export shall be so arranged and shipped as to minimize the risk of injury, damage to health or cruel treatment;

(iv) The import of any specimen listed in Appendix I of Schedule VII shall not be used for primarily commercial purposes;

(v) The re-export of living specimens of species listed in Appendix I or Appendix II of Schedule VII shall require the prior grant and submission of a re-export certificate issued as per the provisions of the Convention;

(vi) The proposed recipient of a living specimen shall be suitably equipped to house and take care of it;

(vii) The import of any specimen of a species included in Appendix I or Appendix II of Schedule VII shall require the submission of either an export permit or a re-export certificate;

(viii) The introduction from the sea of any specimen of a species included in Appendix I or Appendix II of Schedule VII shall require the grant of a certificate from the Management Authority of the country of introduction issued under the provision of the Convention;

(ix) The export of any specimen of a species included in Appendix III of Schedule VII from any country which has included that species in Appendix III shall require grant and submission of an export permit which shall only be granted when conditions (ii)and (iii) have been met;

(x) The import of any specimen of a species included in Appendix III of Schedule VII shall require:

(a) the submission of certificate of origin; and

(b) where the import is from a country which has included that species in Appendix III, an export permit; or

(c) in the case of re-export, a certificate granted by the Management Authority of the country of re-export that the specimen was processed in that country or is being re-exported, may be accepted by the country of import as evidence that the provisions of the Convention have been complied with in respect of the specimen concerned."

(5) The following shall be inserted as the new Section 49H: "49H.-

(1) The Central Government may, by notification in the Official Gazette, designate any institute established by it and engaged in research in wildlife, as the Scientific Authority for the purposes of this Chapter.

(2) The designated Scientific Authority shall advise the Management Authority in such matters as may be referred to it by the Management Authority.

(3) Whenever the Scientific Authority determines that the export of specimens of any such species should be limited in order to maintain that species throughout its range at a level consistent with its role in the ecosystems in which it occurs and well above the level at which that species might become eligible for inclusion in Appendix I of Schedule VII, the Scientific Authority shall advise the Management Authority of suitable measures to be taken to limit the grant of export permits for specimens of that species.

(4) The Scientific Authority while advising the Management Authority shall be guided by the following principles, namely:

(a) that such export or import shall not be detrimental to the survival of that species; and

(b) proposed recipient of a living specimen is suitably equipped to house and take care for it.

(5) The Scientific Authority shall monitor the export permits granted by the Management Authority for specimens of species included in Appendix II of Schedule VII.

(6) The Scientific Authority shall identify and inform the Management Authority of exotic species of animals and plants that are not covered by the Convention and that require regulation:

(i) to protect the indigenous gene pool of the wildlife found in India ; or

(ii) because such species are invasive in nature and may pose a threat to the wildlife or ecosystems of India; or

(iii) because such species are, in the opinion of the Scientific Authority,critically endangered in the habitats in which they occur naturally."

(7) The following shall be inserted as the new Section 49I: "49I.- In the performance of the duties and exercise of the powers by or under this Chapter, the Management Authority and the Scientific Authority shall be subject to such general or special directions, as the Central Government may, from time to time,give in writing."

The following shall be inserted as the new Section 49J: "49J.-

(1) The Central Government may, by notification in the Official Gazette, constitute a co-ordination committee, for the purpose of ensuring co-ordination between the Management Authority and Scientific Authority, State Chief Wildlife Wardens and various other enforcement agencies dealing with trade in wild life.

(2) The co-ordination committee shall meet at such time and place and shall observe such rules of procedure in regard to the transaction of business at its meetings, including the quorum at its meetings, as may be prescribed."

(8) The following shall be inserted as the new Section 49K. 49K.-

(1) The Central Government may, by notification in the Official Gazette, and consultation with the Management Authority and the Scientific Authority, amend, ary or modify Schedule VII annexed to this Act.

(2) Nothing contained in this Chapter and Schedule VII, shall affect anything contained in other provisions of the Act and the Schedules I, II, III, IV, V and VI to this Act.

(3) Notwithstanding anything contained elsewhere in this Act, where the same species is listed under Schedules I, II, III, IV, V, or VI to this Act as well as Schedule VII to this Act, the provisions of the Act relevant to Schedules I to VI shall apply."

(9) The following shall be inserted as the new Section 49L."49L. International Trade in Scheduled Specimens.-

(1) No person shall enter into any trade in scheduled specimens included in Appendix I of Schedule VII to this Act. Provided that the Scheduled specimens included in Appendix I of Schedule VII bred in captivity for commercial purposes, except those which cannot be released in the wild, or of a plant species included in Appendix I thereof artificially propagated for commercial purposes shall be deemed to be scheduled specimen included in Appendix II of Schedule VII.

(2) Subject to the provisions contained in sub-section (1), no person shall enter into any trade in any scheduled specimen except in accordance with the required permits granted by the Management Authority or the officer authorised by it in such manner as may be prescribed.

(3) Every person trading in any scheduled specimen shall report the details of the scheduled specimen(s) and the transaction to the Management Authority or the officer authorised by it in such manner as may be prescribed.

(4) Every person, desirous of trading in a scheduled specimen, shall present it for clearance to the Management Authority or the officer authorised by it or a custom officer only at the ports of exit and entry specified for the purpose.

(5) Every person breeding in captivity or artificially propagating any Scheduled specimen listed in Appendix I of Schedule VII shall apply to the Management Authority for registration and a license to conduct such activity in the time and manner prescribed by the Central Government in this behalf. The Management Authority shall grant such licences only in accordance with rules made in this behalf by the Central Government."

(10) The following shall be inserted as the new Section 49M. "49M. Possession, Breeding and domestic trade of Scheduled Specimens or Exotic Species.-

(1) Every person possessing an exotic species or scheduled specimen shall report the details of such specimen or specimens to the Management Authority or the officer authorised by it in such manner as may be prescribed.

(2) The Management Authority or the officer authorised by it may, on being satisfied that the exotic species or scheduled specimen was in his possession prior to the coming into force of this Chapter, or was obtained in conformity with the Convention, this Act and any rules made hereunder, register the details of such scheduled specimen or exotic species and issue a registration certificate in the prescribed manner allowing the owner to retain such specimen.

(3) Any person who transfers possession, by any means whatsoever, of any scheduled specimen or exotic species shall report the details of such transfer to the Management Authority or the officer authorised by it in the manner prescribed,

(4) The Management Authority or the officer authorised by it shall register all transfers of scheduled specimens or exotic species and issue the transferee with a registration certificate in the manner prescribed.

(5) Any person in possession of any live scheduled specimen or exotic species which bears any offspring shall report the birth of such offspring to the Management Authority or the officer authorised by it in the time and manner prescribed.

(6) The Management Authority or the officer authorised by it shall register any offspring born to any scheduled specimen or exotic species and issue the owner with a registration certificate in the manner prescribed.

(7) No person shall possess, transfer or breed any scheduled specimen or exotic species except in conformity with this section and the rules prescribed in this behalf.

(8) The owner of an exotic species or scheduled specimen shall take all necessary precautions to ensure that it does not contaminate the indigenous gene pool of the wildlife found in India in the country in any manner."

(11) The following shall be inserted as the new Section 49N. "49N.- No person shall alter, deface, erase or remove a mark of identification affixed upon the exotic species or scheduled specimen or its package."

(12) The following shall be inserted as the new Section 49O. "49O.-

(1) Every exotic species or scheduled specimen, in respect of which any offence against this Act or any rule made there under has been committed, shall be the property of the Central Government.

(2) The provisions of section 39 shall, so far as may be, apply in relation to the exotic species or scheduled specimen as they apply in relation to wild animals and animal articles referred to in sub-section (1) of that section.

(3) Where such specimen is a live animal, the Central Government shall ensure that it is housed and cared for by a recognized zoo or rescue centre when it cannot be returned to its natural habitat."

16. Amendment of Section 50 of the Act

(1) In sub-section (1) of Section 50 of the principal Act the words "or any officer authorised by the Management Authority" shall be inserted after the words "any Forest Officer" and the words "or any custom officer not below the rank of an inspector or any officer of the coast guard not below the rank of an Assistant Commandant" shall be inserted after the words "a sub-inspector"

(2) In sub-section (8) clause (b) of Section 50 of the principal Act the words "and accused persons" shall be inserted after the words "to enforce the attendance of witnesses"

(3) The following shall be inserted after sub-section 4 of section 50 of the new Act as subsection 4A: "During any enquiry or trial of an offence under this Act, where it appears to the Judge or Magistrate that there is prima facie case that any property including vehicles and vessels, seized under clause (c) of sub-section (1) of section 50 of this Act was involved in anyway in the commission of an offence under this Act, the Judge or Magistrate shall not order the return of such property to its rightful owner until the conclusion of the trial of the offence. Section 451 of the Code of Criminal Procedure 2 of 1974 shall stand so modified in its application to offences under this Act."

(4) The following sub-sections shall be inserted after sub-section (9) of Section 50 of the principal Act: "(10) Power to undertake controlled delivery- The Director Wildlife Crime Control Bureau constituted in sub-section (Y)(1)(a) of Section 38

or any other officer authorized by him in this behalf, may, notwithstanding anything contained in this Act, undertake controlled delivery of any consignment to-(a) any destination in India; (b) a foreign country, in consultation with the competent authority of such foreign country to which such a consignment is destined, in such manner as may be prescribed.

(11) Police to take charge of article seized and delivered- An officer-in-charge of a police station as and when so requested in writing by an officer of the several departments mentioned in Section 50(1), shall take charge of and keep in safe custody, pending the order of the Magistrate, all articles seized under this Act within the local area of that police station and which may be delivered to him, and shall allow any officer who may accompany such articles to the police station or who may be deputed for the purpose, to affix his seal to such articles or to take samples of and from them and all samples so taken shall also be sealed with a seal of the officer-in charge of the police station.

(12) Obligation of officers to assist each other- All officers of the several departments mentioned in Section 50(1) shall, upon notice given or request made, be legally bound to assist each other in carrying out the provisions of this Act."

17. Insertion of new Section 50A : After Section 50 of the principal Act, Section 50A shall be inserted as follows: "50A. Government Scientific Expert Body.-

(1) The Central Government or State Government may appoint a Scientific Expert Body to examine, analyse or identify the species in question, as may be required in respect of any proceeding under this Act.

(2) The Scientific Expert Body shall comprise of such experts, on such terms, as may be notified in the Rules to this Act, as prescribed by the Central Government from time to time.

(3) Where the Scientific Expert Body has been asked to examine, analyse or identify any animal article, species or any other material in the course of any proceeding under this Act, the resulting report of this Body may be used as evidence in any enquiry, trial or other proceedings under this Act."

18. Amendment of Section 51 of the Act : (1) Section 51 of the principal Act shall be substituted with the following Section: "51. Penalties.-

(1) Any person who contravenes any provision of this Act or any rule or order made there under, or the breach of any of the conditions of any license or permit granted under this Act, shall be guilty of an offence against this Act, and shall, on conviction, be punishable as follows:

(a) Offences relating to certain Species- Where such offence relates to any animal specified in Schedule I or part II of Schedule II or the meat of such animal or animal article, trophy, or uncured trophy derived from such animal, such offence shall be punishable with imprisonment for a term which shall not be less than five years but may extend to seven years or with a fine which shall not be less than one lakh rupees but which may extend to twenty-five lakh rupees or both.

Provided that in case of a second or subsequent offence of the nature mentioned in this sub-section, such offence shall be punishable with a term of imprisonment which shall not be less than seven years and with a fine which shall not be less than five lakh rupees but which may extend to fifty lakh rupees.

(b) Offences relating to Trade in certain Species- Where such offence relates to the sale or purchase or transfer or offer for sale or trade for any other mode of consideration of any animal specified in schedule I or part II of schedule II or the meat of such animal or animal article, trophy, or uncured trophy derived from such animal or any violation of Chapter VA, such offence shall be punishable with imprisonment for a term which shall not be less than seven years or with a fine not less than fifteen lakh rupees or both.

Provided that in case of a second or subsequent offence of the nature mentioned in this sub-section, such offence shall be punishable with a term of imprisonment which shall not be less than seven years and with a fine which shall not be less than thirty lakh rupees.

(c) Offences relating to Trade in other Species- Where such offence relates to the sale or purchase or transfer or offer for sale or trade for any other mode of consideration of any animal specified in Part I of Schedule II, Schedule III and Schedule IV, or the meat of such animal or animal article, trophy, or uncured trophy derived from such animal, such offence shall be punishable with imprisonment for a term which shall not be less than three years or with a fine which shall not be less than one lakh rupees or both.

Provided that in case of a second or subsequent offence of the nature mentioned in this sub-section, such offence shall be punishable with a term of imprisonment which shall not be less than five years and with a fine which shall not be less than three lakh rupees.

(d) Offences relating to National Parks and Sanctuaries- Where such offence relates to hunting in a sanctuary or a National Park or altering the boundaries of a sanctuary or a National Park, such offence shall be punishable with imprisonment for a term which shall not be less than five years but which may extend to seven years or with a fine which shall not be less than five lakh rupees but which may extend to twenty-five lakh rupees or both.

Provided that in case of a second or subsequent offence of the nature mentioned in this sub-section, such offence shall be punishable with a term of imprisonment which shall not be less than seven years and with a fine which shall not be less than thirty lakh rupees.

(e) Offences relating to Tiger Reserves- Where such offence relates to hunting in a tiger reserve or altering the boundaries of a tiger reserve, such offence shall be punishable with imprisonment for a term which shall not be less than seven years or with a fine which shall not be less than five lakh rupees but which may extend to thirty lakh rupees or both.

Provided that in case of a second or subsequent offence of the nature mentioned in this sub-section, such offence shall be punishable with a term of imprisonment which shall not be less than seven years and with a fine which shall not be less than fifty lakh rupees.

(f) Offences relating to Teasing of Animals- Where such offence relates to a contravention of the provisions of Section 38(j), such offence shall be punishable with imprisonment for a term which may extend to six months or with a fine which may extend to five thousand rupees or with both.

Provided that in case of a second or subsequent offence of the nature mentioned in this sub-section, such offence shall be punishable with a term of imprisonment which may extend to one year or with a fine which may extend to ten thousand rupees or with both.

(g) Other Offences- (1) Where such offence relates to any other contravention of any provision of this Act or any rule or order made there under, or the breach of any of the conditions of any license or permit granted under this Act, such offence shall be punishable with imprisonment for a term which may extend to three years or with a fine which shall not be less than twenty five thousand rupees or with both.

Provided that in case of a second or subsequent offence of the nature mentioned in this sub-section, such offence shall be punishable with a term of imprisonment which shall not be less than three years but which may extend to seven years and also with a fine which shall not be less than fifty thousand rupees.

(2) When any person is convicted of an offence against this Act, the court trying the offence may order that any captive animal, wild animal, animal article, trophy, uncured trophy, meat, ivory imported into India or an article made from such ivory, any specified plant, or part or derivative thereof in respect of which the offence has been committed, and any trap, tool, vehicle, vessel or weapon, used in the commission of the said offence be forfeited to the State Government and that any licence or permit, held by such person under the provisions of this Act, be cancelled.

(3) Such cancellation of licence or permit or such forfeiture shall be in addition to any other punishment that may be awarded for such offence.

(4) Where any person is convicted of an offence against this Act, the court may directthat the licence, if any, granted to such person under the Arms Act, 1959 (54 of 1959), for possession of any arm with which an offence against this Act has been committed,shall be cancelled and that such person shall not be eligible for a licence under the Arms Act, 1959 (54 of 1959), for a period of five years from the date of conviction.

(5) Nothing contained in section 360 of the Code of Criminal Procedure, 1973 (2 of 1974) or in the Probation of Offenders Act, 1958 (20 of 1958) shall apply to a person convicted of an offence with respect to hunting in a sanctuary or a National Park or of an offence against any provision of Chapter VA unless such person is under eighteen years of age.

(6) The offences punishable under clauses (a), (b), (d), and (e) of sub-section (1) of this Section shall be tried by a Sessions Judge and any other offence under this Act shall be tried by the Court of a Chief Judicial Magistrate.

(7) A Judge or Magistrate passing a sentence of three years or more against an accused may order that proceedings under Chapter VIA of this Act (Forfeiture of Property Derived from Illegal Hunting and Trade) are initiated against the accused"

19. Amendment of Section 51A of the Act : Section 51 of the principal Act shall be substituted with the following Section:

(1) Notwithstanding anything contained in the Code of Criminal Procedure, 1973(2 of 1974), -

(a) Every offence punishable with a term of imprisonment for three years or more under this Act shall be cognizable offence

(b) No person accused of an offence under section 51 (1) (a), (b), (d) and (e) shall be released on bail or on his own bond unless-

(i) The public prosecutor has been given the opportunity to oppose the application for such release, and

(ii) Where the public prosecutor opposes the application, the court is satisfied that there are reasonable grounds for believing that he is not guilty of such offence and that he is not likely to commit any offence while on bail.

Provided that the Magistrate may authorise the detention of the accused person if heis satisfied that adequate grounds exist for doing so, but total period of detention shall not exceed: (i) ninety days, where the investigation relates to an offence punishable with imprisonment for a term not less that seven years (ii) sixty days, where the investigation relates to any other offence

Provided further that on the expiry of sixty or ninety days, as the case may be, the accused person shall be released on bail if he furnishes bail, and every person released on bail under this sub-section shall be deemed to be so released under the provisions of Chapter XXXIII of the Code of Criminal Procedure 1973(2 of 1974).

20. Insertion of New Section 52A : After Section 52 of the principal Act, the following shall be inserted as a new Section: "52A. Punishment for allowing premises, etc., to be used for commission of an offence- Whoever, being the owner or occupier or having control or use of any house, room, enclosure, space, place, animal or conveyance, knowingly permits it to be used for the commission by any other person of an offence punishable under any provision of this Act, shall be punishable with the punishment provided for that offence."

21. Insertion of New Section 52B : "52 B: Causing disappearance of evidence of offence, or giving false information of offence- Whoever, knowing or having reason to believe that an offence under this Act has been committed, causes any evidence of the commission of that offence to disappear, or gives any information with respect to the offence which he knows or believes to be false, shall be punishable with imprisonment for a term which may extend to three years or with a fine which shall not be less than twenty five thousand rupees or with both."

22. Amendment of Section 55 of the Act

(1) The following clause shall be inserted after sub-clause (ac) of Section 55 of the principal Act: "(ad) the Management Authority or any Officer, including an Officer of the Wildlife Crime Control Bureau, authorised in this behalf by the Central Government; or"

(2) The following shall be inserted as a proviso to Section 55 of the principal Act, i.e., after sub-clause (c) of Section 55 of the principal Act:" Provided that a Court may also take cognizance of any offence under this Act without the accused being committed to it for trial, upon perusal of a police report under section 173 of the Code of Criminal Procedure 1973 of the facts constituting an offence under this Act."

23. Amendment of Section 58J of the Act : The following shall be inserted as a separate paragraph in Section 58J of the principal Act, after the words "person affected." "It is hereby clarified that the burden of proving that the property in question is exempted under the proviso to Section 58C(2) of this Act shall also lie with the person affected."

24. Amendment of Section 63 of the Act

(1) The following clause shall be inserted after clause

(ai) of sub-section (1) of Section 63 of the principal Act:"

(aii) The rules, standards or procedures and any other matter pertaining to Scientific Research;

(aiii) Any matter relating to Leg hold Traps;"

(2) The following clauses shall be inserted after clause (j) of sub-section (1) of Section 63: "

(ji) the terms and conditions of service including salaries and allowances of the officers and employees of the Management Authority under sub-section (5) of section49F;

(jii) the rules of procedure for transaction of business at meetings of the cordination committee including quorum under sub-section (2) of section 49J;

(jiii) the manner of granting permits for possessing or trading in scheduled specimens under sub-section (2) of Section 49L; and the manner of furnishing reports of such specimens to the Management Authority under sub-section (3) of Section 49L;

(jiv) the rules to regulate persons breeding in captivity or artificially propagating specimens listed in Appendix I of Schedule VII to this Act;

(jv) any matter referred to in Section 49M of this Act;

(jvi) any other matter relating to Scheduled specimens or exotic species;"

(3) The following shall be inserted as sub-section (1A) of Section 63 of the principal Act, i.e., after clause (l) of sub-section (1) of Section 63 of the principal Act: "(1A) Notwithstanding anything contained above or elsewhere in this Act, the Central Government shall have the power to makes rules on any subject contained in the Act."

25. Insertion of new Schedule VII : The following schedule shall be inserted after Schedule VI of the principal Act: "Schedule VII

THE DRAFT WILD LIFE (PROTECTION) AMENDMENT BILL, 2010

Statement of Objects and Reasons : The Wild Life (Protection) Act 1972 establishes the legal framework for the protection and conservation of various species of plants and animals and the proper management of their habitats. The Act includes but is not limited to, the regulation and control of trade in parts and products derived from such species. Despite the penalties already provided for, there seems to be no reduction in the instances of wildlife crime that continue to be reported across the country. Many of these are perpetrated by organized, international gangs of criminals who have sohphisticated networks spread across the country and abroad. The current penalties have failed to act as deterrents with few convictions having taken place since the Act came into force. Furthermore, there is no strong financial disincentive that prevents poachers and smugglers from engaging in illicit wildlife trade, which, in terms of value, ranks only behind the illegal trade in arms and narcotics. To address these and other issues, the Wild life (Protection) Amendment Act seeks to make the following broad changes:

[A] **Enhance Penalties and Strengthen Officials :** The present legislation aims to further strengthen the deterrent powers of the said Act by increasing the penalties for offences, especially those involving unlawful trade in wildlife products. It also seeks to increase the efficiency of the procedure governing the prosecution of offences and empower officers who are crucial to the enforcement of the said Act.

[B] **Greater Harmony with CITES provisions :** The Convention on the International Trade in Endangered Species of Wild Fauna and Flora (CITES) aims to regulate the international trade in endangered species of animals and plants. India became a party to the said Convention in July, 1976. The Amendment Act aims to fully implement India's international obligations under the said convention. This includes the establishment of a Management Authority to regulate the trade in exotic species of animals and plants that are alien to India, which may negatively affect the eco-systems of India if introduced.

[C] **Ban on Trade in Peacock Feathers :** So far, the said Act has allowed the possession and domestic trade in naturally shed peacock feathers. However, the demand within India for peacock tail feathers outstrips the supply. This problem has lead to the rampant poaching of the national bird not only for its tail feathers but also for its meat. The present legislation bans the trade in peacock tail feathers and articles made from them allowing reasonable exceptions for their use in religious ceremonies.

[D] **Wildlife Research :** Good science is essential for the proper conservation and management of India's wildlife. Encouraging wildlife research is a crucial part of the endeavor to save the nations dwindling wildlife. The present legislation aims to promote independent scientific research and imposes obligations upon the Central Government to frame comprehensive rules and procedures governing the same.

[E] **Provisions for Leg Hold Traps :** Leg hold-Traps are the main method used by organized poachers to bring down animals like the tiger and the leopard. These specific traps are dangerous devices that can cause grave damage to even human beings, and are not regulated by any law at present. The use of these traps, in general, is already a violation of the Prevention of Cruelty (Capture of Animals) Rules, 1979. The present legislation bans the sale, manufacture and use of these traps and regulates their possession as well.

---

* Here are some of the major amendment of WLP act, the reader is advised to refer the manual for the WLP act for detailed information.

CHAPTER 32

SCHEDULED WILDLIFE OF INDIA

SCHEDULE I

(See Secs. 2, 8, 9, 11, 40, 41, 43, 48, 51, 61 and 62)

PART I

Mammals

1. Andaman wild pig (*Sus andamanensis*)
1-A. Bharal (*Ovis nahura*)
1-B. Binturong (*Arctictis binturong*)
2. Blackbuck (*Antilope cervicapra*)
2-A. *
3. Brow-antlered deer or thamin (*Cervus eldi*)
3-A. Himalayan brown bear (*Ursus arctos*)
3-B. Capped langur (*Presbytis pileatus*)
4. Caracal (*Felis caracal*)
4-A. Catacean spp.
5. Cheetah (*Acinonyx jubatus*)
5-A. Chinese pangolin (*Manis pentadactyla*)
5-B. Chinkara or Indian gazella (*Gazella gazella bennetti*)
6. Clouded leopard (*Neofelis nebulosa*)
6-A Crab-eating macaque (M*acaca irus umbrosa*)
6-B. Desert cat (*Felis libyca*)
6-C. Desert fox (*Vulpes bucopus*)
7. Dugong (*Dugong dugon*)
7-A. Ermine (*Mustela erminea*)
8. Fishing cat (*Felis viverrina*)
8-A. Four-horned antelope (*Tetraceros quadricornis*)

8-B. *

8-C. *

8-D. Gangetic dolphin (*Platanista gangetica*)

8-E. Gaur or Indian bison (*Bos gaurus*)

9. Golden cat (*Felis temmincki*)

10. Golden langur (*Presbytis geei*)

10-A. Giant squirrel (*Ratufa macroura*)

10-B. Himalayan ibex (*Capra ibex*)

10-C. Himalayan tahr (*Hemitragus jemlahicus*)

11. Hispid hare (*Caprolagus hispidus*)

11-A. Hog badger (*Arctonyx collaris*)

12. Hoolock gibbon (H*ylobates hoolock*)

12-A. *

12-B. Indian elephant (*Elephas maximus*)

13. Indian lion (*Panthera leo persica*)

14. Indian wild ass (*Equus hemionus khur*)

15. Indian wolf (*Canis lupus pallipes*)

16. Kashmir stag (*Cervus elaphus hanglu*)

16-A. Leaf monkey (*Presbytis phayrei*)

16-B. Leopard or panther (*Panthera pardus*)

17. Leopard cat (*Felis bengalensis*)

18. Lesser or red panda (*Ailurus fulgens*)

19. Lion-tailed macaque (*Macaca silenus*)

20. Loris (*Loris tardigradus*)

20-A. Little Indian porpoise (*Neomeris phocaenoides*)

21. Lynx (*Felis lynx isabellinus*)

22. Malabar civet (*Viverra megaspila*)

22-A. Malay or sun bear (*Helarctos malayanus*)

23. Marbled cat (*Felis marmorata*)

24. Markhor (*Capra falconeri*)

24-A. Mouse deer (*Tragulus meminna*)

25. Musk deer (*Moschus moschiferus*)

25-A. Nilgiri langur (*Presbytis johni*)

25-B. Nilgiri tahr (*Hemitragus hylocrius*)

26. Nayan or great Tibetan sheep (*Ovis ammon hodgsoni*)
27. Pallas's cat (*Felis manul*)
28. Pangolin (*Manis crassicaudata*)
29. Pygmy hog (*Sus salvanius*)
29-A. Ratel (*Mellivora capensis*)
30. Indian one-horned rhinoceros (*Rhinoceros unicornis*)
31. Rusty-spotted cat (*Felis rubiginosa*)
31-A. Serow (*Capricornis sumatraensis*)
31-B. Clawless otter (*Aonyx cinerea*)
31-C. Sloth bear (*Melursus ursinus*)
32. Slow loris (*Nyeticebus coveang*)
32-A. Small Travancore flying squirrel (*Petinomys fuscopapillus*)
33. Snow leopard (*Panthera uncia*)
33-A. Snubfin dolphin (*Orcaella brevirostris*)
34. Spotted linsang (P *rionodon pardicolor*)
35. Swamp deer (all sub-species of *Cervus duvauceli*)
36. Takin or Mishmi takin (*Budorcas taxicolor*)
36-A. Tibetan antelope or chiru (*Panthelops hodgsoni*)
36-B. Tibetan fox (*Vulpes ferrilatus*)
37. Tibetan gazella (*Procapra picticaudata*)
38. Tibetan wild ass (*Equus hemionus kiang*)
39. Tiger (*Panthera tigris*)
40. Urial or shapu (*Ovis vignei*)
41. Wild buffalo (*Bubalus bubalis*)
41-A. Wild yak (*Bos grunniens*)
41-B. Tibetan wolf (*Canis lupus chancol*)

PART II

Amphibians and reptiles

1. *
1-A. *
1-B. Audithia turtle (P*elochelys bibroni*)
1-C. Barred, oval, or yellow monitor lizard (V*aranus flavescens*)
1-D. Crocodiles (including the estuarine or saltwater crocodile) (*Crocodilus porosus* and *Crocodilus palustris*)

1-E. Terrapin (*Batagur baska*)
1-F. Eastern hill terrapin (*Melanochelys tricarinata*)
2. Gharial (*Gavialis gangeticus*)
3. Ganges soft-shelled turtle (*Trionyx gangeticus*)
3-A. Golden gecko (*Caloductyloides aureus*)
4. Green sea turtle (*Chelonia mydas*)
5. Hawksbill turtle (*Eretmochelys imbricata imbricata*)
6. *
7. Indian egg-eating snake (*Elachistodon westermanni*)
8. Indian soft-shelled turtle (*Lissemys punctata*)
9. Indian tent turtle (*Kachuga tecta tecta*)
9-A. Kerala Forest Terrapin (*Hoesemys sylratiea*)
11. Leathery turtle (*Dermochelys coriacea*)
12. Loggerhead turtle (*Caretta caretta*)
13. Oliverback loggerhead turtle (*Lepidochelys olivacea*)
14. Peacock-marked soft-shelled turtle (*Trionyx hurum*)
14-A. Pythons (*Genus python*)
14-B. Sail terrapin (*Kachuga kachuga*)
14-C. Spotted black terrapin (*Geoclemys hamiltoni*)
15. *
16. *
17. *
17-A. *

PART II

Birds

1. Whale shark (*Rhincodon typus*)
2. Shark and ray
   (i) *Anoxypristis cuspidata*
   (ii) *Carcharhinus hemiodon*
   (iii) *Glyphis gangeticus*
   (iv) *Glyphis glyphius*
   (v) *Himantura fluviatilis*
   (vi) *Pristis microdon*
   (vii) *Pristis zijsron*

(viii) *Rhynchobatus djiddensis*

(ix) *Urogymnus asperrimus*

3. Sea horse (*All sygnathidians*)

4. Giant grouper (*Epinephelus lanceolatus*)

PART III

Birds

1. Andaman teal (*Anas gibberifrons albogularis*)

1-A. Assam bamboo partridge (*Bambusicola fytchii*)

1-B. Bazas (*Aviceda jerdoni* and *Aviceda leuphotes*)

1-C. Bengal florican (*Eupodotis bengalensis*)

1-D. Black- necked crane (*Grus nigricollis*)

1-E. Blood pheasants (*Ithaginis cruentus tibetanus, I.c.kuseri*)

1-F.

2. Cheer pheasant (*Catreus wallichi*)

2-A. Eastern white stork (*Ciconia ciconia boyciana*)

2-B. Forest spotted owlet (*Athena blewitti*)

2-C. Frogmouths (*Genus batrachostomus*)

3. Great Indian bustard (*Chloriotis nigriceps*)

4. Great Indian hornbill (*Buceros bicornis*)

4-A. Hawks (*fam. Accipitridae*)

4-B. Hooded crane (*Grus monacha*)

4-C. Hornbills (*Ptiloaemus tickelli austeni, Aceros nipalensis, Rhyticeros undulatus ticehursti*)

4-D. Houbara bustard (*Chlamydotis undulata*)

4-E. Hume's bar-backed pheasant (*Syrmaticus humiae*)

4-F. Indian pied hornbill (*Anthracoceros malabaricus*)

5. Jerdon's courser (*Cursorius bitorquatus*)

6. Lammergeier (*Gypaetus barbatus*)

7. Large falcons (*Falco peregrinus, F.biarmicus, F.chicquera*)

7-A. Large whistling teal (*Dendrocygna bicolour*)

7-B. Lesser florican (*Sypheotides indica*)

7-C. Monal pheasants (*Lophophorus impejanus, L. sclateri*)

8. Mountain quail (*Ophrysia superciliosa*)

9. Narcondam hornbill (*Rhyticeros (undulatus) narcondami*)

9-A. *

10. Nicobar megapode (*Megapodius freycinet*)

10-A. Nicobar pigeon (*Caloenas nicobarica pelewensis*)

10-B. Osprey or Fish eating eagle (*Pandion haliaetus*)

10-C. Peacock pheasants (P *olyplectron bicalcaratum*)

11. Peafowl (*Pavo cristatus*)

12. Pink-headed duck (*Rhodonessa carryophyllacea*)

13. Scalater's monal (*Lophophorus sclateri*)

14. Siberian white crane (*Grus leucogeranus*)

14-A. *

14-B. Tibetan snow cock (*Tetraogallus tibetanus*)

15. Tragopan pheasants (*Tragopan melanocephalus, T. blythii, T. satyra, T. temminckii*)

16. White-bellied sea eagle (*Haliaetus leucogaster*)

17. White-eared pheasants (*Crossoptilon crossoptilon*)

17-A. White spoonbill (*Platalea leucorodia*)

18. White-winged wood duck (*Cairina scutalata*)

PART IV

Crustacea and Insects

1. Butterflies and Moths

Family Amathusidae	Common English name
Discophora deo deo	Duffer, banded
Discophora sondaica muscina	Duffer, common
Faunis faunula faunoloides	Pallid fauna
Family Danaidae	
Danaus gautama gautamoides	Tigers
Euploea crameri nicevillei	Crow, spotted black
Euploea midamus roepstorfti	Crow, blue-spotted
Family Lycaenidae	
Allotinus drumila	Darkie, crenulate/great
Allotinus fabius penormis	Angled darkie
Amblopala avidiena	Hairstreak, Chinese
Amblypodia ace arata	Leaf blue
Amblypodia alea constanceae	Rosy oakblue

Amblypodia ammon arial	Mala yan bush blue
Amblypodia arvina ardea	Purple brown tailless oakblue
Amblypodia asopia	Plain tailless oakblue
Amblypodia comica	Comic oakblue
Amblypodia opalina	Opal oakblue
Amblypodia zeta	Andaman tailless oakblue
Biduanda melisa cyana	Blue posy
Callophyrs leechii	Hairstreak, ferruginous
Castalius rosimon alarbus	Pierrot, common
Charana cepheis	Mandarin blue, Cachar
Chlioria othona	Tit, orchid
Deudoryx epijarbas amatius	Comelian, scarce
Everes moorei	Cupid, Moore's
Gerydus biggsii	Bigg's brownie
Gerydus symethus diopeithes	Great brownie
Heliophorus hybrida	Sapphires
Horaga albimacula	Onyxes
Jamides ferrari	Caeruleans
Liphyra brassolis	Butterfly, moth
Listeria dudgeni	Lister's hairstreak
Logania Watsoniana subfasciate	Mottle, Watson's
Lycaenopsis binghami	Hedge blue
Lycaenopsis haraldus ananga	Hedge blue, Felder's
Lycaenopsis purpa prominens	Common hedge blue
Lycaenopsis quadriplaga dohertyi	Naga hedge blue
Nacaduba noreia hampsonii	Lineblue, white-tipped
Polymmatus orbitulus leela	Greenis h mountain blue
Pratapa icetas mishmia	Royal, dark blue
Simiskina phalena harterti	Brilliant, boardlanded
Sinthusa virgo	Spark, pale
Spindasis elwesi	Silverline, Elwes's
Spindasis rukmini	Silverline, khaki
Strymonidia mackwoodi	Hairstreak, Mackwood's
Tajuria ister	Royal, uncertain
Tajuria luculentus nela	Royal, Chinese

Tajuria yajna yajna	Royal, chestnut and black
Thecla ataxus zulla	Wonderful hairstreak
Thecla bieti menlera	Indian purple hairstreak
Thecla letha	Watson's hairstreak
Thecla paona	Paona hairstreak
Thacla pavo	Peacock hairstreak
Virachola smiles	Guava blues
Family Nymphalidae	
Apatura ulupi ulupi	Emperor, tawny
Argynnis hegamone	Silver-washed fritillary
Calinaga Buddha	Freak
Charaxes durnfordi nicholi	Rajah, chestnut
Cirrochroa fasciata	Yeomen
Diagora nicevillei	Siren, scarce
Dilipa morgiana	Emperor, golden
Doleschallia bisaltide andamana	Autumn leaf
Eriboea moori sandakanus	Malayan nawab
Eriboea schreiberi	Blue nawab
Eulaceura manipurensis	Emperor, Tytler's
Euthalia durga splendens	Barons/Counts/Dutchesses
Euthalia iva	Duke, grand
Euthalia khama curvifascia	Duke, Naga
Euthalia telchinia	Baron, blue
Helcyra hemina	Emperor, white
Hypolimnas missipus	Eggfly, danaid
Limenitis austenia purpurascens	Commodorre, grey
Limenitis zulema	Admirals
Melitaea shandura	Fritillaries/Silverstripes
Neptis antilope	Sailer, variegated
Neptis aspasia	Sailer, great hockeystick
Neptis columella kankena	Sailer, short-banded
Neptis cydippe kirbariensis	Sailer, Chinese yellow
Neptis ebusa	Sailer, lascar
Neptis jumbah binghami	Sailer, chestnut-streaked
Neptis manasa	Sailer, pale hockeystick

Neptis nyctens	Sailer, hockeystick
Neptis Poona	Lascar, Tytler's
Neptis sankara	Sailer, broad-banded
Panthoporia jina jina	Bhutan sergeant
Panthoporia reta moorei	Malay staff sergeant
Prothoe franckii regalis	Begum, blue
Sasakia funebris	Empress
Sephisa chandra	Courtier, eastern
Symbrenthia silana	Jester, scarce
Vanessa antiopa yednula	Admirables
Family Papilionidae	
Chilasa clytia clytia f. commixtus	Common mime
Papilio elephenor	Spangle, yellow-crested
Papilio liomedon	Swallowtail, Malabar banded
Parnassius aeco geminifer	Apollo
Parnassius delphius	Banded apollo
Parnassius hannyngtoni	Hannyngton's apollo
Parnassius imperator augustus	Imperial apollo
Parnassius stoliczkanus	Ladakh banded apollo
Polydorus coonsambilanga	Common clubtail
Polydorus crassipes	Black windmill
Polydorus hector	Crimson rose
Polydorus nevilli	Nevill's windmill
Polydorus plutonius pembertoni	Chinese windmill
Polydorus polla	Deniceylle's windmill
Family Pieridae	
Aporia harrietae harrietae	Black veins
Baltia butleri sikkima	White butterfly
Colias colias thrasibulus	Clouded yellows
Colias dubi	Dwarf clouded yellow
Delias sanaea	Jezebel, pale
Pieris krueperi devta	Butterfly cabbage/ White II
Family Satyriidae	

Coelites nothis adamsoni Cyllogenes janetae

Elymnias peali

Elymnias penanga philansis Erabia annada annada

Erabia narasingha narasingha Lethe distans

Lethe dura gammiee Lethe europa tamuna Lethe gemina gafuri Lethe guluihal guluihal Lethe margaritae

Lethe ocellata lyncus Lethe ramadeva

Lethe satyabati

Mycalesis orseis nawtilus

Parargemenava maeroides

Yothima dohertyi persimilis

1-A. Coconut or Rubber crab (*Birgus latro)*

2. Dragonfly (*Epioplebia laidlawi*)

[PART IV A]

Coelenterates

1. Reef building coral (*All Scleractinians)*
2. Black coral (*All Antipatharians)*
3. Organ pipe coral (*Tubipora musica)*
4. Fire coral (*All Millipora species)*
5. Sea fan (*All Gorgonians)*

[PART IV B]

Mollusca

1. *Cassis cornuta*
2. *Charonia tritonis*
3. *Conus milneedwardsi*

[PART IV C]

Echinodermata

Sea cucumber(AllHolothurians)

SCHEDULE II

(See Secs. 2, 8, 9, 10, 11, 40, 41, 43, 48, 51, 61 and 62)

PART I

1.

1-A. Assamese macaque (*Macaca assamensis*)

2. Bengal porcupine (*Atheruras mecrourus assamensis*)

3.

3-A. Bonnet macaque (*Macaca radiata*)

3-B. * *

3-C. Cetatean spp.(other than those listed in Sch.I and Sch.II Part II)

4. * *

4-A. Common langur (P*resbytis entellus*)

5.

6 * *

7. Ferret badgers (*Melogale moschata, M. persenata*)

8. * *

9. * *

10.

11. Himalayan crestless porcupine (*Hystrix hodgsoni*)

11-A. Himalayan newt or salamander (*Tyletotriton verrucosus*)

12. *

13. * *

14.

15. *

16. Pig-tailed macaque (*Macaca nemestrina*)

17.

17-A. Rhesus macaque (*Macaca mulatta*)

18.

19. Stump-tailed macaque (*Macaca speciosa*)

20. * *

21. * *

22. Wild dog or dhole (*Cuon alpinus*)

23. * *

24. Chameleon (*Chameleo calcaratus*)

25. Spiny-tailed lizard or sanda (*Uromastix hardwickii*)

PART II

1. Beetles,

Family Amathusidae

Aemona amathusia amathusia

Amathusia philippus andamanicus

Amathusia amythaonam

Family Cucujidae

Carinophloeus raffrayi

Cucujus bicolor

Cucujus grouvelle

Discophora deo deodoides
Discophora lepida lepida
Discophora timora andamanensis
Enispe cycnus
Faunis sumeus assama
Sticopthalma nourmahal
Thauria aliris amplifascia

Family Carabidae
Agonotrechus andrewesi
Amara brucei
Amara eleganfula
Brachinus atripennis
Brososoma gracile
Brosopus bipillifer
Broter ovicollis
Calathus amaroides
Callistominus belli
Chalenius championi
Chalenius kanarae
Chalenius masoni

Family Chrysomelidae
Acrocrypta rotundata
Bimala indica
Clitea indica
Gopala pita
Griva cyanipennis
Nisotra cardoni
Nisotra madurensis
Nisotra nigripennis
Nisotra semicoerulea
Nisotra striatipennis
Nonarthra patkaia
Psylliodes plana
Cucujus imperialis
Heterojinus semilacteneus
Laemophloeus belli
Laemophloeus incertus
Pediacus rufipes

Family Danaidae
Euploea melanoleuca
Euploea midamus rogenhoferi

Family Erycinidae
Abisara kausambi
Dodona adonira
Dodona dipoea
Dodona egeon
Libythea lepita

Family Hesperiidae
Baoris philippina
Bebasa sena
Halpe homolea

Family Inopeplidae
Inopeplus albonotalus

Family Lycaenidae
Allotinus subviolaceous manychus
Amblypodia aberrans
Amblypodia aenea
Amblypodia agaba aurelia
Amblypodia agrata
Amblypodia alesia
Amblypodia apidanus ahamus
Amblypodia aresta aresta
Amblypodia bazaloides

Psylliodes shira
Sebaethe cervina
Sebaethe patkaia
Sphaeroderma brevicorne
Amblypodia paraganesa zephpreeta
Amblypodia paralea
Amblypodia silhetensis
Amblypodia suffusa suffusa
Amblypodia yendava
Apharitis tilacinus
Araotes lapithis
Artipe eryx
Bindahara phocides
Bothrinia chennellii
Castalius rox us manluena
Catapoecilma elegans myositina
Charana jalindra
Cheriterlla truncipennis
Chliaria kina
Deudoryx hypargyria gaetulia
Enchrysops onejus
Everes kalaroi
Heliphorus androcles moorei
Horaga onyx
Horaga viola
Hypolycaena nilgirica
Hypolycaena thecloides nicobarica
Iraota rochana boswelliana
Jamides alectokandulana
Jamides celeodus pura
Jamides kankena
Lampides boeticus
Lilacea albocaerulea
Lilacea atroguttata
Lilacea lilacea
Amblypodia camdeo
Amblypodia ellisi
Amblypodia fulla ignara
Amblypodia ganesa watsoni
Orthomiella pontis
Pithecops fulgens
Polymmatus devanica devanica
Polymmatus metallica metallica
Polymmatus orbitulus jaloka
Polymmatius yeonghusbandi
Poritia erycinoides elisei
Poritia hewitsoni
Poritia plusrata geta
Pratapa bhotes
Pratapa blanka
Pratapa deva
Pratapa icetas
Rapala buxaria
Rapala chandrana chandrana
Rapala nasala
Rapala ref ulgens
Rapala rubida
Rapala scintilla
Rapala ophinx ophinx
Rapala varuna
Spindasis elima elima
Spindasis lohita
Spindasis nipalicus
Suasa lisides
Surendra todara
Tajuria albiplaga
Tajuria cippus cippus
Tajuria culta
Tajuria diaeus
Tajuria illurgoodes

Lilacea melaena
Lilacea minims
Logania massalia
Lycaenesthes lycaenina
Mahathala ameria
Mahathala atkinsoni
Magisba malaya presbyter
Nacaduba aluta coelestis
Nacaduba ancyra aberrans
Nacaduba dubiosa fulva
Nacaduba helicon
Nacaduba hermus major
Nacaduba pactolus
Neucheritra febronia
Niphanda cymbia
Thecla kirbariensis
Thecla suroia
Thecla syla assamica
Thecla vittata
Thecla ziba
Thecla zoa
Una usta
Yasoda tripunctata

Family Nymphalidae

Adolias cyanipardus
Adolius dirtea
Adolius khasiana
Apatura chevana
Apatura parvata
Apatura sordida
Apatura ulupi florenciae
Argynnis adippe pallida
Neptis nandina hamsoni
Argynnis altissima

Tajuria illurgis
Tajuria jangala andamanica
Tajuria melastigma
Tajuria sebonga
Tajuria thydia
Tajuria yajna istroides
Tarucus callinara
Tarucus dharta
Thaduka multicaudata kanara
Thecla ataxus ataxus
Thecla bitei
Thecla icana
Thecla jakamensis
Thecla kabrea
Thecla khasia
Euthalia merta eriphylea
Euthalia nara nara
Euthalia patala taoana
Euthalia teuta
Herona marathus andamana
Hypolimnas missipus
Hypolimnas polynice birmana
Kallima albofasciata
Kallima alompora
Kallima philarchus horsfieldii
Limenitis austenia austenia
Limenitis damava
Limenitis dudu
Melitaea robertsi lutko
Neptis ananta
Neptis anjana nashona
Neptis aurelia
Neptis magadha khasiana

Neptis narayana

Argynnis clara clara
Argynnis pales horla
Atella iscippe
Calinaga buddha brahaman
Charaxes aristogiton
Charaxes fabius sulphurous
Charaxes nabruba
Charaxes marmax
Charaxes polyxena heman
Chersonesia rahria arahrioides
Cyrestis cocles
Diagora persimilis
Doleschallia bisaltide malabarica
Eriboea athamas andamanicus
Eriboea delphis
Eriboea dolon
Eriboea lissainei
Euripus consimilis
Euripus halitherses
Euthalia anosia
Euthalia cocytus
Euthalia duda
Euthalia durga durga
Euthalia evalina landabilis
Euthalia franciae
Euthalia gauda acontius
Graphium evemon albociliates
Graphium gyas gyas
Graphium megarus megarus
Papilio bootes
Papilio Buddha
Papilio fuscus andamanicus
Papilio machaon verityi
Papilio mayo
Parnassius charltonius charltonius
Neptis radha radha
Neptis soma
Neptis zaida
Neurosigma doubledayi doubledayi
Pantoporia asura asura
Pantoporia kanwa phorkys
Pantoporia larymna siamensis
Pantoporia pravara acutipemnis
Pantoporia ranga
Parthenos sylvia
Penthema lisarda
Symbrenthia niphanda
Vanesa egea agnicula
Vanesa lalbum
Venesa polychloros fervida
Venesa prarsoides dohertyi
Venesa urticoe rizama

Family Papilionidae
Bhutanitis liderdalii
Chilasa epycides epycides
Chilasa paradoxa telearchus
Chilasa slateri slateri
Graphium aristeus anticrates
Graphium arycles arycles
Graphium eurypylus macronius
Erebia manii manii
Erebia seanda opima
Erites falcipennis
Hipparchis heydenreichi shandura
Lethe atkinsoni
Lethe baladeva
Lethe brisanda
Lethe goalpara goalpara
Lethe insana insana

Parnessius epaphus hillensis
Parnessius jacquemonti jacquemonti
Polydorus latreillei kabrua
Polydorus plutonius tytleri
Teinopalpus imperialis imperialis
Lethe nicetella

Family Pieridae

Aporia nabellica
Appias albina darada
Appias indra shiva
Appias lyncida latifasciata
Appias wardica
Baltia butleri butleri
Cepora nadian remba
Cepora nerissa dapha
Colias eocandica hinducucica
Colias eogene
Colias ladakensis
Colias stoliczkana Miranda
Delias lativitta
Dercas lycorias
Euchloe charlonia lucilla
Eurema andersoni ormistoni
Metaporia agathon
Pieris deota
Pontia chloridice alpina
Saletara panda chrysaea
Valeria avatar avatar
Neorina patria westwoodii

Family Satyridae

Aulocera brahminus
Cyllogenes suradeva
Elymnias melilas milamba
Elymnias vasudeva
Erebia annada suroia
Lethe jalaurida
Lethe kabrua
Lethe latiaris latiaris
Lethe moelleri moelleri
Lethe naga naga
Lethe pulaha
Lethe scanda
Lethe serbonis
Lethe siderea
Lethe sinorix
Lethe tristigmata
Lethe violaceodicta kanjupkula
Lethe visrava
Lethe yama
Maniola davendra davendra
Melanitis zitanius
Mycalesis adamsoni
Mycalesis anaxias
Mycalesis botama chamba
Mycalesis heri
Mycalesis lepcha bethami
Mycalesis malsarida
Mycalesis misenus
Mycalesis mestra
Mycalesis mystes
Mycalesis suavolens
Neorina hilda
Oeneis buddha gurhwalica
Parantirrhoea marshali
Paraga maerula maefula
Ragadia crisilda crito
Rhapicera sttricus kabrua
Ypthima bolanica

Erebia hygriva *Ypthima lycus lycus*
Erebia kalinda kalinda *Ypthima mathora mathora*
Ypthima similes affectata Zipotis saitis

1-A. Civets (all species of *Viverridae* except *Malabar civet*)
1-B. Common fox (*Vulpes bengalensis*)
1-C. Flying squirrels (all species of the *genera Bulopetes, Petaurista, Pelomys* and *Eupetaurus*)
1-D. Giant squirrels (*Ratufa indica,* and *R. bicolour*)
2.
2-A. Himalayan black bear (*Selenarctos thibetanus*)
2-B. Jackal (*Canis aureus*)
2-C. Jungle cat (*Felis chaus*)
2-D. Marmots (*Marmota bobak himalayana, M. caudata*)
2-E. Martens (*Martes foina intermedia, M.flavigula, M.watkinsii*)
3. * *
4. *
4. Otters (*Luthra, L.perspicillata*)
4-A. Pole cats (V*ermela peregusna, M.putorius*)
4-B. Red fox (*Vulpes vulpes, V.montana, V.griffithi, V.pusilla*)
5. *
5-A. Sperm whale (*Physter macrocephalus*)
6. *
7. Weasels (*Mustela sibirica, M.kathian, M.altaica*)
8. Checkered keelback snake (*Xeno chrophis piscator*)
9. Dhaman or rat snake (*Ptyas mucosus*)
10. Dog-faced water snake (*Cerberus rhynchopi*)
11. Indian cobras (all sub -species of the *genus Naja*)
12. King cobra (*Ophiophagus hannah*)
13. Oliveceous keelback snake (*Artretium schistosum*)
14. Russel's viper (*Vipera ruselli*)
15. Varanus species (excluding yellow monitor liazard)

SCHEDULE III

(See Secs. 2, 8, 9, 11 and 61)

1. *
2. Barking deer or muntjac (*Muntiacus muntjak*)

3. *
4. *
5. Chital or spotted deer (*Axis axis*)
6. * *
7. Gorals (*Nemorhaedus goral, N.hodgsoni*)
8. * *
9. * *
10. * *
11. Hog deer (*Axis porcinus*)
12. Hyaena (*Hyaena hyaena*)
13. * *
14. Nilgai (*Boselaphus tragocamelus*)
15. * *
16. Sambar (*Cervus unicolor*)
17. * *
18. * *
19. Wild pig (*Sus scrofa*)
20. Sponges (All *Calcareans*)

SCHEDULE IV

(See Secs. 2, 8, 9, 11 and 61)

1. * *

1-A. *

2. *

3. *

3-A. Five-striped palm squirrel (*Funambulus pennanti*)

4. Hares (Black Naped, Common Indian, Desert and Himalayan mouse hare)

4-A. Hedgehog (*Hemiechinus auritus*)

4-B. * *

4-C. *

4-D. *

4-E. Indian porcupine (*Hystrix indica*)

5. *

6. *

6-A. Mongooses (all species of *genus Herpestes*)

6-B. *

7. * *

7-A. * *

7-B. *

8. *

8-A.

9. * *

9-A. *

10. *

11. Birds (other than those, which appear in other Schedules):

1. Avadavat (*Estrildinae*)
2. Avocet (*Recurvirostridae*)
3. Babblers (*Timaliinae*)
4. Barbets (*Capitonidae*)
5. Barnowls (*Tytoninae*)
6. Bitterns (*Ardeidae*)
7. Brown-headed gull (*Larus brunnicephalus*)
8. Bulbuls (*Pycnonotidae*)
9. Buntings (*Emberizidae*)
10. Bustards (*Otididae*)
11. Bustard-quails (*Turnicidae*)
12. Chloropsis (*Irenidae*)
13. Comb duck (*Sarkidiornis melanotos*)
14. Coots (*Rallidae*)
15. Cormorants (*Phalacrocoracidae*)
16. Cranes (*Gruidae*)
17. Cuckoos (*Cuculidae*)

17-A. Curlews (*Scolopacinae*)

18. Darters (*Phalacrocoracidae*)
19. Doves including the Emerald dove (*Columbidae*)
20. Drongos (*Dicruridae*)
21. Ducks (*Anatidae*)

22. Egrets (*Ardeidae*)
23. Fairy bluebirds (*Irenidae*)
24. Falcons (*Falconidae*), except the shaheen and peregrine falcons (*Falco peregrinus*), the saker and laggar falcons (*F. biarmicus*), and the redheaded merlin (*F. chicquera*)
25. Finches including the chaffinch (*Fringillidae*)
26. Flamingos (*Phoenicopteridae*)
27. Flowerpeckers (*Dicaeidae*)
28. Flycatchers (*Muscicapidae*)
29. Geese (*Anatidae*)
30. Goldfinches and allies (*Carduelinae*)
31. Grebes (*Podicipitidae*)
32. Herons (*Ardeidae*)
33. Ibises (*Threskiornithidae*)
34. Ioras (*Irenidae*)
35. Jays (*Corvidae*)
36. Jacanas (*Jacanidae*)

36-A. Junglefowl (*Phasianidae*)

37. Kingfishers (*Alcedinidae*)
38. Larks (*Alaudidae*)
39. Lorikeets (*Psittacidae*)
40. Magpies including the Hunting magpie (*Corvidae*)
41. Mannikins (*Estrildinae*)
42. Megapodes (*Megapodidae*)
43. Minivets (*Campephagidae*)
44. Munias (*Estrildinae*)
45. Mynas (*Sturnidae*)
46. Nightjars (*Caprimulgidae*)
47. Orioles (*Oriolidae*)
48. Owls (*Strigidae*)
49. Oystercatchers (*Haematopodidae*)
50. Parakeets (*Psittacidae*)
51. Partridges (*Phasianidae*)
52. Pelicans (*Pelecanidae*)
53. Pheasants (*Phasianidae*)

54. Pigeons (*Columbidae*) except the Blue rock pigeon (*Columba livia*)
55. Pipits (*Motacillidae*)
55-A. Pittas (*Pittidae*)
56. Plovers (*Charadriinae*)
57. Quails (*Phasianidae*)
58. Rails (*Rallidae*)
59. Rollers or Blue jays (*Coraciidae*)
60. Sandgrouses (*Pteroclididae*)
61. Sandpipers (*Scolopacinae*)
62. Snipes (*Scolopacinae*)
63. Spurfowls (*Phasianidae*)
64. Starlings (*Sturnidae*)
65. Stone curlews (*Burhinidae*)
66. Storks (*Ciconiidae*)
67. Stilts (*Recurvirostridae*)
68. Sunbirds (*Nectariniidae*)
69. Swans (sic) (*Anatidae*)
70. Teals (*Anatidae*)
71. Thrushes (*Turdinae*)
72. Tits (*Paridae*)
73. Tree pies (*Corvidae*)
74. Trogons (*Trogonidae*)
75. Vultures (*Accipitridae*)
76. Waxbills (*Estrildinae*)
77. Weaverbirds or bayas (*Ploceidae*)
78. White-eyes (*Zosteropidae*)
79. Woodpeckers (*Picidae*)
80. Wrens (*Troglodytidae*)

12. Snakes [other than those species listed in Sch.I, Part II; and Sch.II, Part II]
    (i) *Amblycayhalidae*
    (ii) *Amilidae*
    (iii) *Boidae*
    (iv) *Colubridae*
    (v) *Dasypeptidae*
    (vi) *Elapidae* (cobras, kraits and coral snakes)

(vii) *Glauconidae*

(viii) *Hydrophidae* (freshwater and sea snakes)

(ix) *Ilysidae*

(x) *Leptotyphlopidae*

(xi) *Typhlopidae*

(xii) *Uropeltidae*

(xiii) *Viperidae*

(xiv) *Xenopeltidae*

13. Freshwater frogs (*Rana spp.*)
14. Three-keeled turtle (*Geomyda tricarinata*)
15. Tortoises (*Testudinidae, Tryonichidae*)
16. Viviparous toads (*Nectophyrynoides spp.*)
17. Voles
18. Butterflies and moths:

Family Danaidae

Euploea core simulatrix

Euploea crassa

Euploea dioeletianus ramsahai Euploea mulciber

Family Hesperiidae

Baoris farri

Hasaro vitta

Hyarotis adrastus

Oriens concinna

Pelopidas assamensis

Pelopidas sinensis

Polytrema discreta

Polytrema rubricans

Thoressa horiorei

Family Lycaenidae

Tarucus ananda

Family Nymphalidae

Eiuthalia lubentina

Family Pigeridae

Appias agathon ariaca

Appias libythea

Appias nero galba

Prioneris sita

19. Mollusca:

(i) *Cypraea lamacina*

(ii) *Cypraea mappa*

(iii) *Cypraea talpa*

(iv) *Fasciolaria trapezium*

(v) *Harpulina arausiaca*

(vi) *Lambis chiragra*

(vii) *Lambis chiragra arthritica*

(viii *Lambis crocea*

(ix) *Lambis millepeda*

(x) *Lambis scorpius*

(xi) *Lambis truncata*

(xii) *Placenta placenta*

(xiii) *Strombus plicatus sibbaldi*

(xiv) *Trochus niloticus*

(xv) *Turbo marmoratus*

SCHEDULE V

(See Secs. 2, 8, 61 and 62)

Vermin

1. Common crow
2. *
3. Fruit bats
4. *
5. Mice
6. Rats
7. *

SCHEDULE VI

(See Section 2)

1. Beddomes cycad (*Cycas beddomei*)
2. Blue vanda (*Vanda coerulea*)
3. Kuth (*Saussurea lappa*)

4. Ladies slipper orchid (*Paphiopedilium*)
5. Pitcher plant (*Nepenthes khasiana*)
6. Red vanda (*Renanthera imschootiana*)

THE WILDLIFE (TRANSACTION AND TAXIDERMY) RULES, 1973

G.S.R. 198(E), Dated the 9th April, 1973

In exercise of the powers conferred by Cl. (b) of sub-section (1) of Sec. 63 of the Wild Life (Protection) Act, 1972 (53 of 1972), the Central Government hereby makes the following rules.

1. Short title, extent and commencement

(1) These rules may be called the Wildlife (Transactions and Taxidermy) Rules, 1973.

(2) They extend to the whole of the State of Bihar, Gujarat, Haryana, Himachal Pradesh, Madhya Pradesh and Uttar Pradesh.

(3) They sha ll come into force on the 9th April, 1973.

2. Definition

In these rules, unless the context otherwise requires

(a) "Act" means the Wildlife (protection) Act, 1972 (53 of 1972);

(b) "Form" means a Form appended to these rules;

(c) "Licensee" means a licensee under Chapter V of the Act;

(d) "Officer" means the Chief Wildlife Warden or any other officer whom the State

Government may, for the purposes of these rules, by notification in the official Gazette, appoint;

(e) "specified animal" means any animal which is specified in Sch.I or Part II of Sch. II to the Act and which is

(i) captured or kept or bred in captivity, or

(ii) found wild in nature.

3. Acquiring, receiving or keeping specified animal, etc. in control, custody or possession or put under process of taxidermy or make articles, etc.,

(1) No person shall

(i) acquire, receive, keep in his control, custody or possession, any specified animal or any animal articles, trophy, uncured trophy, or meat derived therefrom, or

(ii) put under a process of taxidermy or make animal articles containing part of whole of such animal, except with the previous permission of the officer.

(2) Every application for such permission shall be made in Form I.

(3) On receipt of an application made under sub-rule (2), the officer may, after making such inquiry as he may think fit and within a period of fifteen days from the date of receipt of the application, either grant or refuse to grant the permission;

Provided that no such permission shall be granted unless the officer is satisfied that the specified animal or animal article, trophy, uncured trophy or meat, referred to in sub-rule (1) has been lawfully acquired.

(4) Where the officer refuses to grant the permission, he shall record the reason for so doing and a copy of the reason so recorded shall be communicated to the licensee applying for the permission.

(5) Every permission granted under sub-rule (3) shall be in Form II.

4. Submission of report of stocks

(1) Every licensee to whom permission has been granted under sub -rule (3), of Rule 3 shall submit, to the officer who has granted the said permission, report regarding the stocks of specified animal or animal article, trophy, uncured trophy or meat, referred to in sub-rule (1) of Rule 3, in Form III within a period of [thirty days] of the acquisition, receipt or keeping of the same in his control, custody or possession.

(2) The officer, after receiving such report, may arrange to affix identification marks on such stocks.

5. Sale of specified animal, etc.

(1) No licensed dealer shall sell or offer for sale any specified animal or any animal article, trophy or uncured trophy derived therefrom, except to a person authorised to purchase by a permission granted by the officer and where the sale is effected the purchaser shall surrender the permission to the licensed dealer.

(2) Every application for permission to purchase shall be made in Form IV.

(3) On receipt of an application made under sub-rule (2), the officer may, after making such inquiry as he may think fit, and within a period of ten days from receipt of the application, either grant or refuse to grant the permission.

(4) Where the officer refuses to grant the permission, he shall record the reasons for so doing and a copy of the reasons so recorded shall be communicated to the person applying for the permission.

(5) Every permission granted under sub-rule (3) shall be valid up to a period of one month from the date of issue of the same.

(6) Every licensed dealer shall, at the time of each sale, issue a voucher in relation to the specified animal or animal article, trophy or uncured trophy referred to in sub-rule (1), to the person authorised to purchase.

(7) Each voucher shall contain the following particulars, namely

(a) date of issue of the voucher;

(b) the amount of price realised or to be realised;

(c) name and address of the licensed dealer issuing the voucher;

(d) name and address of the person to whom the voucher is issued;

(e) permission number of the person authorised to purchase;

(f) description of the specified animal/ animal article/ trophy/ unc ured trophy derived therefrom and number;

(g) whether such specified animal/ animal article/ trophy/ uncured trophy was/were required to be declared under Sec.44 of the Wild Life (Protection) Act, 1972 (53 of 1972), and if so, whether it/they has/have been declared; signature of the licensed dealer issuing the voucher;

(h) signature of the person to whom the voucher is issued.

6. Taxidermy or making animal article

(1) Every licensed taxidermist or licensed manufacturer shall, at the time of returning the trophy or animal article, issue a voucher to the owner of the said trophy or animal article.

(2) Each voucher shall contain the following particulars, namely:

(a) date of issue of voucher;

(b) charges realised or to be realised;

(c) name and address of the licensed taxidermist/ manufacturer issuing the voucher;

(d) name and address of the person to whom the voucher is issued;

(e) whether uncured trophy/ trophy/ animal article was required to be declared under Sec.40 or Sec.44 of the Wild Life (Protection) Act, 1972 (53 of 1972), and if so whether it/ they has/have been declared;

(f) signature of the licensed taxidermist/ manufacturer issuing the voucher.

7. Maintenance of vouchers

(1) The voucher referred to in Rules 5 or 6 shall be in triplicate and serially numbered.

(2) The duplicate and the triplicate copy of the voucher shall be retained by the licensed dealer, licensed taxidermist or licensed manufacturer, and the original copy of the voucher shall be given to the person referred to in sub -rule (7) of Rule 5 or sub -rule (1) of rule 6.

(3) Every book containing blank vouchers shall be presented to the officer for affixing his initials to stamps on such book before it is brought into use.

(4) (a) Every licensed dealer, licensed taxidermist or licensed manufacturer shall send in monthly batches, not later than the seventh day of every month, the duplicate copies of vouchers retained by him, to the officer.

(b) Every permission surrendered to a licensed dealer at the time of sale shall also be enclosed along with the duplicate copies aforesaid.

8. Transport of specified animal, etc.

(1) No licensee shall transport from one place to another within the State any specified animal, animal articles, trophy or uncured trophy derived therefrom, except with the previous permission of the officer.

(2) Every application for such permission shall be made in Form VI.

(3) On receipt of an application made under sub-rule (2), the officer may, after making such inquiry as he may think fit, and within a period of seven days from the date of receipt of the application, either grant or refuse to grant the permission; Provided that no such permission shall be granted unless the officer is satisfied that the specified animal or animal article, trophy or uncured trophy, referred to in sub-rule (1) has been lawfully acquired.

(4) Where the officer refuses to grant the permission, he shall record the reasons for so doing and a copy of the reasons so recorded shall be communicated to the licensee applying for the permission.

(5) Every permission granted under sub-rule (3) shall be in Form VII.

9. Appeal

(1) Any licensee or a person aggrieved by an order made by the Chief Wildlife Warden or any officer granting the permission under sub-rule (3) of rule 3, sub-rule (3) of rule 5 or sub-rule (3) of Rule 8, ma y prefer an appeal, if

(i) the order is made by an officer other than the Chief Wildlife Warden, to the Chief Wildlife Warden, or

(ii) the order is made by the Chief Wildlife Warden, to the State Government.

(2) In the case of an order passed in appeal by the Chief Wildlife Warden under Cl.

(i) of sub -rule (1), a second appeal shall lie in the State Government

(3) No appeal shall be entertained unless it is preferred within fifteen days from the date of the communication to the applicant of the order appealed against.

Provided that the appellate authority may admit any appeal after the expiry of the period aforesaid, if it is satisfied that the appellant had sufficient cause for not preferring the appeal in time.

FORM I

(See sub-rule (2) of Rule 3)

Application for permission to acquire, receive, keep specified animal, animal article, etc. or put under process of taxidermy or make animal article

To,

The

Sir, ...

....

....

1. I......................resident of ....Taluk.....................District .... and holding License No. .... Granted under Sec.44 (4) of the Wild Life (Protection) Act, 1972 (53 of 1972), request that I may be granted permission to acquire/ receive/ keep in any control/ custody/ possession of specified animal/ animal article/ trophy/ uncured trophy/ meat derived from specified animal and/ or put under process of taxidermy/ make animal article containing part/ whole of such animal.

2. I furnish below the particulars in relation to such specified animal/ animal article/ trophy/ uncured trophy/ meat:

   (1) Species of animal

   (2) Number

   (3) Description(including sex, if possible)

   (4) Source from which to be obtained

      (i) Address and License No. if any

      (ii) Whether declaration made/ permission/ licence obtained under Secs. 40, 43 or 44 of the Wild Life

      (Protection) Act, 1972, and if so the particulars:

   (5) Particulars of certificate of ownership

   (6) Identification mark, if any

   (7) Premises in which intended to be kept

   (8) Purpose for which to be acquired/ received/ kept in control/ custody/ possession

   (9) If to be put under p rocess of taxidermy or to make animal article,

      (a) No. of trophies/ articles to be made

      (b) Description of such trophies/ articles

      (c) To whom will they be returned

      (d) Probable date by which they will be returned

3. I hereby declare that to the best of my knowledge and belief the information furnished herein is true

and complete.

....

Signature of the applicant

Strike out whichever is not applicable

FORM II

(See sub -rule (2) of Rule 5)

Possession to acquire, receive, keep in control, custody or permission of specified animal, animal article, etc. or put under process of taxidermy or make animal article.

Shri …………………… Holding Licence No. ……………….. granted under Sec.44(4) of the Wild Life (Protection) Act, 1972 (53 of 1972) is hereby permitted to acquire/ to keep under his control/ custody/ possession of specified animal/ animal article/ trophy/ uncured trophy/ meat derived from specified animal of the following description, or put under process of taxidermy or make animal article containing part or whole of such animal:

(1) Species of animal

(2) Description (including sex, if given in the application)

(3) Number

(4) Source from which to be obtained

(5) Licence/ Permission No. of the source from which to be obtained

(6) Particulars of the Cer tificate of Ownership

(7) Identification mark, if any

(8) Premises in which intended to be kept

(9) Purpose for which permitted to be acquired/ received/ kept in control/ custody/ possession

(10) If permitted to be put under process of taxidermy or to make animal article,

 (a) No. of trophies/ articles to be made

 (b) Description of such trophies/ articles

 (c) To whom they should be returned

 (d) Probable date by which they would be returned

Issued by me this …………………………day of……………………………….

Seal:

Place:

………………………………………

Signature and Designation

Date:

Strike out whichever is not applicable

FORM III

(See sub-rule (1) of Rule 4)

Report of stocks

To,

The ________________________

____

____

1. Full name, address and Licence No. of the Licensee
2. Stock held on the date of report in specified animals:
   (a) Species and sex
   (b) Number
   (c) Adult or juvenile
   (d) Premises where kept
3. Stock held on the date of report in animal articles:
   (a) Description, including species of animal from which derived
   (b) Number
   (c) Dimension or weight
   (d) Premises where kept
4. Stock held on the date of report in trophies:
   (a) Description, including species of animal from which derived
   (b) Number
   (c) Dimension or weight
   (d) Premises where kept
5. Stock held on the date of report in uncured trophies:
   (a) Description, including species of animal from which derived
   (b) Number
   (c) Dimension or weight
   (d) Premises where kept
6. Remarks, if any

   I do hereby declare that the information given above is true to the best of my knowledge and belief.

Place: ....

Date: Signature of the person making declaration

Strike out whichever is not applicable

FORM IV

(See sub-rule (2) of Rule 5)

Application for permission to purchase specified animal, etc.

To

The ___________________________

____

____

Sir,

I/ We,................residing at..................Taluk...............District................ Request that I/ We may be granted permission to purchase specified animal/ animal article/ trophy/ uncured trophy derived from specified animal of the following description, from a Licence:

(1) Number and description of
  - (a) specified animal
  - (b) animal article
  - (c) trophy
  - (d) uncured trophy

(2) Purpose for which the purchase is to be made

(3) I/ We hereby declare that to the best of my/ our knowledge and belief the information furnished herein is true and complete.

....

Signature(s) of the applicant(s)

Place:

Date:

Strike out whichever is not applicable

FORM V

(See sub -rule (5) of Rule 5)

Permission to purchase specified animal etc.

S/ Shri .... is/ are hereby permitted to purchase specified animal/ animal article/ trophy/ uncured trophy derived from specified animal of the following description, from.................................for the purpose of ....

Number and description of

(a) Specified animal

(b) Animal article

(c) Trophy

(d) Uncured trophy

Issued by me this .... day of ....

....

Signature and Designation

Seal:

Place:

Date:

Note: This permission shall be valid up to a period of one month from the date of issue.

Strike out whichever is not applicable

FORM VI

(See sub -rule (2) of Rule 8)

Application for permission to transport specified animal etc.

To

The ____________________________

____

____

Sir,

I, ...residing at...................Taluk...................District .... holding Licence No.....................granted under Sec.44 (4) of the Wild Life (Protection) Act, 1972 (53 of 1972), request that I may be granted permission to transport the following :

(1) Species of specified animal or from which the animal article/ cured trophy/ uncured trophy is derived ____________________________

(2) Number ____________________________

(3) Description (including sex if possible) ____________________________

(4) Identification mark, if any ____________________________

(5) Source of procurement and the Licence/ Permission No. ______________

(6) Certificate of ownership, if any ____________________________

(7) Mode of transport ____________________________

(8) Route ____________________________

(9) Period required for transport ____________________________

(10) Destination ____________________________

I hereby declare that to the best of my knowledge and belief the information furnished herein is true and complete.

....

Signature of the applicant

Place:

Date:

Strike out whichever is not applicable

FORM VII

(See sub -rule (5) of Rule 8)

Permission to transport specified animal etc.

Shri .... Holding Licence No. .... granted under Sec.44(4) of the Wild Life (Protection) Act, 1972 (53 of 1972), is hereby permitted to transport in the manner prescribed below specified animal/ animal article/ cured trophy/ uncured trophy derived from specified animal, from .... to..................................

(i) Mode of transport

(ii) Route

(iii) Period allowed for transport

(iv) Remarks

Issued by me this .... day of ....

....

Signature and Designation

Seal:

Place:

Date:

Strike out whichever is not applicable

THE WILDLIFE (STOCK DECLARATION) CENTRAL RULES, 1973

G.S.R. 29(E)

In exercise of the powers conferred by Cl. (a) of sub-section (1) of Sec.63 of the Wild Life (Protection) Act, 1972 (53 of 1972), the Central Government hereby makes the following rules, namely

1. Short title and commencement

(1) These rules may be called the Wildlife (Stock Declaration) Central Rules, 1973.

(2) They shall come into force in the State of Madhya Pradesh on the 25th January, 1973 and in other States and Union Territories on such date as the Central Government may, by notification appoint, and different dates may be appointed for for different States and union Territories.

2. Declaration by manufacturer or dealer or taxidermist in, animal article, etc.

Every manufacturer of, or dealer in, animal article or every dealer in captive animals, trophies, or uncured trophies, or every taxidermist shall, within fifteen days from the commencement of Wild Life (Protection) Act, 1972, declare his stock of animal article, captive animal, trophies, and uncured trophies, as the case may be, as on the date of such declaration to the Chief Wildlife Warden in the form given below.

Form of Declaration

(See sub-section (2) of Sec. 44)

To

The Chief Wildlife Warden

State or Union territory of ....

1. Full name and address of the manufacturer/ dealer/ taxidermist making the declaration ________________________________

2. Actual stock held on the date of declaration in animal articles:

   (i) Description including name of animal from which derived ________________

   (ii) Number________________________________

   (iii) Dimensions or weight________________________________

   (iv) Premises where kept________________________________

3. Actual stock held on the date of declaration in captive animals:

   (i) Species and sex________________________________

   (ii) Number ________________________________

(iii) Adult or juvenile ____________________

(iv) Premises where kept ____________________

4. Actual stock held on the date of declaration in trophies:

(i) Description including name of animal from which derived ____________________

(ii) Number ____________________

(iii) Dimensions or weight ____________________

(iv) Premises where kept ____________________

5. Actual stock held on the date of declaration in uncured trophies:

(i) Description including name of animal from which derived ____________________

(ii) Number ____________________

(iii) Dimensions or weight ____________________

(iv) Premises where kept ____________________

6. Remarks, if any

I do hereby declare that the information given above is true to the best of my knowledge and belief.

....

Signature of the person making the declaration

Place:

Date:

THE WILDLIFE (PROTECTION) LICENSING (ADDITIONAL MATTERS FOR CONSIDERATION) RULES, 1983

G.S.R. 328(E), dated 13th April, 1983

In exercise of the powers conferred by Cl.(a) of sub-section (1) of Sec.63, read with Cl. (b) of sub-section (4) of Sec. 44 of Wild Life (Protection) Act, 1972 (53 of 1972), the Central Government hereby makes the following rules, namely

1. Short title, extent and commencement
   (1) These rules may be called the Wild Life (Protection) Licensing (Additional Matters for Consideration) Rules, 1983
   (2) They shall extend to the whole of India except the State of Jammu and Kashmir.
   (3) They shall come into force on the date of their publication in the official Gazette.
2. Definition

   In these Rules, unless the context otherwise, requires, "Act" means the Wild Life (Protection) Act, 1972 (53 of 1972).
3. Additional matters for consideration for grant of licence under Sec.44 of the Act.

   For the purposes of granting a licence referred to in sub-section (1) of Sec.44 of the Act, the Chief Wildlife Warden or the authorised officer, as the case may be, shall in addition to the matters specified in Cl. (b) of sub-section (4) of that section, have regard to the following other matters, namely -
   (i) capacity of the applicant to handle the business concerned with referred to facilities, equipment and suitability of the premises for such business;
   (ii) the source and the manner in which the supplies for the business concerned would be obtained;
   (iii) number of licences for the relevant business already in existence in the area concerned;
   (iv) implications which the grant of such licence would have on the hunting or trade of the wild animals concerned.

Provided that no s uch shall be granted if the said implications relate to any wild animal specified in Sch.I or Part II of Sch.II to the Act, except with the previous consultation of the Central Government.

WILDLIFE (PROTECTION) RULES, 1995

G.S.R. 348(E)

In exercise of powers conferred by clause (k) of sub-section (1) of Sec.63 of the Wild Life (Protection) Act, 1972 (53 of 1972), the Central Government hereby makes the following rules, namely:

1. Short title and commencement -
   (1) These Rules may be called Wildlife (Protection) Rules, 1995.
   (2) They shall come into force from the date of their publication in the Official Gazette.
2. In these rules, unless the context otherwise requires -
   (a) "Act" means the Wild Life (Protection) Act, 1972.
   (b) "Section" means the Section of the Act.
3. The manner of the notice under clause (c) of Sec.55 -
   (1) The notice to the Central Government or the State Government or any authorised officer, as the case may be, shall be given in form "A" annexed to these rules.
   (2) The person giving notice to the Central Government or the State Government or any authorised officer shall send the notice by registered post to:
      (a) The Director of Wildlife Preservation, Government of India in the Ministry of Environment and Forests, New Delhi; and
      (b) (i) The Secretary to the concerned State Govt./ Union Territory in charge Wildlife, or
         (ii) The Chief Wildlife Warden of the concerned State Govt./ Union Territory, or
         (iii) Any authorised officer of State Govt./ Union Territory.

FORM "A"

(See sub-rule (1) of Rule 3)

From:

____

____

____

To:

____

____

____

Notice under Sec. 55 of the Wild Life (Protection) Act, 1972.

Whereas an offence under the Wild Life (Protection) Act, 1972 has been committed/ is being committed by [Full name(s) and complete address(es)] ....
....
....

And whereas the brief facts of the offence(s) are enclosed;

I/ We hereby gives notice of 60 days under Sec. 55 of the Wild Life (Protection) Act, 1972, my/ our intention to file a complaint in the court of ....for violation of section(s)................................ of the Wild Life (Protection) Act, 1972.

I am/We are enclosing the following documents as evidence of proof of the violation of the said Act. (Documentary evidence may include photographs/ reports/ statements of witness(es) for enabling enquiry into the alleged violation/ offence).

[No.1-2/ 91/ WLI]

SARWESHWAR JHA, Jt. Secy.

WILDLIFE (SPECIFIED PLANTS - CONDITIONS FOR POSSESSION BY LICENSEE) RULES, 1995

G.S.R. 349(E)

In exercise of powers conferred by Clause (a) of sub-section (1) of Sec. 63 of Wild Life (Protection) Act, 1972 (53 of 1972), the Central Government hereby makes the following rules, namely:

1. Short title, extent and commencement -

(1) These Rules may be called Wildlife (Specified Plants - Conditions for possession by licensee) Rules, 1995.

(2) These rules shall come into force from the date of commencement of provisions of Chapter IIIA of the Wild Life (Protection) Act, 1972.

2. Definition -

In these Rules, unless the context otherwise requires, "Act" means the Wild Life (Protectio n) Act, 1972 (53 of 1972).

3. Conditions and other matters subject to which the licensee may keep any specified plants in his custody or possession -

(1) No licensee shall acquire or receive or keep in his control, custody or possession any specified pla nt or part or derivative thereof in respect of which a declaration under Sec. 17E of the Act has not been made.

(2) No licensee shall acquire, purchase or receive any specified plant or part or derivative thereof from any person other than a licensed dealer in specified plants or a cultivator having a license for cultivation of specified plants under the Act.

(3) Licensee shall keep the stock of specified plants so purchased by him only in the premises approved by the Chief Wildlife Warden of the State.

[No.1-2/ 91/ WLI]

SARWESHWAR JHA, Jt. Secy.

WILDLIFE (SPECIFIED PLANT STOCK DECLARATION) CENTRAL RULES, 1995.

G.S.R. 350(E)

In exercise of powers conferred by Clause (h) of Sec. 63 read with Sec. 17E of Wild Life (Protection) Act, 1972 (53 of 1972), the Central Government hereby makes the following rules, namely:

1. Short title and commencement -
   (1) These rules may be called the Wildlife (Specified Plant Stock Declaration) Central Rules, 1995.
   (2) These rules shall come into force from the date of commencement of provisions of Chapter IIIA of the Wild Life (Protection) Act, 1972.
2. Declaration of stocks by a cultivator or dealer in specified plants, parts and derivatives thereof -

Every cultivator of specified plants and the dealer in specified and derivatives thereof shall, within 30 days from the commencement of provisions of Chapter IIIA of Wild Life (Protection) Act, 1972, declare his stocks of specified plants, parts and derivatives thereof, as the case may be, as on the date of such declaration to the Chief Wildlife Warden in the form given below:

Form of Declaration

(See Sec. 17E and sub -section (2) of Sec. 44)

To

The Chief Wildlife Warden,

State/ Union Territory of ....

1. Full name and address of the cultivator or dealer in specified plants, parts and derivatives thereof making the declaration..
2. Actual stock held on the date of declaration:

Name of the specified plant (including scientific name)	Known Uses	Description of stock	Quantity held in stock Kgs number	Premises where stock are kept	Date of procure-ment	Source and specific area of procurement	Documentary proof, if any
1	2	3	4	5	6	7	8

3. Remarks, if any ....
   I do hereby declare that the information given above is true to the best of my knowledge and belief.

....

Signature of the person making the declaration

Place:

Date:

[No.1-2/ 91/ WLI]

SARWESHWAR JHA, Jt. Secy.

DECLARATION OF WILD LIFE STOCK RULES,2003

REGISTERED NO. DL-33004/99

The Gazette of India

EXTRAORDINARY

PART II - Section 3 - Sub-section (ii)

PUBLISHED BY AUTHORITY

NO. 365] NEW DELHI, FRIDAY,APRIL 18,2003/ CHAITRA 28, 1925

MINISTRY OF ENVIRONMENT AND FORESTS

NOTIFICATION

New Delhi, the 18th April,2003

S.O. 445(E). - In exercise of the powers conferred by sub-section (1) and (3) of Sec. 40A read with Sec. 63 of the wild life (Protection) Act, 1972 (53 of 1972), the Central Government hereby marks the following rules, namely:-

1. Short title and commencement -

1. These rules may be called the Declaration of wild life stock Rules, 2003.
2. They shall come into force on the date of their publication in the Official Gazette

2. Definitions. -

In these rules, unless the context otherwise requires,-

(a) "Act" meansthe wild life (Protection) Act, 1972 (53 of 1972);

(b) "From" means the form annexed to these rules;

(c) all other words and expressions used in these rules shall have the meanings respectively assigned them in the Act

3. Publicity of intent of notification and Assistance in making application

i. The Chief Wildlife The chief wildlife warden or the officer authorized by the state Government in this regard shall cause to give wide publicity to the intent of this notification in the regional languge through eletronic or print media or such other means.

ii. The Chief wildlife Warden of the officer authorized by the state Government in this regard shall take necessary action to assist the local communities and individuals especially the poor and illiterate in the declaration of their possession, filling up the specified from and any other matter connected there with and shallmake every attempt to ensure that no individual or community associated with animals is deprived os this opportunity.

4. Procedure fro filling application. -

a. Warden or the officer authorized by the state Government in this regard shall be presented in the Form annexed to these rules by the application either in person or by an agent or by duly authorized legal practitioner or sent by registered post address to the Chief Wild Life Warden or the officer authorized by the State Government in this regard of the concerned State or the Union territory.

b. The application under sub rule (1) shall be present in four complete sets within a period of one hundered of one eight days from the date of publication of these rules.

c. The applicant may attach to and present with his application an acknowledgement slip as is given in the From, Which shall be signed by the official receiving the application on behalf of the Chief Wild Life Warden or he officer authorized by the State Government in this regard in ackonwledgement of the receipt of the application.

5. **Presentation and scrutiny of application**

   1. The chief Wildlife warden or the officer authorized by the State Government in this regard shall endsore on every application the date on which it is presented or deemed to have been presented under that rule and shall sign the endorsement.
   2. If on scrutiny the application is found to be in order, it shall be duly registered and given serial number
   3. If the application, on scrutiny, is found to be defective, the same shall be returned to the application within fifteen days for rectifying the defects and resubmitting the corrected application within fifteen days from the date of its receipt.
   4. If the applicant fails to rectify the defect within the time allowed under sub-rule (3), the Chief Wild Life Warden or thr officer authorized by the State Government in this regard may, by order and for the reasons to be recorded in writing decling to register the application.

6. **Place of filling application** - The application shall file application with the Chief Wild Life Warden of the officer authorized by the State Government in this regard

7. **Date and palce of hearing to be notified** - The Chief wild life Warden of the officer authorized by the State Government in this regard shall notify to the parties the date, palce and time of hearing of each application if required.

8. **Decision on application.** -

   a. The Chief Wild Life Warden of the officer authorized by the State Government in this regard shall verify the facts mentioned in the application and make such inqury as may be required.

   b. The Chief Wild Life Warden shall, as far as possible, decide the application within six months of the dates of its presentation and communicate the same to the applicant in writing under his own signature by register post

9. **Hearing on application ex- parte.** - Where on the date fixed for hearing the application, the application fails to appear without intimation, the Chief Wild Life Warden or the officer authorized by the State Government in this regard may at their discretion adjourn or decide the application ex-parte.

10. Inquiry by the Chief Wild Life Warden or Authorized officer.-

(1) The Chief Wild Life Warden or the officer authorized by the State Government n this regard shall conduct a detailed inquiry and take all actions as provided in Sec.41 of the Act.

(2) A copy of the report pertaining to sub -rule (1) of this rule, shall be provided to the application

11. Certificate of ownership

(1) The Chief Wild Life Warden shall provide a certificate ownership to the applicant whose claim is found valid.

(2) The cert ificate of ownership shall be provided as per the provisions of Sec. 42 of the Act

(3) The certificate of ownership shall contain the facsimile of the identification mark and case of live animals the identification number of the transponder (microchip) implanted shall be mentioned in the certicate.

12. Dealing with declared object.- Any captive animal, animal article trophy or uncutred trophy under sub-section (1) Sec. 40A and in respect of which certificate of ownership has not been granted or obtained, sha ll be treated as government property

13. Order to be signed and dated.- Every order of the Chief Wild Life Warden shall be in writing and shall be signed and dated by the Chief Wild Life Warden.

14. Commnication of order to parties.- Every order passed on the application shall be communicated to the application either in person or by registered post free of cost.

[F. No. 1-1/2003 WL-I]

FORM

APPLICATION UNDER SECTION 40A OF WILD LIFE PROTECTION ACT, 1972 FOR CERTIFICATION OF OWNERSHIP

To

The Chief Wild Life Warden or the Authorized Officer

State or Union Territory of ....

....

(i) I...................... ....

(Surname) (First name) (Middle Name)

son/daughter of ....

(Surname) (First name) (Middle Name)

presently residing at House

Number..................... Taluk...................................

District...................................

State... Pin Code....................................and having permanent residence at House

Number..................... Taluk ....

District...................................

State...................................... Pin Code.................................. Here by declare that I am in control or possession of captive animal and/or its offspring bred in captivity/ animal article/ trophy/ uncured trophy/ derived from animal (strike out whichever is not applicable) specified in Schedule I or part II of Schedule II o the Wildlife (Protection) Act, 1972 having following description

1. Common name of the animal species
2. Zoological name (Mention sub-species if any):
3. Description of the item
4. State the condition of the item (provide four colour photographs of size 8" x 6" covering front, left and right profiles and full photograph):
5. Number of item:
6. Method of procurement: Purchase/gift/inheritance/any other modes specify:
7. Date of procurement:
8. Name of person/ institute from whom obtain:
9. Address of person/ institute referred to in (6) above:
10. Size (in meters/cms):

(i) Length

(ii) Width

(iii) Height

11 Weight (in Kgs/gms):

12 any specific mark that can help in ident ification of the item

13 Mention the age and sex in case of live animals

(ii) I hereby declare that the above referred captive animal / item has been kept, store or maintained at the following address

....

....

....

I hereby declare that the above referred captive animal / item was acquire by me through legal means but no declaration has been made by me under sub -section (1) or sub-section (4) of sec. 40 of the Wild Life (Protection) Act. 1972

I further declare that I have read and understood the provisions contain in Sec. 40A, 42 and 43 of wild Life (Protection 0 Act, 1972 and state that the above shall not be transferred to anyone by any mode except by way of inheritance.

I hereby give my consent for fixing an identification mark to each item and transponder in case of captive animal and assure that mark of transponder will not be erased, altered or damaged and in the event of any damage, alternation or change of the mark, I shall inform the competent within twenty- four hours.

I do hereby declare that the information given above is true to the best my knowledge and belief.

Place :

Date :

Signature of the person making the declaration

(Name)

Acknowledge Slip

Receipt of the application filed by Shri/ Smt .... present residing at .... (Ful Address and Telephone Number) .... in the office of the .... Is hereby acknowledged.

Official Seal

Signature

GUIDELINES FOR APPOINTMENT OF HONORARY WILDLIFE WARDENS

The Need

1. People's participation and support is crucial for nature and Wildlife conservation. One of the important ways of enlisting such support is by involving the community leaders and other persons of standing, who have the interest as well as the capacity to render assistance for this cause. Such assistance can be very useful in control over poaching for this clandestine trade in wild animals or their articles, identification of relatively less known wildlife refuges needing protection, carrying the message of conservation to the people living in and around the sanctuaries and national parks, and related matters. This objective can be accomplished if really suitable public men are identified, duties and Honorary Wildlife Wardens, with their responsibilities and powers clearly defined.

Legal Status

2. Sec.4 of the Wild Life (Protection) Act, 1972 empowers the State Government to appoint
   (a) a Chief Wildlife Warden;
   (b) Wildlife Wardens; and
   (c) Such other officers and employees as may be necessary for the purpose of the Act.

   Honorary Wildlife Wardens can be appointed under sub-section (c) of Sec.4 of the Act. Under Sec.59 of the aforesaid Act, such Honorary Wildlife Wardens shall be deemed to be public servants within the meaning of Sec.21 of the Indian Penal Code.

Criteria for Selection

3. It is very important that the right persons are selected for appointment as Honorary Wildlife Wardens. Eve ry State has a Chief Wildlife Warden and it is mainly his duty to recommend the names of suitable person for this purpose. However, in order to assist him in this regard as well as to introduce a measure of wider participation, the members of the State Wildlife Advisory Board should be requested to suggest suitable names, especially from their own areas.
4. The following criteria should be kept in mind while assessing the suitability of a person as an Honorary Wildlife warden:
   (a) Genuine concern for Wildlife conservation.
   (b) Personal record free of involvement in any activity detrimental to the interest of nature and Wildlife conservation.

   Any person involved in commercial exploitation of Wildlife should not be considered.

   (c) Capacity to render help to the official machinery.
   (d) Local standing which make him/ her effective, especially in conveying the conservation message.

5. An important point to bear in mind is the identification of areas particularly prone to poaching, e.g. forests in the vicinity of urban centres and cantonments or close to sanctuaries and national parks. Likewise, centres of clandestine trade in wildlife and products thereof should be identified; so also areas where damage to the people or their property from wild animals is heavy. Selection of persons as honorary Wildlife Wardens must be related to such problem areas because it is these areas, which need priority attention and where public participation is needed most.

Procedure and Appointment

6. Under Sec.6 of the Wild Life (Protection) Act, 1972, every State and Union Territory has a Wildlife Advisory Board to aid and advise the Government in matters connected with the protection of wildlife. The appointment of an Honorary Wildlife Warden should be generally with the recommendation of this Advisory Board. The Chief Wildlife Warden should submit the proposals for this purpose at the meeting(s) of the Board and then seek the orders of the Government.
7. While recommending any person for such appointment, the criteria led down in paras 4 and 5 above must be kept in mind by the Board.
8. The appointment of an Honorary Wildlife Warden should, in the first instance, be generally for a period of one year. Thereafter on the recommendation of the Wildlife Advisory Board, it may be renewed for a period not exceeding 2-3 years at a time.
9. The Wildlife Advisory Board of each State/ Union Territory should review the functioning of the scheme of Honorary Wildlife Wardens at least once every year.
10. The appointment order of an Honorary Wildlife Warden should clearly specify the jurisdiction, which should normally be a district or a few districts, in the area where the person resides. However, there is no objection to making members of the State Wildlife Advisory Board Honorary Wardens for larger areas.
11. Each Honorary Warden should be issued an identity card having his signature and photograph duly attested by the Chief Wildlife Warden. The Chief Wildlife Warden should also give each Honorary Warden a small booklet containing the Wild Life (Protection) Act and the Rules made there under as well as the duties, responsibilities, and power of an Honorary Wildlife warden.
12. The State Government may, at its discretion, terminate the appointment of an Honorary Wildlife Warden at any time, without assigning reasons.

Duties and Responsibilities

13. The main duty and responsibility of an Honorary Wildlife Warden is to assist whole heartedly the State organization responsible for Wildlife conservation work, especially with regard to the following matters:
    (a) Control of poaching and clandestine trade in wild animals and products/ articles thereof.

(b) Detection and prosecution of offences under the Wild Life (Protection) Act and the Rules made thereunder.

(c) Preventing damage to the habitat of Wildlife.

(d) Identification and selection of areas suitable to be declared as sanctuaries, national parks, closed areas, etc; as well as measures for their proper protection.

(e) Measures for dealing with the problem of damage by wild animals to life and property, including the assessment and payment of compensation, etc.

(f) Carrying the message of conservation to the people and enlisting public support for nature and Wildlife conservation. The effort should be especially directed to the communities living in or near the declared Wildlife reserves.

(g) Any other matter connected with the protection of Wildlife, which may be entrusted by the Wildlife Advisory Board or the Chief Wildlife Warden of the State, from time to time.

Powers

14. In accordance with sub-section (3) of Sec.4 of the Wild Life (Protection) Act, 1972, an Honorary Wildlife Warden appointed under sub-section (2) (c) of Sec.4 shall be subordinate to the Chief Wildlife Warden of the State and under Sec.59 of aforesaid Act, he shall be deemed to be a public servant within the meaning of Sec.21 of the Indian Penal Code. Protection for action taken in good faith is provided under Sec.60 of the Act.

15. With a view of making the Honorary Wildlife Wardens useful and effective it is necessary that the following specific powers under the Wild Life (Protection) Act, 1972 should be delegated to them:

(a) Power to inspect records of licences under Sec.47 (b) of the Act;

(b) Powers of entry, search seizure and detention under Sec.50 for prevention and detection of offences under the Act.

16. Suitable Honorary Wildlife Wardens could be authorised also to file complaints in courts in accordance with Sec.55 of the Wild Life (Protection) Act, 1972. Normally, however, an Honorary Wildlife Warden should bring the offence detected by him to the notice of the Wildlife Warden having jurisdiction for making proper investigation and lodging a complaint in the court as laid down in Sec.55 of the Act.

17. Apart from the above, the State Government may delegate any other power under the aforesaid Act, as it may cons ider necessary.

General

18. Just as it is expected that the Honorary Wildlife Wardens should assist the State Wildlife organization, it is equally essential that the Chief Wildlife Warden and the whole State Machinery responsible for the protection of Wildlife should take all possible steps to associate the Honorary Wildlife Wardens in their work. This can be achieved best by fostering a spirit of mutual trust and confidence.

19. No staff or vehicle support can be provided to Honorary Wardens as a matter of course. However, if the circumstances warrant, the departmental staff should provide all possible help and assistance. Instructions to this effect should be issued by the State Government to all concerned officers in the field.
20. It is also appropriate that the actual expenses incurred by an Honorary Warden on travel by public transport for carrying out the duties assigned to him should be reimbursed by the State Government. In addition, all actual expenses incurred in the detection of an offence under the Wild Life (Protection) Act, 1972, which leads to successful prosecution may be reimbursed after due verification.
21. The State Government should recognise outstanding work or service rendered by any Honorary Warden. Such recognition can be by way of a letter of commendation, or a certificate signed by the Minister in charge of the Department, or the membership of the State Wildlife Advisory Board. Cash grants could also be considered in suitable cases.

MINISTRY OF ENVIRONMENT AND FORESTS

NOTIFICATION

New Delhi, the 10th July, 2001

RECOGNITION OF ZOO (AMENDMENT) RULES, 2001

G.S.R. 520(E) - In exercise of the power conferred by Clauses (g) of sub-section (1) of Sec.63 of the Wild Life (Protection) act, 1972 (53 of 1972), the Central Government hereby makes following rules to amend the 'Recognition of Zoo Rules, 1992' namely: -

1. **Short title and commencement:**
   (1) These rules may be called the Recognition of Zoo (Amendment) Rules, 2001.
   (2) They shall come into force on the date of their publication in the Official Gazette.
2. **Definition:** In these rules, unless the context otherwise requires,
   (a) "Act" means the Wild Life (Protection) Act, 1972 (53 of 1972);
   (b) "Enclosure" means any accommodation provided for Zoo animals;
   (c) "Enclosure barrier" means a physical barrier to contain an animal within an enclosure;
   (d) ["Endangered species" means species included in Sch. I and Sch. II of the Act except Black buck;]
   (e) ["Critically endangered species" means indigenous species whose total number, in all the zoos put together doesn't exceed 200 but shall include tiger, Asiatic lion and panther;]
   (f) "Form" means form set forth in Appendix A to these rules;

(g) "Performing purposes" means any effort to force the animal to carry out unnatural act including performance of circus tricks;

(h) "Stand-of barrier" means a physical barrier set back from the outer edge of an enclosure barrier;

(h) "Zoo operator" means the person who has unlimited control over the affairs of the Zoo provided that -

(i) in the case of firm or other association of individuals, any one of the individual partners or members thereof shall be deemed to be the Zoo operator;

(ii) in the case of a company, any director, manager, secretary or other officer, who is in-charge of and responsible to the company for the affairs of the Zoo shall be deemed to be the Zoo operator;

(iii) in the case of zoo owned or controlled by the Central Government or any State Government or any local authority, the person or person appointed to manage the affairs of the zoo by the Central Government, the State Government or the local authority, as the case may be shall be deemed to be the Zoo operator.

3. **Application for recognition:** An application under Sec.38H of the Act for recognition of a zoo shall be made to the Central Zoo Authority in Form A.

4. **Fees for application:**

(a) There shall be paid in respect of every application under rule 3 a fee of rupees five hundred.

(b) The amount of the fee shall be paid through Demand Draft/ Postal Order (s) in favour of the Central Zoo Authority, New Delhi.

5. **Documents to be filled along with the application and particulars it should contain:** Every application shall be accompanied by the prescribed fee and shall contain clear particulars as to the matters specified in Form A.

6. **Power to make inquiries and call for information:** Before granting recognition to a zoo under Sec.38H of the Act, the Central Zoo Authority may make such inquiries and require such further information to be furnished, as it deems necessary, relating to the information furnished by the zoo in its application in Form A.

7. **Form of recognition:** The recognition granted to a zoo shall be subject to the following conditions, namely:

(a) that the recognition unless granted on a permanent basis, shall be for such period not less than one year as may be specified in the recognition;

(b) that the zoo shall comply with such standards and norms as are or may be prescribed or imposed under the provisions of the Act and these rules from time to time.

8. **Renewal of recognition:**

(a) Three months before the expiry of the period of recognition, a recognised zoo desirous of renewal of such recognition may make an application to the Central Zoo Authority in Form A.

(b) The provisions of rules 3, 4,5,6 and 7 shall apply in relation to renewal of recognition as they apply in relation to grant of recognition except that, the fee payable in respect of an application for renewal of recognition shall be rupees two hundred.

9. **Classification of Zoos:** For the purposes of deciding standards and norms for recognition of zoos and monitoring and evaluating their performance, the zoos, on the basis of number of animals, species, endangered species and number of animals of endangered species exhibited, shall classified into four categories as specified below:

Category of Zoo	Large	Medium	Small	Mini
Number of animal exhibited	More than 750	Less than 500-750	200-499	200
Number of species exhibited	More than 75	Less than 50-75	20-49	20
Number of endangered species exhibited	More than 15	10-15	5-9	——
Number of animals of endangered species exhibited	More than 150	100-149	50-99	——

[(9A) Central Zoo Authority may allow a mini zoo to keep animals of endangered species subject to the condition prescribed by it with regard to health, care, facilities and upkeep of animals including deployment of supervisory level staff including veterinarian.]

10. Standards and norms subject to which recognition under Sec. 38H of the Act shall be granted:

The Central Zoo Authority shall grant recognition with due regard to the interests of protection and conservation of wild life and such standards, norms and other matters as are specified below:

(1) The primary objective of operating any zoo shall be the conservation of wildlife and no zoo shall take up any activity that is inconsistent with the objective.

(2) No zoo shall acquire any animal in violation of the Act or rules made thereunder.

(3) No zoo shall allow any animal to be subjected to the cruelties as defined under the Prevention of Cruelty to Animals Act, 1960, (59 of 1960) or permit any activity that exposes the animals to unnecessary pain, stress or provocation, including use of animals for performing purposes.

(4) No zoo shall use any animal, other than the elephant in plains and yak in hilly areas for riding purposes or draughting any vehicle.

(5) No zoo shall keep any animal chained or tethered unless doing so is essential for its own well being.

(6) No zoo shall exhibit any animal that is seriously sick, injured or infirm.

(7) Each zoo shall be closed to visitors at least once a week.

(8) Each zoo shall be encompassed by a perimeter wall at least two metres high from the ground level. The existing zoos in the nature of safaris and deer parks will continue to have chain link fence of appropriate design and dimensions.

(9) The zoo operators shall provide a clean and healthy environment in the zoo by planting trees, creating green belts and providing lawns and flower beds, etc.

(10) The built up area in any zoo shall not exceed twenty five percent of the total area of the zoo. The built up area includes administrative buildings, stores, hospitals, restaurants, kiosks and visitor rest sheds, etc., animal houses and 'pucca' roads.

(11) No zoo shall have the residential complexes for the staff within the main campus of the zoo. Such complex, if any, shall be separated from the main campus of the zoo by a boundary wall with a minimum height of two metres from the ground level.

[(11A) Every zoo shall prepare a collection plan of animals to be housed and displayed in the zoo, keeping due regard to the availability of land, water, electricity and climatic condition of the area.]

Administrative and Staffing Pattern:

(12) [Every zoo shall have one full-time officer in-charge of the zoo. The said officer shall be delegated adequate administrative and financial powers to purchase, feed and medicine and carry out emergency repair of animal enclosures, as may be necessary for proper upkeep and care of zoo animals.]

(13) [Every large, medium and small zoo shall have an official with Master Degree in Wildlife Science/ Zoology as a full time curator solely responsible for looking after the upkeep of animals and maintenance of animal enclosures.]

(14) Each large zoo shall have at least two full-time veterinarians and medium and small zoo shall have at least one veterinarian. The mini zoo may at least have arrangement with any outside veterinarian for visiting the zoo every day to look after the animals.

[(14A) Every zoo shall have veterinarians of following description and qualification: m®0p6

Category	Sr. Vet.	Jr. Vet.
Large zoo	1	1
Medium zoo	1	0
Small zoo	1	0

Sr. Veterinary Officer shall have B.V.Sc. and A.H having experience of working in a zoo recognised by Central Zoo Authority for at least by five years.

Veterinary officer shall have B.V.Sc. and A.H with Diploma in Zoo and Wildlife Animal Health Care Mana gement or Masters degree in Wildlife Diseases and Management from a recognised University.]

Animal Enclosures - Design, Dimensions and other Essential Features:

(15) All animal enclosures in a zoo shall be so designed as to fully ensure the safety of animals, caretakers and the visitors. Stand of barriers and adequate warning signs shall be provided for keeping the visitors at a safe distance from the animals.

(16) [All animal enclosures in a zoo shall be so designed as to meet the full biological requirements of the animal housed therein. The enclosures shall be of such size as to ensure that the animals get space for their free movement and exercise and the animals within herds and groups are not unduly dominated by individual in case of species, which cannot be kept in groups for behavioural or biological reasons, separate enclosures will be provided for each animal. The enclosures will not be smaller than the dimensions given in Appendix II to these rules.]

[(16A) Zoo operators shall provide appropriate screening between the adjacent enclosures to safeguard against the animals getting excited or stressed because of the visibility of animals in other enclosures.]

(17) [The zoo operators shall endeavour to simulate the conditions of the natural habitat of the animal in the enclosures as closely as possible. Planting of appropriate species of trees for providing shade and shelters, which merge in the overall environment of the enclosures, shall also be provided. Depending upon the availability of land and technical feasibility, most shall be provided as enclosure barrier.]

(18) [Every mammal in the zoo shall be provided food inside a feeding cell/ retiring cubicle or feeding kraal. The number and size of feeding cells or kraals will also be such that the dominant animals do not deprive other animals from getting adequate food. The endangered mammalian species shall be provided individual feeding cells/ night shelters of the dimensions as specified in Appendix I to these rules. Each cubicle/ cell have resting, feeding, drinking water and exercising facilities according to the biological needs of the species. Proper ventilation and lighting for the comfort and well being of animals shall be provided in each cell/ cubicle/ enclosures.]

(19) Proper arrangement of drainage of excess of water and arrangements for removal of excreta and residual water from each cell/ cubicle/ enclosures shall be made.

(20) Designing of any new enclosure for endangered species shall be finalized [with the approval of] the Central Zoo Authority.

Hygiene, Feeding and Upkeep:

(21) Every zoo shall ensure timely supply of wholesome and unadulterated food in sufficient quantity to each animal according to the requirement of the individual animals, so that no animal remains undernourished.

(22) Every zoo shall provide for a proper waste disposal system for treating both the solid and liquid wastes generated in the zoos.

(23) All left over food items, animal excreta and rubbish shall be removed from each enclosure regularly and disposed of in a manner congenial to the general cleanliness of the zoo.

(24) The zoo operators shall make available round the clock supply of potable water for drinking purposes in each cell/ enclosure/ cubicle.

(25) Periodic application of disinfectants in each enclosure shall be made according to the directions of the authorised veterinary officer of the zoo.

Animal Care, Health and Treatment:

(26) The animals shall be handled only by the staff having experience and training in handling the individual animals. Every care shall be taken to avoid discomfort, behavorial stress or physical harm to any animal.

(27) The condition and health of all animals in the zoo shall be checked every day by the person in-charge of their care. If any animal is found sick, injured, or unduly stressed, the matter shall be reported to the veterinary officer for providing treatment expeditiously.

(28) Routine examination including parasites checks shall be carried out regularly and preventive medicines including vaccination be administered at such intervals as may be decided by the authorised veterinary officers.

(29) The zoo operators shall arrange for medical check-ups of the staff responsible for upkeep of animals at least once in every six months to ensure that they do not have infections of such diseases that can infect the zoo animals.

(30) Each zoo shall maintain animal history sheets and treatment cards in respect of each animal of endangered species, identified by the Central Zoo Authority.

Veterinary Facilities:

(31) Every large and medium zoo shall have full- fledged veterinary facilities including a properly equipped veterinary hospital, basic diagnostic facilities and comprehensive range of drugs. Each veterinary hospital shall have isolation and quarantine wards for newly arriving animals and sick animals. These wards should be so located as to minimise the chances of infections spreading to other animals of the zoo.

[(31A) Every zoo operator shall provide one qualified lab assistant/ compounder for assisting the veterinarian in health care of the zoo animals.]

(32) Each veterinary hospital shall have facilities for restraining and handling sick animals including tranquillising equipments and syringe projector. The hospital shall also have a reference library on animal health care and upkeep.

(33) The small and mini zoos, where full- fledged veterinary hospital is not available, shall have at least a treatment room in the premises of the zoo where routine examinations of animals can be undertaken and immediate treatment can be provided.

(34) Every zoo shall have a post-mortem room. Any animal that dies in a zoo shall be subjected to a detailed post- mortem and the findings recorded and maintained for a period of at least six years.

(35) [Each zoo shall have proper facility for disposal of carcasses without affecting the hygiene of the zoo. However, carcasses of large cats shall be disposed off only by burning in presence of director or an officer not below the rank of a curator duly authorised by the director.]

Breeding of Animals:

(36) [Every zoo shall keep in its collection only such number of animals and such species for which appropriate housing facility exists. The zoo operators shall be responsible for ensuring that the number of animals of any species does not go beyond the holding capacity of the enclosures available in the zoo and housing standards are not compromised for keeping the excessive numbers.]

(37) Every zoo shall keep the animal in viable, social groups. No animal will be kept without a mate for a period exceeding one year unless there is a legitimate reason for doing so or if the animal has already passed its prime and is of no use for breeding purposes. In the event of a zoo failing to find a mate for any single animal within this period, the animal shall be shifted to some other place according to the directions of the

Central Zoo Authority.

(38) No zoo shall be allowed to acquire a single animal of any variety except when doing so is essential either for finding a mate for the single animal housed in the said zoo or for exchange of blood in a captive breeding group.

(39) [All zoos shall participate in planned breeding programme of endangered species approved by Central Zoo Authority in consultation with the Chief Wildlife Warden of the State. For this purpose, they shall exchange animals between zoo by way of breeding loans, gifts, etc. as per the direction of Central Zoo Authority.]

(40) To safeguard against uncontrolled growth in the population of prolifically breeding animals, every zoo shall implement appropriate population control measures like separation of sexes, sterilization, vasectomy, tubectomy and implanting of pallets, etc.

(41) No zoo shall permit hybridisation either between different species of animals or different races of the same species of animals. Maintenance of records and Submission of Inventory to The Central Zoo Authority:

(42) Every zoo shall keep a record of the birth, acquisitions, sales, disposals and deaths of all animals. The inventory of the animal housed in each zoo as on 31st March of every year shall be submitted to the Central Zoo Authority by 30th April of the same year.

(43) [Every zoo shall also submit a brief summary of the death of animals in the zoo for every financial year, along with the reasons of death identified on the basis of post-mortem reports and other diagnostic tests, by 30th April of the following year. In case of death of critically endangered species, a report along with details specified above shall be submitted to Central Zoo Authority within twenty four hours.]

(44) [Every zoo shall submit an annual report of the activities of the zoo in respect of each financial year to the Central Zoo Authority. With respect to mini zoos, a consolidated report may be submitted by the Chief Wildlife Warden of the respective State/ Union Territories.]

Education and Research:

(45) Every enclosure in a zoo shall bear a sign board displaying scientific information regarding the animals exhibited in it.

(46) Every zoo shall publish leaflets, brochures and guidebooks and make the same available to the visitors, either free of cost or at a reasonable price.

(47) Every large and medium zoo shall make arrangements for recording, in writing, the detailed observations about the biological behaviour, population dynamics and veterinary care of the animals exhibited as per directions of the Central Zoo Authority so that a detailed database could be developed. The database shall be exchanged with other zoos as well as the Central Zoo Authority.

Visitor Facilities:

(48) The zoo operators shall provide adequate civic facilities like toilets, visitor sheds and drinking water points at convenient places in the zoo for visitors.

(49) First-aid equipments including anti- venom shall be readily available in the premises of the zoo.

(50) Arrangements shall be made to provide access to the zoo to disabled visitors including those in the wheel chair.

Development and Planning:

(51) Each zoo shall prepare a long-term master plan for its development. The zoo shall also prepare a management plan, giving details of the proposal and activities of development for next six years. The copies of the said plans shall be sent to the Central Zoo Authority.

APPENDIX A

APPLICATION FOR GETTING RECOGNITION FROM THE CENTRAL ZOO AUTHORITY UNDER SECTION 38H (Sub-section 2)

FORM - A

To

The Member-Secretary,

Central Zoo Authority of India, New Delhi.

We want to get recognition under section 38H of the Wild Life (Protection) Act, 1972 in respect of ___________________________ Bank Draft/ Postal Order for Rs.500/- drawn in favour of Central Zoo Authority is also enc losed. The required information in respect of ___________________________ is as under:

1. Name of the Zoo:
2. Location of the Zoo and Area:
3. Date of establishment:
4. Name of controlling authority/ operator:
5. Total number of visitors to the Zoo during the last three years (Year wise):
6. Total number of days on which Zoo is open to visitors during a calendar year:
7. Number of animals exhibited by the Zoo:

Stock position during the current financial year

Number of species exhibited	Stock Position on the close of preceding year	Births	Acquisitions	Deaths	Disposals	Stock as on the date of application
MAMMALS						
BIRDS						
REPTILES						
AMPHIBIANS						
FISHES AND OTHERS						
INVERTEBRATES						

8. Total number of enclosures:
   (i) Open air moated enclosures:
   (ii) Closed cages/ aviaries:

9. List of endangered species bred during last 3 years:
10. Veterinary facilities:
    (i) Whole time veterinarian available or not:
    (ii) Facilities available in the Veterinary Hospital:
        (a) Operation theatre/ Surgical room
        (b) X-ray facility
        (c) Squeeze cages
        (d) In-door patient ward
        (e) Quarantine ward
        (f) Dispensary
        (g) Nursery for hand-rearing animal babies
        (h) Pathological laboratory
        (i) Tranquillising equipments/ drugs
11. Whether the following facilities exist in the zoo:
    (i) Kitchen
    (ii) Food store
    (iii) Deep freeze
    (iv) Potable water facility
    (v) Food distribution van/ rickshaw etc.
12. Sanitary care and disease control:
    Whether -
    (i) Pollution free water to animals for drinking is available?
    (ii) Proper drainage system exists in enclosures?
    (iii) Regular disposal of refuse material is done?
    (iv) Programme for control of pests and predators exist?
    (v) Preventive measures like deworming and vaccination are being provided?
13. Amenities to visitors:
    Whether -
    (a) Public facilities like toilets/ bathrooms exist?
    (b) Sufficient number of drinking water taps available?
    (c) Visitor information centre and nature interpretation centre exist?
    (d) Zoo education facilities have been provided?
    (e) Public telephone booths are available?
    (f) Kiosks and restaurants are available at the zoo?

14. Safety measures for visitors:

    Whether -

    (a) Effective stand-of barriers have been provided around enclosures?

    (b) Adequate number of warning sign boards exist?

    (c) First-aid measures are available?

15. Budget of the Zoo for last 3 years:

    Revenue Grants Total expenditure

16. Annual report, Guide books, Brochure or any other publication (copies enclosed):

17. Master plan of the Zoo (copy enclosed):

____

Signature of the applicant

APPENDIX - I

MINIMUM PRESCRIBED SIZE FOR FEEDING/ RETIRING CUBICLE/ ENCLOSURES FOR IMPORTANT MAMMALIAN SPECIES OF CAPTIVE ANIMALS.

Name of the Species.

[Size of feeding Cubicle/ Night shelter (L x B x H) -in meters]

FAMILY - *Felidae*:	Length	Breadth	Height
Tiger and lions			
Panther	2.75	1.80	3.00
	2.00	1.50	2.00
Clouded leopard & snow leopard	2.00	1.50	2.00
Small cats	1.80	1.50	1.50
FAMILY - *Elephantidae*:			
Elephant	8.00	6.00	5.50
FAMILY - Rhinocerotidae:			
One-horned Indian Rhinoceros	5.00	3.00	2.50
FAMILY - *Cervidae:*			
Brow antlered deer	3.00	2.00	2.50
Hangul	3.00	2.00	2.50
Swamp deer	3.00	2.00	2.50
Musk deer	2.50	1.50	2.00
Mouse deer	1.50	1.00	1.50
FAMILY - *Bovidae:*			
Nilgiri tahr	2.50	1.50	2.00
Chinkara	2.50	1.50	2.00
Four horned antelope	2.50	1.50	2.00
Wild buffalo	3.00	1.50	2.00
Indian bison	3.00	2.00	2.50
Yak	4.00	2.00	2.50
Bharal, goral, wild sheep and markhor	2.50	1.50	2.00
FAMILY - *Equidae:*			
Wild Ass	4.00	2.00	2.50
FAMILY - *Ursidae:*			
All types of Indian bears	2.50	1.80	2.00
FAMILY - *Canidae:*			
Jackal, wolf & wild dog	2.00	1.50	1.50
FAMILY -*Vivirridae:*			
Palm civet	2.00	1.00	1.00
Large Indian civet & binturong	2.00	1.50	1.00
FAMILY - Mustellidae:			
Otters all types	2.50	1.50	1.00
Ratel/ Hogbadger	2.50	1.50	1.00
Martens	2.00	1.50	1.00
FAMILY - *Procyonidae*:			
Red Panda	3.00	1.50	1.00
FAMILY - *Lorisidae:*			
Slow loris and slender loris	1.00	1.00	1.50
FAMILY - *Cercopithecidae:*			
Monkeys and langurs	2.00	1.00	1.50

APPENDIX - II

MINIMUM PRESCRIBED SIZE FOR OUTDOOR OPEN ENCLOSURE FOR IMPORTANT MAMMALIAN SPECIES OF CAPTIVE ANIMALS

Sl. No.	Name of the Species	Minimum size of outdoor enclosures (per pair)	Minimum area extra per additional animal
		Square meter	
FAMILY - *Felidae:*			
1.	Tiger and lions	1000	250
2.	Panther	500	60
3.	Clouded leopard	400	40
4.	Snow leopard	450	50
FAMILY - *Rhinocerotidae:*			
5.	One-horned Indian Rhinoceros	2000	375
FAMILY - *Cervidae:*			
6.	Brow antlered deer	1500	125
7.	Hangul	1500	125
8.	Swamp deer	1500	125
FAMILY - *Bovidae:*			
9.	Wild buffalo	1500	200
10.	Indian bison	1500	200
11.	Bharal, Goral, Wild sheep and Serow	350	75
FAMILY - *Equidae:*			
12.	Wild Ass	1500	200
FAMILY - *Ursidae:*			
13.	All types of Indian bears	1000	100
FAMILY - *Canidae:*			
14.	Jackal, Wolf & Wild dog	400	50
FAMILY - *Procyonidae:*			
15.	Red Panda	300	30
FAMILY - *Cercopithecidae:*			
16.	Monkeys and langurs	500	20

Note:

1. The dimensions have been given only in respect of the species, which are commonly displayed in zoos.
2. No dimensions for outdoor enclosures have been prescribed for Chinkara and Chowsingha because of the problem of infighting injuries. These animals may be kept in battery type enclosures of the dimensions suggested by the Central Zoo Authority.
3. The designs of enclosures for Sch. I species, not covered by this Appendix, should be finalized only after approval of the Central Zoo Authority.]

(S.C SHARMA)

Addl. DGF (Wildlife) and Director, Wildlife Preservation

[F. No. 7-4/ 99 (8A)]

NUMBER OF ZOOS AND CAPTIVE WILDLIFE FACILITIES IN STATES AND UNION TERRITIRIES OF INDIA

STATE/ UT s.	ZOOS/ NATURE PARKS	DEER AQUARIUMS PARKS		SAFARI PARKS	TOTAL EDUCATION BREEDING CENTRES		SNAKE
Andaman and Nicobar Islands	1	0	0	0	0	0	1
Andhra Pradesh	3	14	3	0	1	1	22
Arunachal Pradesh	3	0	0	0	1	0	4
Assam	1	0	0	0	3	0	4
Bihar	5	1	0	0	0	0	6
Delhi	1	1	0	0	0	0	2
Goa	1	0	0	0	0	0	1
Dadra and Nagar Haveli(UT)	1	2	0	0	0	0	3
Gujarat	8	4	0	0	2	2	16
Haryana	5	2	0	0	1	0	8
Himachal Pradesh	4	1	2	0	3	0	10
Jammu & Kashmir	2	1	0	0	0	0	3
Karnataka	19	3	4	1	0	0	27
Kerala	3	1	1	1	2	0	8
Madhya Pradesh	5	0	0	0	1	1	7
Maharastra	10	1	1	2	2	1	17
Manipur	1	0	0	0	0	0	1
Meghalaya	2	0	0	0	0	0	2
Mizoram	1	0	0	0	0	0	1
Nagaland	1	0	0	0	0	0	1
Orissa	2	7	3	0	1	0	13
Pondicherry	1	0	0	0	0	0	1
Punjab	5	3	1	0	0	0	9
Rajasthan	6	1	0	0	1	0	8
Tamil Nadu	8	1	0	1	1	2	13
Tripura	1	0	0	0	0	0	1
Sikkim	1	1	0	0	0	0	2
Uttar Pradesh	3	7	0	0	6	0	16
West Bengal	3	0	1	1	0	1	6
TOTAL	**107**	***51**	**16**	**6**	**25**	**8**	**213**

* *The actual number could be much more but State-wise details are not available.*

CHAPTER - 2

IMPORT-EXPORT POLICY (APRIL 1992- MARCH 1997)

Extracts from the Export & Import Policy (1st April 1992- 31st March 1997)

As application from 01.04.94 to 31.03.96

Chapter III

Para 7. Definitions

(2) "Act" means the Foreign trade (Development & Regulation) Act 1992 (No. 22 of 1992).

(24) "Manufacture" means to make, produce, fabricate, assemble, process or bring into existence, by hand or by machine, a new produce having a distinctive name, character or use and shall include process such as refrigeration, repacking, polishing, labelling and segregation. Manufacture for the purpose of this policy, shall also include agriculture, aquaculture, animal husbandry, floriculture, pisciculture, poultry and sericulture.

(31) "Policy" means the Export & Import Policy 1992-97 as amended from time to time.

Chapter IV

Para 8. Exports & Imports free unless regulated

Exports and Imports may be done freely, except to the extent they are regulated by the provisions of this Policy or any other law for the time being in force.

Para 9. Form of Regulation

The Central Gove rnment may, in public interest, regulate the import or export of goods by means of a Negative list of Imports or a Negative list of Exports, as the case may be.

Para 10. Negative Lists

The Negative lists may consist of goods that Import or Export of whic h is prohibited, restricted through licensing or otherwise, or canalised. The Negative lists of Exports and the Negative lists of Imports shall be as contained in this Policy.

Para 11. Prohibited goods

Prohibited goods shall not be imported or exported.

Para 16. Procedure

The Director General of Foreign Trade may, in any case or class of cases, specify the procedure to be followed by an exporter or importer or by any licensing, competent or the authority for the purpose of implementing the provisions of the Act, the Rules and Orders made thereunder this Policy. Such procedures shall be included in the Handbook of Procedures and published by means of a Public Notice. Such procedures may, in like manner, be amended from time to time.

Chapter XV

NEGATIVE LISTS OF IMPORTS

155 PROHIBITED ITEMS

3. Wild animals including their parts and products and ivory.

156 RESTRICTED ITEMS

D.SEEDS, PLANTS AND ANIMALS

1. Animals, Birds & Reptiles (including their parts & products)

   Import permitted against a licence to zoos and zoological parks, recognised scientific/ research institutions, circus companies, private individuals, on the recommendation of Chief Wildlife Warden of a State Government subject to the provisions of the Convention on International Trade in Endangered Species of Wild Fauna & Flora (CITES)

4. Plants, Fruits & Seeds

   (a) Import of seeds of wheat, paddy coarse cereals, pulses, oilseeds and fodder for sowing is permitted without a licence subject to fulfilment of the provisions of the New Policy on Seed Development 1988 and in accordance with a permit for import granted under the Plants, Fruits and Seeds. (Regulation of Import into India) Order 1989.

   (b) Import of seeds of vegetable flowers, fruits and plants, tubers and bulbs of flowers, cutting, sapling, budwood, etc. of flowers and fruits for sowing or planting is permitted without a licence I accordance with a permit for import granted under the Plants, Fruits and Seeds. (Regulation of Import into India) Order 1989.

   (c) Import of Seeds, Fruits and Plants for consumption or other purposes is permitted against a licence or in accordance with a Public Notice in this behalf.

   (d) Import of plants, their products and derivatives shall also be subject to the provisions of the Convention of International Trade in Endangered Species of Wild fauna & Flora (CITES).

Chapter XVI

NEGATIVE LIST OF EXPORTS

PART I

158 PROHIBITED ITEMS

1. All forms of wild animals including their parts and products except Peacock tails including handicrafts made thereof and manufactured Articles and Shavings of Shed Antlers or Chital and Shambhar subject to condition as specified in Annexure to Public Notice No. 15-ETC (N)/ 92-97 date 31st March, 1993

issued by the Director General of Foreign Trade and reproduced in the Handbook of Procedures (Vol. 1)

2. Exotic Birds.
3. All items of plants included in Appendix I of the Convention of International Trade in Endangered Species (CITES), wild orchids as well as plants as specified in Public Notice No. 47 (PN)/ 92-97 dated 30th March, 1994 issued by the Director General of Foreign Trade and reproduced in the Handbook of Procedures (Vol. 1)
7. Wood and wood products in the form of logs, timber, stumps, roots, barks, chips, powder, flakes, dust, pulp and charcoal except sawn timber ma de exclusively out of imported teak, logs/ timber subject to conditions as specified in Annexure to Public Notice No. 15-ETC (PN)/ 92-97 date 31st March, 1993 issued by the Director General of Foreign Trade and reproduced in the Handbook of Procedures (Vol. 1)
9. Sandalwood.
10. Red Sanders wood in any form whether raw, processed or unprocessed as well as any product made thereof.

PART II

159 RESTRICTED ITEMS

(EXPORTS PERMITTED UNDER LICENCE)

8. Fur of domestic animals, excluding lamb fur skin.
10. Hides and skins namely:
    - (i) Cutting and fleshing of hides and skins used as raw materials for manufacture of animal glue gelatine.
    - (ii) Raw hides and skins, all types excluding lamb fur skin.
    - (iii) All categories of semi-processed hides and skins including E.I. tanned and wet blue hides and skins and crust leather.

APPENDICES

COMMON VETERINARY TERMINOLOGY/DEFINATIONS

Veterinary Preventive Medicine : Refers to attempts to keep animal healthy when the disease is imminent. It deals with the communicable diseases. Prevention can be done with help of vaccine, sera and sanitary measures such as isolation, quarantine, disinfection, notification etc. includes all the practice that strengthen genetic and immunological resistance to the diseases, provide sound nutrition and minimize exposure to disease.

State Medicine : Part of preventive medicine dealing with the safeguarding of the animal as well as human health. Inspection of meat, milk etc.

Bacteraemia : A state in which micro organisms invade the blood stream but do not multiply. (Passive)

Carrier : An individual who harbors specific organisms of a disease in the body without manifesting symptoms but still serving as a means of conveying infection. It may be Incubated carrier, convalescent carrier or healthy true carrier.

Chemotherapy : The treatment of disease by chemical agents.

Contagious disease : Is one, which is caused by organisms that are rapidly transferred from one individual to another by direct or indirect contact. **Contagious diseases are all infectious, but all infectious diseases are not contagious.**

Disease : Deviation from the health. Disharmony within the body, between the body and mind and between the man and environment. The result of collision between a pathogenic agent and susceptible individual. Disease can be classified

I. According to the mode of origin (*A. Hereditary*: epilepsy, *B. Congenital*: Intra uterine, atresia ani, *C.Acquired*: on exposure to pathogen).

II. According to etiology : Specific (Bacterial,Viral etc.) and non specific etiology.

III. According to clinical occurrence: ***A. Primary /Idiopathic*:** Distemper, ***B. Intercurrent:*** Sequalae – Chorea, ***C. Secondary*:** Babesiosis-distemper

IV. According to severity/duration: ***A. Per acute***: RP, ***B. Acute***: FMD, ***C. Subacute***: Anthrax, ***D. Chronic***: TB, Glanders.

V. According to organs affected: *A.General*: Influenza, *B.Local:* Actino bacillosis.

VI. According to changes of organs: *A.Functional*: epilepsy, *B. Structural*: B.Q.

VII. According to distributions:. Sporadic,.Endemic,.Epidemic, Pandemic.

VIII. According to population affected: Exotic & indigenous.

Disinfectant : An agent that prevents infection by destruction of pathogenic microorganisms usually applied to inanimate objects.

Endemic : A disease is said to be endemic when it is prevalent in a geographic area continuously at a relatively low level without importation from outside. The virulence of the organisms is usually low and cases are limited.

Epidemics : Occurrence of a disease in a population clearly in excess of normal expectation.

Epidemiology : Science deals with incidence, distribution and control of disease in a population. Patterns of occurrence disease.

Epizootic : A disease of animals that spread rapidly and are widely diffused.

Epizootiology : A science that deals with epizootic and the factors involved in the occurrence and spread of the disease of animals.

Immunity : power to resist infection the action of certain poisons. It may be inherited, acquired naturally or artificially. E.g.: Snake venom - mongoose, pigeon—morphine, Fowl –anthrax, tetanus, Horse –FMD, Rat –TB, OX-Glanders. Active –Vaccination, Passive –Serum.

Incubation Period : It is the period from the time of entry of infection in to the body up to the production of the first symptom.

Infectious disease : One that is caused by a living agent such as bacteria, protozoa, viruses or fungi and may or may not be contagious.e.g. Blue tongue.

Incidence rate : It is a measure of attack. Incidence rate of the disease is number of new cases of a particular disease occurring within specified period in a given population. It is dynamic measures of the risk of getting sick.

$$\text{Incidence rate (\%)} = \frac{\text{Number of new cases reported during period}}{\text{Av. population during that period}} \times 100$$

Pandemic : Wide geographical distribution of a disease on a large scale. From one country to another country in a short time or occurs at the same time in the different countries.

Prognosis : Forecasting of a probable outcome and the course of a disease with regards to the prospect of early or late recovery. It is based on the nature of the disease, severity of the symptoms and the condition of the animal.

Prophylaxis : The prevention of disease by various measures.

Prevalence rate : It is static measure of the no. of total cases prevailing at a certain time.

$$\text{Prevalence rate (\%)} = \frac{\text{Total No. of cases existing at that period of time (New \& old)}}{\text{Average population during that period}} \times 100$$

Pyemia : If a pyogenic organism gains access to the lymph or blood stream and gets carried to the other parts of the body and produces abscesses the condition is called as pyaemia.

Quarantine : It is the isolation of healthy man or animal who have come in contact with an infectious disease for a period of time equal to the longest incubation period of the disease to prevent contact with those who are not exposed.

Septicemia : A condition in which pathogenic bacteria and their associated proteins are present in the blood and the organisms multiply there (toxaemia +hyperthemia + microorganism).

Sero diagnosis : Diagnosis made by means of the reactions taking place with blood serum.

Sporadic disease : It refers to the occasional or infrequent occurrence of cases. The cases are so few and separated in space and time that they show little or no connection with each other.

Surveillance : It is continued watchfulness over the distribution and trends of incidence of a disease through systematic collection, consolidation and evaluation of morbidity and mortality reports.

Toxaemia : A general intoxication due to the absorption of bacterial products usually toxins.

Vector : Vector is arthropod or invertebrate host which transmit the infection by inoculation in to the skin or mucous membrane by biting or by depositing the infective materials on the skin /feed/objects.

Vehicle : It is non living thing or substance such as milk, water, meat, pus, serum, dust, veg. Fruits etc. by which an infectious agent passes from an individual to a susceptible one.

Table 1. Value of Physiological Parameter/ vital Parameters in Wild Animals.

Sr.No	Name	Body temperature	Respiration rate per minute	Heat rate or pulse rate per minute
1	Tiger	37.8-39.9°C	10-12	40-50
2	Lion	37.06-38.6°C	10-12	40-50
3	Leopard (Panther)	38.0-39.0°C	12-16	42-45
4	Cheetah	38.0-38.5°C	10-16	43-45
5	Wolf	35.90-36.5 °c	10-15	28-35
6	Elephant	36.0-37.0°C	10-16	25-30
7	Rhinoceros	37.0-39.0 ° C	20-40	70-140

[Table Contd...

Contd. Table]

Sr.No	Name	Body temperature	Respiration rate per minute	Heat rate or pulse rate per minute
8	Giraffe	38.0-38.8°C	12-20	40-50
9	Kangaroo	37.0-39.0 ° C	20-40	70-140
10	Bear	37.5-38.3°C	15-20	60-90
11	Antlers	38.0-39.0°C	18-25	45-60
12	Yak	38.0-38.5°C	12-16	40-50
13	Alligator	80-90°F		
14	Chameleons	55-75°F		
15	Turtles	75-85°F		
16	Snakes	75-90°F		
17	Scorpians	18-32°C		
18	Hare	38-40°C	32-60	135-325
19	Avifauna	106-108°F		233 22

Table 2. Site of recording pulse in different specieis of wild animals.

S. No.	Name	Site of pulse recording
1.	Tiger & Lion	femoral / tail artery
2.	Rhinoceros	ear / tail artery
3.	Elephantear	artery
4.	Giraffe	facialj coccygeal artery
5.	Bear	femoral artery
6.	Antlers	femoral artery
7.	Yak	facial artery
8.	Cheetah	femoral/ tail artery
9.	Wolf	femoral artery
10.	Leopard (Panther)	femoral artery
11.	Nilgai	middle coccygeal artery

Table 3. Site of i/m and s/c injection in wild animals.

S.NO.	Name	Subcutaneous	Intramuscular
1	Tortoise	corapecae	Plastem
2	Snake	ventral site of abdominal	region abdominal inter between scales abdominal inter scu tes
3	Rhino	between skin fold	shoulder & rump
4	Boar	neck	hind quarter

[Table Contd...

Contd. Table]

S.NO.	Name	Subcutaneous	Intramuscular
5	Porcupine	ventral site of abdominal	neck
6	Elephant	between skin fold	hind quarter/ shoulder
7	Nilgai	neck	hind quarter/shoulder
8	Snow leopard	neck skin fold	rump
9	Tiger & musk deer	skin fold	rump and shoulder
10	Lizard and crocodile		Lateral aspect of the arm and forearm of the front limb
11	Turtle		Lateral aspect of front limb
12	Avifauna	Wing web, interscapular region & gron	Superficial pectoral muscles on either side of neck bone

Table 4. Site of collection of blood in wild animals.

S.No	Name	Site of collection
1.	Snake	Dorsal buccal vein
2.	Turtle	Jugular vein/Occipital sinus
3.	Lizard	Ventral tail vein
4.	Hare	Lateral saphenous vein
5	Avifauna	Jugular vein(small birds under 100 gnns) Cutaneous ulnavein/ Caudal tibial vein(Layer bird)
6.	Dear	Jugular vein
7.	Samber	Jugular vein
8.	Leopard	tarsal vein/Saphenous vein
9.	Lion	Tarsal vein
10.	Monkey	Jugular vein
11.	Nilgai	Jugular vein
12.	Tiger	Tarsal vein
13.	Elephant	Ear vein
14.	Snow Leopard	Tarsal vein/Saphenous vein
15.	Boar	Ear vein/ Anterior Vena cava

Table 5. REPRODUCTIVE BASIC DATA

Title	Sexual Maturity	Estrus	Gestation Period
Asian Lion	24-36 months	4-16 days (Polyestrous)	100-103 days
African Lion	24-36 months	4-16 days (Polyestrous)	92-113 days
Cheetah	9-16 months	10-14 days (Seasonal Polyestrous)	84-85 days
Leopard	9-16 months	6-7 days (Seasonal Polyestrous)	95-105 days
Tiger	36-48 months	3-6 days (Seasonal Polyestrous)	100-112 days
Jackal	24-36 months	1-14 days (Seasonal Polyestrous)	60-63 days
Indian Fox	24-36 months	1-14 days (Seasonal Polyestrous)	60-63 days
Stripped Hyena	About 2-3 year	45-50 days, polyestrous	88 -92days
Caracal caracal	14-15 months	3-6 days (Polyestrous)	68-81 days
Sloth Bear	36-48 months	18-21 days (Spring season)	210 days
Asian Black bear	About 3 -4 year	12 -35 days (seasonal)	200-240 days (Delayed implantation)
Samber	1-2 year	Approx 16-18 days, Polyestrous	180 days
Spotted Deer	1-2 year	Approx 18-21 days , Polyestrous	210-225 days
Barking Deer	About 1- 2 year	14-21 days, polyestrous	180-210 days
Musk Deer	About 18 months	Seasonal	160-180 days
Hog Deer	8-12 months	Seasonal	80 days
Chinkara	About 2 year	Approx 21-28 days , Polyestrous	150-180 days
Black buck	2-3 years	Approx 16-18 days , Polyestrous	180 days
Four Horned Antelop	About 2 year		225-240 days
Nilgai	About 2-3 year	4 day	240-258 days
Gaur	2-3 year	Polyestrus	275 days
Asiatic wild ass	About 4.5 to 5 year	Seasonal	11 – 11.5 months
Elephant	Male-10-14 years, Female-8-9 years	22 days	18-22 months
Giraff	4-6 years	14-15 days	450 days
Rhinoceroyus	4 years (Female) & 9 years	45 days (Polyestrous)	480 days
Hippopotamus	Male (7.5 year) & Female (3-4 year)	28 days	6-7 months
Wild pig	8-24 months	21 days	115 days
Comon Langur	Male: 6-7 years, Female: 3-4 years	24-20 days	190-210 days
Bonnet Macaque	32-40 months	29 days approximately	168 days
Rhesus Macaque	2.5- 4 years	29 days	135-190 days

Table 6. Common medicine used in birds (rescued, injured,diseased)

Generic Name	Dosage Route Frequency	Indication
ACYCLOVIR	330 mg/kg PO BID 4-7 days. 1000 mg/cup food	Pacheco's Viral Outbreak
AMIKACIN	15 mg/kg IM, SQ BID 10 mg/kg IM, SQ BID	Gram negatives, Cranes
AMOXICILLIN	100 mg/kg IM BID 100 mg/kg PO TID 200 mg/kg IM once	Prophylactic during Surgery
AMPICILLIN	50-100 mg/kg IM QID 15 mg/kg IM BID 150-200 mg/kg PO BID-TID	Cranes
BACITRACIN	15 mg/100 birds PO as feed additive30 mg/ 100 birds PO as feed additive	Antidiarrhoeal in chickens, Prevention of necrotic enteritis Antidiarrhoeal in turkey.
	200 gm/ ton of feed PO daily	Prevent ulcerative enteritis in young quail
CARBENACILLIN	100 mg/kg IV, IM, Intratracheal BID-TID 1 ml/10 ml for nebulization 100-200 mg/kg PO BID	Bad taste
CEFALOTHIN	100 mg/kg IM QID	
CEFAZOLIN	25-30 mg/kg IV, IM TID	Cranes.
CEFOTAXIME	100 mg/kg IM BID50-100 mg/kg	IV, IM 3-6x day 75 mg/kg IM TID 12 weeks frozen 10 days in refridge very rapid serum clearance
CEFTIOFUR	50-100 mg/kg IM 3-6x day	very rapid serum clearance
CEFTRIAXONE	50-100 mg/kg IV, IM 3-6x day	
CEPHALEXIN	35-50 mg/kg PO QID	
CHLORAMPHENICOL (Succinate) (Palmitate)	50 mg/kg, IV, IM TID 75 mg/kg PO TID	Do not use in raptor
CHLORHEXIDINE	12 tsp/gal drinking water	Antiviral
CHLORTETRACYCLINE	1% large bird, 0.5% in small bird PO 45 days 500 mg/litre water	Impregnanted feed, use antifungal too.
	50·100 mg/ton teed PO	Prevent chronic resp diseases, chickens prevention sinusitis/ hexamita, turkeys treatment of CRD/
	100·200 mg/ton feed PO	Infectious enteritis, chicken treatment of infectious sinusitis/ enteritis and prevention synovitis, turkeys
	200 mg/ ton feed PO	Treat synovitis / CRD, chickens / turkeys

[Table Contd...

Contd. Table]

Generic Name	Dosage Route Frequency	Indication
		Fowl cholera ducks
	200-400 mg/ton feed PO	Treat *Salmonella typhimurium*, turkey poult
	400 mg/ton feed PO up to 21 days	E coli, chickens
	500 mg/kg ton feed PO 5 day	
CIPROFLOXIN	20-40 mg/ kg PO BID	Gram negative
CLOXACILLIN	50-250 mg/kg IM, PO	Raptors
DOXYCYCLINE	25 - 100 mg/kg IM q 4-5 days 20 mg/kg IV once only 25 mg/kg PO BID (syrup) 50 mg/kg PO SID (suspension) 200 ppm impregnated seed PO 100 mg/kg mash PO 1 gr/kg of food	Critical psitticosis (oral forms can cause regurgitation in macaws)
ENROFLOXACIN	15 mg/kg IM BID 30 mg/kg PO BID	Gram negative
	200 ppm in water PO 100 ppm in mash PO 10% liquid: 250 mg/liter drinking water PO	Flock treatment (taste bad) Flock treatment
ERYTHROMYCIN	6 mg/kg IM BID 50-100 mg/kg PO BID 100·200 mg/litre for PO 100 ml/litre saline nebulization 15 min TID 92.5 gm/ton feed PO	
	7-14 days. 185 mg/ton teed PO 5-8 days	Prevention CRD/coryza, layers-chicken/turkeys Prevent CRD breeder chickens/turkey
ETHAMBUTOL	15-25 mg/kg PO SID	avian TB, al00l with Strep and Ritampin
FURADANTIN	1-2 mg PO TID 3-5 days	Enteritis in parakeets
	15 mg/ml: 1-2 drops of this solution lotion PO QID	Enteritis in parakeets
	8-10 mg/kg PO SID	E coli in parrots/ostrich
FURAZOLIDONE (Furo	50 gm/ton feed PO continuously	Typhoid / paratyphoid / pullorum, chicken / turkey
	50-100 mg/ton feed PO continuously	CRD/ infectious snusitis/ synovitis/ enteritidis Quail ulcerative enteritis-chicken turkeys quail
	100 gm/ton feed PO continuously	Prevent histomoniasis/ infectious hepatitis
	200 gm/ton feed PO 2 weeks	Treat severe outbreak any above diseases

Table 7. COMMON REPORTED DISEASES IN ZOO ANIMALS

Sr.No.	Disease	Species						Specific medicine
		Carnivorous	Herbivorous	Primates	Birds	Rodent	Reptile	
1	2	3	4	5	6	7	8	9
1	Tuberculosis	Common	Common	Common	Common	Common	Rare	Isorifazide
2	Enteritis	Common	Common	Common	Common	Common	Rare	Antibiotic, Fluid therapy
3	Gastritis	Common	Common	Common	Common	Common	Rare	Antibiotic, Fluid Therapy
4	Dermatitis	Common	Common	Common	Common	Common	Rare	Anti Fungal Anti biotic
5	Septicemia	Common	Common	Common	Common	Common	Rare	Antibiotic Fluid Therapy Anti pyretic
6	Pyemia	Common	Common	Common	Common	Common	Rare	Antibiotic
7	Canine Distemper*	Common	Nil	Nil	Nil	Nil	Nil	Prophylactic Vaccine
8	Parvo Virus*	Common	Nil	Nil	Nil	Nil	Nil	Prophylactic Vaccine
9	Leptospira*	Common	Nil	Nil	Nil	Nil	Nil	Prophylactic Vaccine
10	Hepatitis	Common	Common	Common	Common	Common	Common	Liver Tonic, Antibiotic
11	Feline pan leukopenia♣	Common	Nil	Nil	Nil	Nil	Nil	Prophylactic Vaccine
12	Rhino tracheitis♣	Common	Nil	Nil	Nil	Nil	Nil	Prophylactic Vaccine
13	Calici♣	Common	Nil	Nil	Nil	Nil	Nil	Prophylactic Vaccine
14	Rabies♦	Common	Common	Common	Unk	Common	Unk	Prophylactic Vaccine
15	Endoparasite	Common	Common	Common	Common	Common	Common	Broad Spectrum Helmenthetic
16	Ectoparasite	Common	Common	Common	Common	Common	Common	Butox Spray and Ivermectin Injection
17	Encephalitis	Common	Unk	Common	Unk	Unk	Unk	No specific treatment, Symptomatic treatment
18	Ranikhet Disease	Nil	Nil	Nil	Common	Nil	Nil	Prophylactic Vaccine

* Common in Canine, ♣ Common in Feline but not noticed in Zoo. ♦ Not noticed in Zoo.

Table 8. COMMON NON INFECTIOUS REPORTED DISEASE IN ZOO ANIMALS

Sr.No.	Disease	Species						Specific medicine	Reptile
		Carnivorous	Herbivorous	Primates	Birds	Rodent	Reptile		
1	2	3	4	5	6	7	8	9	10
1	Heat Stroke	Common	Common	Common	Common	Common	Common	Fluid Therapy	
2	Cold Stroke	Common	Common	Common	Common	Common	Common	Symptomatic	
3	Psychic shocked	Common	Common	Common	Common	Common	Common	Symptomatic	
4	Hemorrhgic Shock	Common	Common	Common	Common	Common	Common	Symptomatic	
5	Emaciation	Common	Common	Common	Common	Common	Common	Symptomatic	
6	Old Age	Common	Common	Common	Common	Common	Common	Symptomatic	
7	General Weakness	Common	Common	Common	Common	Common	Common	Symptomatic	
8	Trauma	Common	Common	Common	Common	Common	Common	Symptomatic	
9	Cardiac Failure	Common	Common	Common	Common	Common	Common	No treatment	
10	Respiratory Failure	Common	Common	Common	Common	Common	Common	No treatment	
11	Snake Bite	Common	Common	Common	Common	Common	Unk	Anti Snake Venum & Fluid Therapy	
12	Debility	Common	Common	Common	Common	Common	Common	Symptomatic	
13	Impection	Common	Common	Common	Common	Common	Common	Symptomatic	
14	Intestinal obstruction	Common	Common	Common	Common	Common	Common	Symptomatic	
15	Fracture	Common	Common	Common	Common	Common	Common	Symptomatic	
16	External injury	Common	Common	Common	Common	Common	Common	Dressing material	
17	Poisoning	Common	Common	Common	Common	Common	Common	Antidote for particular poison	

Table 9. Normal haematological data of zoo and wild animals

Sr.No.	Animal (gm%)	HGB (%)	PCV (106/mm3)	RBC (103/cmm)	WBC	ESR	DLC N%	L%	M%	B%	E%	MCH (pg)	MCV (m3)	MCHC (%)
1	**Monkey (Primates):**													
(a)	Lemur (*Lemur fulvus*)	13-15	40-46	7-9	8-11									
(b)	Gibbon (Hylobatus spp.)		45-50		8-11		55	35	5	5	0			
(c)	Gorilla (*Gorilla gorilla*)	12.5-15.2	39-50	4.56	7.5-13.5		55-80	12-36	3-7	0-1	0-2	27.2	75.64	28.68
(d)	Chimpanzee (Par troglodytes)	15.8	49.5	5.19	10.4		55	38	5	0-1	0-2	26.65	88.85	30
(e)	Orangutan (*Pongo pygmaeus*)	13.4-15.2	41-46	4.42	10.9-14.9		68-83	10-22	3-8	0-1	0-4	25.5	88.13	28.9
2	**Rodents and Logomorphs**													
(a)	Rabbit/Guinea Pig	8-15	30-50	4-7	6-12	2-4	20-60	30-80	2-20	0-1	0-5			
(b)	Hamster	16-18	45-50	1-8	7-10	0-1	18-40	56-80	2-5	0-1	0-1			
(C)	Rat	17-18	35-45	7-10	5-23	1-2	10-50	50-70	0-10	0-1	0-5	57-63	18-19	30-33
3	**Ruminants.**													
(a)	Fallow deer (*Dama dama*)	17.1	42-46	10.3-11.5	4.3-11	4	39	23	1-4	0-1	0-2			
(b)	Giraffe (Giraffe camelopardis)	11.5-13.8	33-35	3.6-4.2	5.8-10.0		68	39	2	0-1	0-1			
(C)	Bison (*Bison bison*)	16.17-16.88	45-47	8-10	7.9-9.10		58-66	23-28	4-6	0-13	2-5			
(d)	Reindeer (*Rangifer tarandus*)	13.5-17	36-59	6.5-11.6	5.0-9.9	5-20	17-74	20-78	0-1	0-6	0-33	35-72	13-22	31-53
4	**Felidae**													
(a)	Domestic cat	8-15	24-45	5-10	5.5-19.5		35-75	15-55	0-7	0-12	0-1			
(b)	African Lion	8-12	35-40	7-8	10.0-15.0	0-5	63	30	5	2	0			
(C)	Bengal tiger/ Jaguar (Dipado)	9-14	35-45	6-8	10.0-15.0	0-5	63	30	5	2	0			
(d)	Cheetah/Leopard	8-13	35-45	6-8	10.0-12.0	0-5	63	30	5	2	0			
5	**Canidae**													
(a)	Fox (Silver, Red, Arctic,Crab)	12-15	35-50	12-15.5	4.6-7.2		21-38, 34-66	45-69, 25-63	,1-3	5,4	1-2, 1-3	47-71	15-19	27-34
(b)	Wolf (Timber, Red, Maned)	12-16	35-45	5.1-8.0	10-12	0-2	39-60	18-25	1-3	2-6		63-69	22-24	35-36
6	**Procyonidae**													
(a)	Racoon	10-11	35-40	11	13-16	1-3	45	49	2	3	0			

[Table Contd...

Contd. Table]

Sr.No.	Animal (gm%)	HGB (%)	PCV (106/mm3)	RBC (103/cmm)	WBC	ESR	DLC N%	L%	M%	B%	E%	MCH (pg)	MCV (m3)	MCHC (%)
7	**Mustalidae**													
(a)	Shunk / Ferret/ Mink	9.5-15.6	38-50	9.5-15.5	9-14	1-3	47	50	1	2	0	54	10	30
8	**Viverridae and Hyaenidae**													
(a)	Small Indian Civet	16.5	22	12.7	7.4	54	32	1-3	2-4					
(b)	Mongoose (slender crab)	13-16	43-52	7.7-12	4.0-5.4	62, 35-47	37, 47-50	1,5-7	0,0-4	0,0				
(c)	Hyena (Strip, spotted, Brown)	8.8-15	34-53	5.6-7.5	6.0-14.0									
9	**Ursidae**													
(a)	Asiatic black bear (*Selenaretod thiobetanus*)	11.8-19.1	33-55	5.40-6.85	7.5-17.9									
(b)	Polar bear (*Thalarctos maritimus*)	15.8-19.3	46-56	5.43-7.69	6.3-13.9		75	12	7	0-1	0-6			
10	**Cetyaceans and Sirenians**													
(a)	Whales (Common trilot/killer)	15.8	45	3.71	10		50-70	20-40	4-10	2-20	0-1	40		
(b)	Dolphines											7	107	38
11	**Perissodactyla, Proboscidae, Hippo**													
(a)	Grant's Zebra	14.3-16.5	44	8.3-8.9	8.3-9.5		50-60	20-25	1-2	0-1		18.7	54	34.7
(b)	(b) White Rhinoceros.	13.1-19.5	36.7-49.5	5.87-8.11	6.0-15.6		40-50	40	0-1	0-2		19-27	55-67	32-42
(c)	Elephant (Asiatic/African)	11.4-12.8	32.4-34.9	2.86-3.50	12.78-16.72		37-43	49-56	4-6	1-3		36-48	113-122	35-36
12	**Raptors(Birds of Prey)**													
(a)	Owl (Wood, eared, tittle)	11	36-40	2.2-2.7	19	2-7	25	62		3.9	8.6			
(b)	King Vulture		40-46	3.2-38	37-46	1								
(C)	Eagle (Bald/Godders)		35-50	1.8-2.2	11-27	0-2	21	7-13	0-5			200	30	

PROFORMA OF HEALTH ASSESSMENT OF ZOO ANIMALS

1	Name of zoo		
2	Type of zoo	Large/Medium/small/mini	
3	Present number of Veterinary Doctors in the zoo		
4	Hospital facilities/treatment room	Yes/No	**Remark**
	i. emergency drugs	Available/not available	
	ii. restraints drugs	Available/not available	
	iii. revival drugs	Available/not available	
	iv. diagnostic equipments	Available/not available	
	v. transportation equipments	Available/not available	
	vi. squeeze cages/physical devices like tranquilizing darts/syringe projectors	Available/not available	
	vii. Staff in hospital	Sufficient/insufficient	
	viii. laboratory	Available/not available Well equipped/not equipped	
	ix. Library/reference books/internet facilities Available/not	AvailableWell equipped/ not equipped	
	x. Isolation wards	Constructed/not constructed	
	xi. quarantine wards	Constructed/not constructed	
5	Post mortem rooms	Available/not available Well equipped/not equipped	
6	Grave yard/incinerators	Available/not available	
7	Hygiene	Good/poor/satisfactory	
8	Disinfection	Good/poor/satisfactory	
9	Feeding/nutrition	Good/poor/satisfactory	
10	Way and means of transportation of nutrition/feed from unit to cages	Manual/tray/lorry/cartGood/ poor/satisfactory	
11	Quarantine procedure	Maintained/not maintained	
12	Marking/Identification system	Name/tattoo/elct.chips	
13	Health standards	Good/poor/satisfactory	
14	Physical condition of animals/birds	Good/poor/satisfactory	
15	Apparent illness /abnormal signs	Found/not found/	
16	Behavioural abnormalities if any	Yes/No	
17	Regular follow up of deworming schedule	Followed/ Not followed Every .... months	
18	Vaccination schedule to felidae	Followed/ Not followed Type of Vaccine	
19	Diagnostic/ treatment support and emergency help sources from	Vety.College/local	
20	Routine parasitic screening	Followed / Not followed	
	i. When?	Every .... month	

	ii. Where?	Local/Vety.College/ ADIO
	iii. Steps for positive cases & retesting	Deworming with
21	Blood testing	Every......... / As and when animal immobilised
22	Dental/general check up/ weighing of animals	Every......... / As and when animal immobilised
23	Enclosure designs and barrier patterns	Satisfactory/ require modification/
24	Enclosure enrichment	Used/ Not used
25	Feeding cubicles & off exhibit facilities for animals	Available/ not available
26	Display and signage of information	
27	Visitor facilities	
28	Whether food /water quality testing is carried out?	Yes/No
29	Do zoo have fencing/compound wall as suggested by CZA?	Yes/No
30	Over all condition of zoo	

PROFORMA OF HEALTH ASSESSMENT OF ZOO ANIMALS

Evaluation Report Date

Over all Remark /suggestions/guidance if any

....

....

....

....

....

....

....

....

....

....

....

....

....

....

....

Date:................ Signature:

1.

2.

3.

4.

5.

Proforma for the Record of Post-Mortem Examination

Animal belong to (Park):______________________________

Place______________________ Ambient temperature______________________

Description of the Animal:

Species:______________ Sex:__________ Age:______________

Date and Time of Death:______________________________

Date and Time of Post-mortem:____________ Conducted by:____________

History:

Clinical Signs and Symptoms (if any):______________________

External examination:

Body condition:______________ Skin coat:______________

Rigor mortis:______________ Mucous membrane:______________

Others:______________________________

Detailed Post-mortem examination:

1. Subcutaneous tissue:______________________________
2. Body cavities:______________________________
3. Digestive system:

   Mouth cavity:______________ Buccal cavity:______________

   Tongue:______________ Pharynx:______________

   Oesophagus:______________________________

   Stomach:______________________________

   Small intestine:______________________________

   Large intestine:______________________________

   Liver and Gall bladder:______________________________

   Lymph node:______________________________
4. Respiratory system:

   Trachea:______________________________

   Lungs:______________________________

   Lymph nodes:______________________________
5. Cardio-vascular system:

   Heart:______________________________

   Pericardial sac:______________________________

   Heart chambers:______________________________

   Large blood vessels:______________________________

   Spleen:______________________________
6. Urogenital system:

   Urinary bladder:______________________________

   Kidneys:______________________________

   Reproductive organs:______________________________

   Mammary glands:______________________________
7. Musculoskeletal system:______________________________
8. Remarks (if any):______________________________

Date:______________ Signature:______________________________

Place:______________ Name and Designation:______________________

Address:______________________________

Proforma for the Dispatch of Material for Laboratory Examination

Proforma to be sent along with the Specimen/ Tissue for Lab. Examination

To,

Address of the Lab.

______________________ **Date:**______________

______________________ **Reference No.**__________

____

Subject: Requisition for laboratory examination

A. Details of the Animal

Animal belongs to______________________

Address______________________

Animal/Species______________ Sex__________ Age__________

Body condition______________________

B. Specimen Details

(a) Tissue submitted for examination

1.__________ 2.__________ 3.__________

4.__________ 5.__________ 6.__________

(b) Preservative used______________________

(c) Date and Time of Death______________________

(d) Date and Time of Collection______________________

C. Symptoms observed (if any)______________________

D. Treatment given (if any)______________________

E. Remark______________________

Signature ______________________

Name______________________

Address______________________

____

____

MATERIALS REQUIRED FOR FIELD LABORATORY

List of Equipments, Chemicals and Glasswares

Equipments

1. Immobilization equipments
   (i) Dis-inject syringe projector complete kit (Model 60 N)
   (ii) Tele-injector – Applicator complete set for range upto 50 m (Model 4V 310)
2. Pant pellet Pistol
3. Binocular microscope
4. Post-mortem set (for large animals)
5. Deep freezer
6. Refrigerator
7. Centrifuge machine
8. Liquid nitrogen container
9. Incubator
10. Hot air oven
11. Spirit lamp
12. Equipment for haematological studies
13. Gluconometer
14. Metal distillation apparatus
15. Mass vaccinator for cattle
16. Protective clothings (Aprons, hand gloves and gum-boots)
17. Disposable syringe (5 ml and 10 ml)
18. Monopan balance

Chemicals

1. Xylene
2. Methonol
3. Alcohol
4. Spirit
5. Acetone
6. Benzene
7. Paraffin wax
8. EDTA
9. Heparin

10. Gram's stain
11. Lugol's iodine
12. Geimsa stain
13. Leishman's stain
14. Methylene blue
15. Formalin 40%
16. Savlon/Dettol
17. D.P.X.
18. Cedar wood oil

Glasswares

1. Vials for collection of faecal sample
2. Vials for blood samples
3. Glass slides
4. Glass coverslips
5. Specimen jars (5 litre and 10 litre capacity)
6. Glass bottles with wide mouth (250 ml and 500 ml)
7. Glass bottles with narrow mouth (500 ml)
8. Beakers (100 ml and 250 ml)
9. Watch glass (5 mm and 10 mm)
10. Petri dishes
11. Funnel
12. Vacutainers with heparin

List of some of the Approachable Disease Diagnostic Centers.

1. College of Veterinary Science, AAU, Anand (Gujarat)
2. Indian Veterinary Research Institute, Izatnagar(U. Khand)
3. College of Veterinary Science, Pantnagar- 263 145 (U. Khand)
4. College of Veterinary Science & A.H., Jabalpur-482 001 M. P.
5. College of Veterinary Science, HAU, Hissar (Haryana)
6. College of Veterinary Science, Chennai (Tamilnadu): Coordinator, Wildlife Health
   (a) Scientist, Wildlife Institute of India, Dehra Dun
   (b) Coordinator, Wildlife Health (WRC), Vety. College, Anand (Gujarat)
   (c) Prof. & Head, Madras Veterinary College, Chennai
   (d) Professor, College of Veterinary Sc. & A.H., Jabalpur (M.P.)

EQUIPMENTS AND SUPPLIES USED IN DISEASE CONTROL PROGRAMME

A. Carcass Collection

i. Large heavy duty plastic bags.

B. Carcass Disposal

i. Equipment for digging trenches or pits.

ii. Incineration

a. Fuel for burning carcasses.

C. Sanitation Procedures

1. Decontamination of environment and structures

i. Chemicals such as Chlorine bleach, Environ etc.

ii. Hand carried spray units

2. Protection of personnel and prevention of mechanical movement of disease agent to other locations by people and equipment.

i. Rubber gloves, foot coverings.

ii. Spray units and chemical disinfectants.

iii. Plastic bags for transportation of field cloths.

D. Surveillance and Observation

i. Monitoring animal population and environmental conditions.

ii. Binoculars and spotting scope.

iii. Maps for tracking the progress of events and animal populations associated with die off.

E. Animal Population and Habitat Manipulation

i. Denying animal use of a specific area.

a. Monitorized means of hazing animal population.

ii. Conservation and maintenance of animal in a specific area.

a. Grain and other source of food.

b. Water

c. 'Area closed' signs to prevent temporary refuge area.

F. Animal Sampling and Monitoring

i. Animal Capture

a. Nest and capture equipments

b. Animal marking.